Biological Effects of Dietary Restriction

W0051365

ILSI Monographs

Carcinogenicity:
The Design, Analysis, and Interpretation of Long-Term Animal Studies
H.C. Grice and J.L. Ciminera, Editors
1988. 279pp. ISBN 3-540-18301-9

Inhalation Toxicology: The Design and Interpretation of Inhalation
Studies and Their Use in Risk Assessment
U. Mohr, Editor-in-Chief
D.L. Dungworth, G. Kimmerle, J. Lewkowski, R.O. McClellan,
W. Stöber, Editors
1988. 318pp. ISBN 3-540-17822-8

Radionuclides in the Food Chain
M.W. Carter, Editor-in-Chief
J.H. Harley, G.D. Schmidt, G. Silini, Editors
1988. 518pp. ISBN 3-540-19511-4

Assessment of Inhalation Hazards:
Integration and Extrapolation Using Diverse Data
U. Mohr, Editor-in-Chief
D.V. Bates, D.L. Dungworth, P.N. Lee, R.O. McClellan, F.J.C. Roe, Editors
1989. 382pp. ISBN 3-540-50952-6

Monitoring Dietary Intakes
I. Macdonald, Editor
1991. 259pp. ISBN 3-540-19645-5

Biological Effects of Dietary Restriction
L. Fishbein, Editor
1991. 349pp. ISBN 3-540-52294-8

ILSI
MONOGRAPHS

Sponsored by
the International
Life Sciences Institute

L. Fishbein
Editor

Biological Effects of Dietary Restriction

With 82 Figures

Springer-Verlag Berlin Heidelberg GmbH

Lawrence Fishbein, PhD
1126 Sixteenth Street, N.W.
Washington, DC 20036
USA

Library of Congress Cataloging-in-Publication Data
Biological effects of dietary restriction / L. Fishbein (ed.).
   p.   cm. — (ILSI monographs)
   Based on a conference convened in Washington, D.C., Mar. 5–7,
1990; cosponsored by the International Life Sciences Institute,
National Center for Toxicological Research, and National Institute
on Aging.
   Includes bibliographical references.
   Includes index.

   ISBN 978-3-642-63494-9   ISBN 978-3-642-58181-6 (eBook)
   DOI 10.1007/978-3-642-58181-6

   1. Low-calorie diet—Congresses.   2. Diet in disease—Congresses.
3. Laboratory animals—Feeding and feeds—Congresses.
4. Nutritionally induced diseases—Congresses.   I. Fishbein, L.
(Lawrence), 1923–   .   II. International Life Sciences Institute.
III. National Center for Toxicological Research.   IV. National
Institute on Aging.   V. Series.
   [DNLM: 1. Diet—congresses.   2. Diet Therapy—congresses.   WB 400
B615   1990]
QP141.A1B56   1991
616.3'9—dc20
DNLM/DLC
for Library of Congress   90-10458

Printed on acid-free paper.

© 1991 by Springer-Verlag Berlin Heidelberg.
Originally published by Springer-Verlag Berlin Heidelberg New York in 1991
Softcover reprint of the hardcover 1st edition 1991

This work is subject to copyright. All rights are reserved, whether the whole or part of the material
is concerned, specifically the rights of translation, reprinting, reuse of illustrations, recitation,
broadcasting, reproduction on microfilms or in other ways, and storage in data banks. Duplication of
this publication or parts thereof is only permitted under the provisions of the German Copyright Law
of September 9, 1965, in its version of June 24, 1985, and a copyright fee must always be paid. Viola-
tions fall under the prosecution act of the German Copyright Law.
The use of general descriptive names, trade marks, etc. in this publication, even if the former are not
especially identified, is not to be taken as a sign that such names, as understood by the Trade Marks
and Merchandise Marks Act, may accordingly be used freely by anyone.

Typeset by Publishers Service of Montana, Bozeman, Montana.

9 8 7 6 5 4 3 2 1

ISBN 978-3-642-63494-9

Series Foreword

The International Life Sciences Institute (ILSI), a nonprofit, public foundation, was established in 1978 to advance the sciences of nutrition, toxicology, and food safety. ILSI promotes the resolution of health and safety issues in these areas by sponsoring research, conferences, publications, and educational programs. Through ILSI's programs, scientists from government, academia, and industry unite their efforts to resolve issues of critical importance to the public.

As part of its commitment to understanding and resolving health and safety issues, ILSI is pleased to sponsor this series of monographs that consolidates new scientific knowledge, defines research needs, and provides a background for the effective application of scientific advances in toxicology and food safety.

Alex Malaspina
President
International Life Sciences Institute

Preface

The influence of quality and quantity of diet regimens (principally, dietary and caloric restriction) on longevity as well as susceptibility to nonneoplastic diseases, cancer, other age-associated diseases, and physiologic processes in numerous animal species has been recognized for at least six decades. The modulation of the onset of human diseases, primarily atherosclerosis, coronary diseases, and cancer, through dietary intervention is receiving increasing attention (partly intensified in recognition of an increased life span in a number of North American, European, and Asian societies). The feasibility of developing dietary guidelines for individuals with known risks for chronic diseases is also being increasingly advocated in some quarters. Optimal diet could result in healthier older populations and decreased debility associated with aging.

Nutritional modification of cancer is a burgeoning research area with the recognition that dietary excess, deficiencies, or imbalances can exert a major role in the etiology or modulation of cancer. More recent advances in assessing the effects of long-term caloric restriction on a host of physiological, biological, molecular, and behavioral parameters in a variety of animal models suggested the opportunity of convening an international conference to more fully assess the progress and future directions to be taken in this important field of research.

The conference "Biological Effects of Dietary Restriction" cosponsored by the International Life Sciences Institute, National Center for Toxicological Research, and National Institute on Aging was convened in Washington, D.C., March 5–7, 1990, to pursue the following major objectives:

Elaborate the complex effects of dietary restriction on longevity and the physiological and biochemical state of laboratory animals;

Describe the different feeding regimens and alterations in dietary composition that are used in dietary restriction research and their effects on the nutritional status of experimental animals;

Evaluate current multidisciplinary research on the underlying mechanisms by which dietary restriction may induce physiological and biochemical changes;

Identify opportunities for further research to elucidate mechanisms and increase understanding of the practical applications of dietary restriction;

Promote discussion of the implications of research on dietary restriction for the use of diet-restricted animals in toxicity testing;

Examine the current status of animal models for dietary restriction and the implications for safety assessment and human health.

The papers and panel discussion presented in this monograph by preeminent researchers in the fields of nutrition, biochemistry, physiology, toxicology, and pathology attest to the overall success in attempting to accomplish our objectives.

Lawrence Fishbein, PhD
Washington, D.C.

Contents

Part I. Feeding Regimens and Diets

Part II. Effects of Dietary Restriction on Toxicological Endpoints

Part VI. Impact of Dietary Restriction on Bioassays and Recommendations for Future Research

Contributors

The complete affiliations for all authors are given as footnotes to the opening pages of their chapters. These page numbers are given in the list below.

R.J. Feuers, National Center for Toxicological Research   198, 207, 245

E. Flescher, University of Texas Health Science Center at San Antonio   172

P. Gao, National Center for Toxicological Research   42

R.A. Good, All Children's Hospital, University of South Florida   147

P. Gray, University of Texas Health Science Center at San Antonio   172

J.R. Harmon, National Center for Toxicological Research   207

D.E. Harrison, The Jackson Laboratory   264

R.W. Hart, National Center for Toxicological Research   185, 198, 207, 245

K. Hashimoto, Food and Drug Safety Center   87

C.J. Henry, ILSI Risk Science Institute   321

A.M. Holehan, Institute of Human Ageing, University of Liverpool   140

R.R. Holson, National Center for Toxicological Research   123

J.D. Hunter, University of Texas at El Paso   198

K. Imai, Food and Drug Safety Center   87

D.K. Ingram, Gerontology Research Center, National Institute on Aging   157, 305

C. Ip, Roswell Park Cancer Institute   65

R. Irwin, National Institute of Environmental Health Sciences   99

H. Iwai, University of Texas Health Science Center at San Antonio   172

J. Johnson, Battelle Columbus Laboratories   99

B. Kent, Mt. Sinai Medical Center   264

J.J. Knapka, National Institutes of Health   12

S. Laganiere, University of Texas Health Science Center at San Antonio   172

J.E.A. Leakey, National Center for Toxicological Research   198, 207, 245

D.-W. Lee, University of Texas Health Science Center at San Antonio   191

E. Lok, Health Protection Branch, Health and Welfare Canada   55

E. Lorenz, All Children's Hospital, University of South Florida   147

E.J. Masoro, University of Texas Health Science Center   115

B.J. Merry, Institute of Human Ageing, University of Liverpool   140

R. Mongeau, Health Protection Branch, Health and Welfare Canada   55

E.A. Nera, Health Protection Branch, Health and Welfare Canada   55

R.A. Pegram, Oak Ridge Associated Universities   27, 42

M. Pollard, Lobund Laboratory, University of Notre Dame   238

G.N. Rao, National Toxicology Program, National Institute of Environmental Health Sciences   16, 321

F.J.C. Roe, Consultant   287, 321

G.S. Roth, Gerontology Research Center, National Institute on Aging   305

F.M. Scalzo, National Center for Toxicological Research   123

L.E. Scheving, University of Arkansas for Medical Sciences   198

F.W. Scott, Health Protection Branch, Health and Welfare Canada   55

J.G. Shaddock, National Center for Toxicological Research   198

W.G. Sheldon, Pathology Associates, Inc.   73

R.J. Scheuplein, Food and Drug Administration   321

M.G. Simic, National Institute of Standards and Technology   217

D.L. Snyder, Center for Gerontological Research, The Medical College of Pennsylvania   238

D.E. Stevenson, Shell Oil Company   321

J.D. Thurman, Pathology Associates, Inc.   73

A. Turturro, National Center for Toxicological Research   185

J.T. Venkatraman, University of Texas Health Science Center at San Antonio   172

R.L. Walford, UCLA School of Medicine   229

R. Weindruch, Department of Medicine, University of Wisconsin   3

W.M. Witt, National Center for Toxicological Research   73

B.S. Wostmann, Lobund Laboratory, University of Notre Dame   238

S. Yoshimura, Food and Drug Safety Center   87

B.P. Yu, University of Texas Health Science Center at San Antonio   191

Cosponsors

International Life Sciences Institute (ILSI)

ILSI Risk Science Institute

ILSI Human Nutrition Institute

ILSI Japan

National Center for Toxicological Research

National Institute on Aging

Scientific Advisory Committee

W.T. Allaben, National Center for Toxicological Research

H.C. Grice, CanTox Inc.

C.J. Henry, ILSI Risk Science Institute

E.E. McConnell, ILSI Risk Science Institute

T.A. Morck, International Life Sciences Institute

G.N. Rao, National Toxicology Program, National Institute of Environmental Health Sciences

F.J.C. Roe, Consultant

A.Turturro, National Center for Toxicological Research

R. Weindruch, National Institute on Aging

Session Chairs

W.T. Allaben, National Center for Toxicological Research

D.E. Harrison, The Jackson Laboratory

C.J. Henry, ILSI Risk Science Institute

J.J. Knapka, National Institutes of Health

E.J. Masoro, University of Texas Health Science Center

A. Turturro, National Center for Toxicological Research

R. Weindruch, National Institute on Aging

Part I
Feeding Regimens and Diets

CHAPTER 1

Terminology and Methods in Dietary Restriction Research

R. Weindruch[1]

Introduction

Dietary restriction (DR) entailing a 30%–70% reduction in caloric intake done without malnutrition favorably alters gerontologic and pathologic outcomes in mice and rats (for reviews see: Holehan and Merry 1986; Masoro 1988; Weindruch and Walford 1988). Over the many years since McCay et al. (1935) first described the life span-increasing action of DR in rats, the model has evolved from an interesting peculiarity to a uniquely powerful tool for investigating the retardation of aging and disease processes. DR stands out from the many putative antiaging interventions tested in rodents because it is the only one which retards a broad spectrum of age-related biologic changes (Weindruch and Walford 1988), reduces mortality rates (Sacher 1988), and extends maximum life span (Ross 1978; and Schneider and Reed 1985).

Life span extension by DR seems to depend on *calorie* restriction (Weindruch and Walford 1988). The limitation of protein, fat, or carbohydrate *without* calorie restriction does not raise the maximum species-specific life span of rodents. Also, maximum life span is not increased by DR limited to the period before weaning, overall vitamin supplementation, or with vitamin E or other antioxidant fortification. Variations in the type of dietary fat, carbohydrate, or protein do not extend maximum life span either.

Most DR studies have involved mice or rats; however, low calorie diets also extend life span in diverse other animals (e.g., rotifers, nematodes, water fleas, spiders, fish). The DR paradigm is not simply a better diet for inactive, overfed rodents in cages.

These low-calorie diets have been described by many terms and studied using several methods. These diverse terms and methods as applied to studies in rodents are the subject of this brief review (for a more comprehensive discussion, see Weindruch and Walford 1988).

[1]Department of Medicine, University of Wisconsin, Madison, WI 53706, USA

Terminology

Terms (and acronyms) used to describe these life span-extending diets include food restriction, dietary restriction (DR), caloric restriction, chronic energy intake restriction (CEIR; Dao et al. 1989), *under*nutrition without *mal*nutrition (Walford et al. 1977), every other day (EOD) feeding (i.e., giving animals free access to food but only every other day; Goodrick et al. 1982), and protein restriction (i.e., ad libitum feeding of a low-protein diet which causes animals to reduce food intake).

The most successful regimens provide essential nutrients in adequate amounts but restrict calorie intake to 30%–70% below the ad libitum level. Effects on longevity seem to depend largely on calorie restriction, since, as just discussed, protocols which restrict fat, protein, or carbohydrate intake *without* calorie restriction do not increase the maximum species-specific life span.

The use of the term "restriction" in describing DR regimens is viewed as undesirable by some due to the negative connotations it may engender. In these times of heightened concern over the proper treatment and very use of experimental animals it is wise to choose labels for important animal models with extreme care. Perhaps phrases such as "low-calorie diets" or "caloric optimization" are more acceptable descriptors. On the other hand, the DR paradigm seems very defendable because the animals subjected to DR do not appear to be in distress and live healthier and longer lives than do their conventionally fed counterparts. In my view, those individuals truly concerned with the health of laboratory animals would be hard-pressed to show any benefits associated with ad libitum feeding.

Methods

Animal Model Selection

The choice of an experimental animal in DR studies depends on the purpose of the investigation. If the goal is to learn about DR's actions on aging, the use of long-lived rodent strains seems most appropriate because it is most significant from a gerontologic standpoint to retard the aging rate of an already long-lived representative of the species. Mice from long-lived F_1 hybrid strains such as C3B10RF$_1$ (Weindruch et al. 1986) or B6CBAF$_1$ (Harrison and Archer 1987) appear quite suitable for gerontologically oriented DR work. Long-lived rat models responding well to DR include the Sprague-Dawley rats studied by Merry and Holehan (1981) and the Wistar rats investigated by Beauchene et al. (1986). To investigate diseases, a long-lived strain is not essential (see Chapter 14 by Good et al.) but still may prove advantageous if one seeks a model for spontaneous diseases which afflict humans in the seventh decade and beyond. For toxicologic studies, as discussed elsewhere in this volume, standardized rodent models are available.

A major and pervasive difficulty affecting aging research involves separating the study of aging from that of diseases. This problem is often a very serious one in DR studies. For example, in a study comparing 2-year-old animals subjected to either control or DR regimens, the collected evidence strongly suggests that the control animals differ from the DR group in more than just diet history. Most 2-year-old controls are disease-afflicted survivors from a cohort that has already exhibited significant mortality. Thus, comparison of old animals in DR studies is often confounded in a serious way by the presence of diseases. Indeed, many of the age-sensitive biologic parameters staying younger longer in DR rodents may do so because of retardation of disease (not aging) processes.

Types of Diets

Most gerontologic studies in rodents have used either commercially available nonpurified diets or defined purified diets. Two types of nonpurified diets exist: *open* formula (which has a defined, fixed composition) and *closed* formula (in which the manufacturer does not disclose the exact composition). The purified diets differ from nonpurified diets in being made of refined proteins (casein is most commonly used), carbohydrate (sucrose, cornstarch), and fat (usually corn oil), with added vitamin and mineral mixtures. A potential disadvantage of nonpurified diets is that they can be contaminated more readily by nonnutritive substances than the purified diets.

The most solidly established and effective way to extend life span in rodents is to feed a restricted amount of a high-quality purified diet. Life span is far more impressively increased by reducing the intake of a nutrient-adequate diet than by either EOD ad libitum feeding or by ad libitum feeding of low protein (typically 4%–10% casein) diets (Weindruch and Walford 1988). The diets fed in restricted amounts are sometimes (but not always) enriched in certain essential nutrients to guard against malnutrition. Diets which provide identical intakes of all known essential nutrients, even though differing in calories, may be referred to as isonutrient diets. A study restricting energy intake via isonutrient diets results in caloric restriction whereas use of nonisonutrient diets is better termed dietary restriction or food restriction.

A 1977 American Institute of Nutrition report suggested that the AIN-76™ diet be the standard purified diet for rodent studies. The widespread use of a single diet is positive in allowing meaningful comparison of data but is negative in discouraging new attempts to improve diets. It is pertinent to note that the AIN Committee ". . . decided that the diet should not contain quantities of vitamins and minerals highly in excess of the requirements for rats and mice as set forth by the Committee on Animal Nutrition, National Research Council (1987)." More recently, an AIN workshop resulted in recommendations for possible changes in the AIN-76™ diet (Reeves 1989). Despite the obvious appeal of using a completely defined diet for many nutritional studies, the AIN-76™ diet itself must be shown to yield maximum life spans as great as less-defined or less widely used diets in order to be considered an adequate standard for gerontologic studies.

Feeding and Housing Strategies

In one study (Nelson and Halberg 1986), the frequency and timing of meals for mice on DR did not influence longevity. Likewise, DR has been successfully carried out by controlled daily or intermittent feeding (see Weindruch and Walford 1988). The lack of an overt effect on longevity of the periodicity of feeding in DR studies stands in sharp contrast to the profound influence of meal timing on many physiologic rhythms (see Duffy et al., Chapter 24).

Individually housing all animals in DR studies is critical because it allows for the food intake of each animal to be accurately determined. To my knowledge, negative sequelae of long-term individual housing in DR studies have not been reported. Because the effects of DR on aging/diseases are so strong it becomes essential for all studies of potential anti-aging interventions to use housing conditions which allow for the accurate measure of food intake. Note that one such candidate intervention, dehydroepiandrosterone (DHEA), has been found to reduce food intake when fed to mice in some studies (Nyce et al. 1984; Weindruch et al. 1984). Therefore, the possibility exists that DR may contribute biologic effects attributed to DHEA. In this regard the extremely long life spans described by Knoll et al. (1989) for rats treated with the monoamine oxidase inhibitor (-) Deprenyl are difficult to interpret because body weights and food intakes were not reported.

One DR method which produced very long-lived mice utilized the purified diets shown in Table 1.1 (Weindruch et al. 1986). Weanling mice from the C3B10RF$_1$ hybrid strain were individually housed in a nonbarrier facility. The mice on DR received on an intermittent basis a diet enriched in protein, vitamins, and minerals such that per week intakes of these nutrients were quite similar for DR and control mice. The DR mice received considerably less carbohydrate

Table 1.1. Composition of diets

Ingredient	Diet 1[a] g/kg of diet	Diet 1[a] g/mouse per week	Diet 2[b] g/kg of diet	Diet 2[b] g/mouse per week
Casein	200.0	4.3	350.0	4.3
Cornstarch	260.8	5.7	157.6	2.0
Sucrose	260.8	5.7	157.6	2.0
Corn oil	135.0	2.9	135.0	1.7
Mineral mixture	60.0	1.4	110.0	1.4
Fiber	56.4	1.2	40.0	0.5
Vitamin mixture	23.0	0.50	42.2	0.52
Brewer's yeast	4.0	0.09	7.4	0.09
Zinc oxide	0.05	1.1×10^{-3}	0.1	1.2×10^{-3}

[a]Diet 1: Diet fed to control mice as seven 3.0- to 3.2-g feedings per week providing ~85 kcal/week. Composition is given as grams of ingredient per kilogram diet. Also indicated is how much of each ingredient a mouse ate in 1 week. For details, see Weindruch et al. (1986).
[b]Diet 2: Diet fed to restricted mice. This diet was enriched in casein, mineral and vitamin mixtures, brewer's yeast, and zinc oxide. Fed as four 2.4- to 3.2-g feedings per week (one feeding on Monday, one on Wednesday, and a double portion on Friday) providing ~40–50 kcal/week to the DR groups.

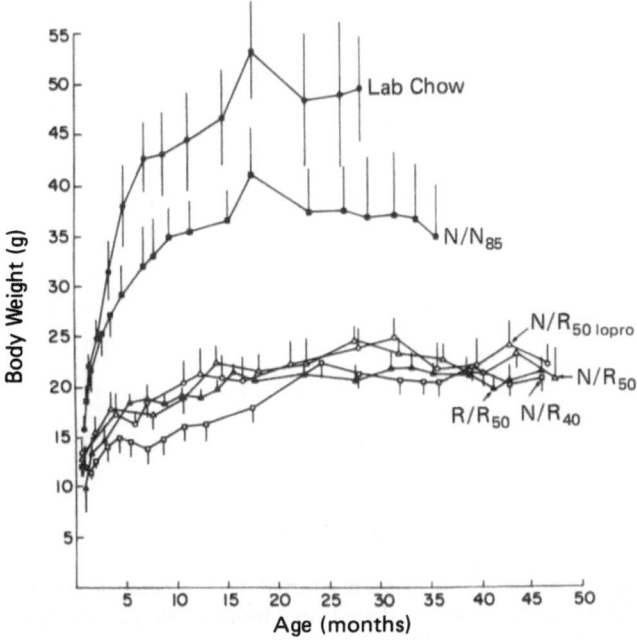

Figure 1.1. Body weights (mean ± SD) of six diet groups of C3B10RF$_1$ female mice. Diet groups: Lab Chow, Purina Lab Chow ad libitum; N/N$_{85}$ fed normally before and after weaning, postweaning diet fed at 85 kcal/week (25% less than ad libitum); N/R$_{50}$, fed normally before weaning, restricted postweaning to 50 kcal/week; R/R$_{50}$, restricted both before and after weaning; N/R$_{50lopro}$ restricted after weaning to 50 kcal/week with a decrease with age in the protein content of the diet; N/R$_{40}$ restricted after weaning to 40 kcal/week. (From Weindruch et al 1986)

(cornstarch and sucrose) and somewhat less fat and fiber per week than the controls. We find that the best results occur when only calorie intake, and not essential nutrient intake, differ between DR and control mice. An important feature of the strategy used in our study was to feed the control group at a level ~25% below the average ad libitum intake level. This produced two desirable outcomes: (1) the control mice were not obese (and therefore not prone to develop obesity-associated diseases) so that one common and important criticism of the DR model was addressed, and (2) the food intakes of control mice were matched and easily determined.

Long-term DR in rats and mice has been done successfully in nonbarrier (Merry and Holehan 1981; Weindruch et al. 1986), specific pathogen-free (Yu et al. 1982) and germ-free (Snyder et al., Chapter 23) facilities. In these particular studies, the rodents raised without barriers lived as long or longer than the environmentally protected animals. The trade off between the safety of a highly controlled microbial environment and its artificiality is an ongoing concern.

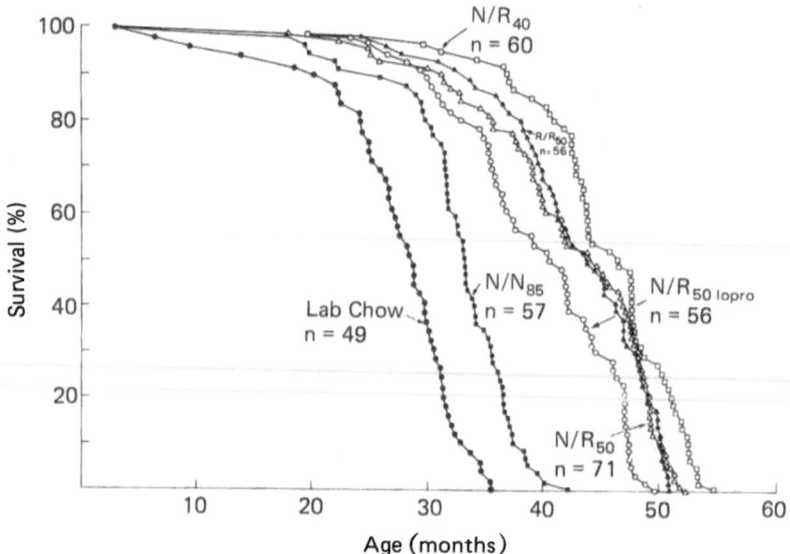

Figure 1.2. Effects of dietary restriction started at 3 weeks of age on survival of C3B10RF₁ female mice. Each symbol depicts one mouse. The diet groups are explained in Fig. 1. (From Weindruch et al. 1986)

Age of Onset

Caloric restriction started early in life (3–6 weeks of age) provides the best available model to study biologic processes which underlie the retardation of aging and chronic diseases. Imposing low calorie diets at or beyond midadulthood is a superior model for possible human use, and although less closely studied than early-onset DR, also retards aging and diseases. A brief illustration of both approaches using data from our studies at UCLA follows.

Figure 1.1 shows body weights of C3B10RF₁ mice subjected to the experimental diets (Table 1.1) from early in life (3 weeks of age; Weindruch et al. 1986). The growth rates and adult body weights approximated the levels of caloric intake. The heaviest mice were allowed free access to Purina Lab Chow and most appeared obese. The other diet groups were fed nutritionally adequate purified diets at 85, 50, or 40 kcal/week.

The survival of these cohorts is shown in Fig. 1.2. A comparison of the survival of mice fed 85 or 40 kcal/week shows that early-onset DR increased the average life span to 45.1 ± 0.9 months (SEM) from 32.7 ± 0.7 months (a 38% increase), and increased by 34% the maximum (10th decile) life span to 53.0 ± 0.3 from 39.7 ± 0.6 months. To my knowledge, this 53-month maximum decile exceeds reported values for mice of any strain.

As Fig. 1.3 shows, mice from medium and long-lived strains which were first subjected to gradual DR in midadulthood (12 months of age) showed 10%–20%

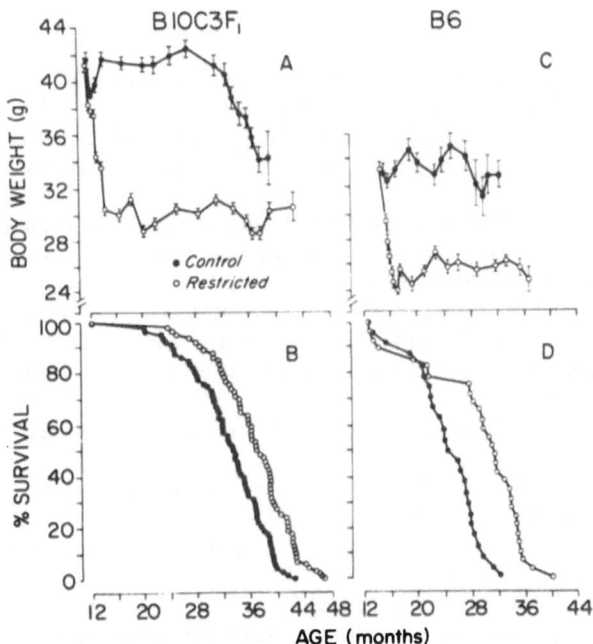

Figure 1.3 A-D. Body weights and survival of B10C3F₁ mice (**A** and **B**) and B6 mice (**C** and **D**) fed control or adult-onset restricted diets. The food restriction started at 12 months of age. Each point in the survival curves represents one male mouse. (From Weindruch and Walford 1982)

increases in average and maximum life span (Weindruch and Walford 1982). This increased longevity was associated with a lower incidence and delayed appearance of lymphoma (a main tumor in these strains). Adult-onset DR was also found to retard aspects of immunologic aging in mice (Weindruch et al 1982). These results suggest that the actions of DR show only a partial and minor dependence on slowing growth and maturation.

Conclusions

The recent increase in interest in the DR paradigm has produced a rapid surge of knowledge on biologic outcomes. This new information has created the opportunity to design increasingly sophisticated experiments aimed at clarifying the mechanisms which underlie DR's retardation of aging and disease processes. On the other hand, relatively little new effort has gone into improving DR methodologies. A significant reason for this situation is the difficulty of convincing peer reviewers of the importance of studies of this sort. Methodologic innovations (e.g., improved formulations for DR diets, environmental optimization, new animal models) could only serve to further stimulate this exciting area of inquiry.

References

American Institute of Nutrition (1977) Ad Hoc Committee on Standards for Nutritional Studies. Report of the Committee. J Nutr 107:1340–1348

Beauchene RE, Bales CW, Bragg CS, Hawkins ST, Mason RL (1986) Effect of age of initiation of feed restriction on growth, body composition, and longevity of rats. J Gerontol 41:13–19

Dao ML, Shao R, Risley J, Good RA (1989) Influence of chronic energy intake restriction on intestinal alkaline phosphatase in C3H/Bi mice and autoimmune-prone MRL/1pr,1pr mice. J Nutr 119:2017–2022

Goodrick CL, Ingram DK, Reynolds MA, Freeman JR, Cider NL (1982) Effects of intermittent feeding upon growth and life span in rats. Gerontology 28:233–241

Harrison DE, Archer JR (1987) Genetic differences in effects of food restriction on aging in mice. J Nutr 117:376–382

Holehan AM, Merry JB (1986) The experimental manipulation of aging by diet. Biol Rev 61:329–368

Knoll J, Dallo J, Yen TT (1989) Striatal dopamine, sexual activity and lifespan of rats treated with (-) Deprenyl. Life Sci 45:525–531

Masoro EJ (1988) Food restriction in rodents: an evaluation of its role in the study of aging. J Gerontol 43:B59–64

McCay CM, Crowell MF, Maynard LA (1935) The effect of retarded growth upon the length of the life span and upon the ultimate body size. J Nutr 10:63–79

Merry JB, Holehan AM (1981) Serum profiles of LH, FSH, testosterone and 5-alpha-DHT from 21 to 1000 days of age in ad libitum fed and dietary restricted rats. Exp Gerontol 16:431–444

National Research Council (NRC) (1978) Nutrient requirements of the mouse. In: Nutrient requirements of laboratory animals, 3rd revised edn. National Academy of Sciences Washington DC, pp 38–53

Nelson W, Halberg F (1986) Meal-timing, circadian rhythms and life span of mice. J Nutr 116:2244–2253

Nyce JW, Magee PN, Hard GC, Schwartz AG (1984) Inhibition of 1,2-dimethylhydrazine-induced colon tumorigenesis in Balb/c mice by dehydroepiandrosterone. Carcinogenesis 5:57–62

Reeves PG (1989) AIN-76 diet: should we change the formulation? J Nutr 119:1081–1082

Ross MH (1978) Nutritional regulation of longevity. In: Behnke JA, Finch CE, Moment GB (eds) The biology of aging. Plenum, New York, pp 173–189

Sacher GA (1977) Life table modification and life prolongation. In: Finch CE, Hayflick L (eds) Handbook of the biology of aging. Van Nostrand Reinhold, New York, pp 582–638

Schneider EL, Reed JD (1985) Modulations of aging processes. In: Finch CE, Schneider EL (eds) Handbook of the biology of aging, 2nd edn. Van Nostrand Reinhold, New York, pp 45–76

Walford RL, Meredith PJ, Cheney KE (1977) Immunoengineering: prospects for correction of age-related immunodeficiency states. In: Makinodan T, Yunis E (eds) Immunology and aging. Plenum, New York, pp 183–201

Weindruch R, Walford RL (1982) Dietary restriction in mice beginning at one year of age: effects on lifespan and spontaneous cancer incidence. Science 215:1415–1418

Weindruch R, Walford RL (1988) The retardation of aging and disease by dietary restriction. Thomas, Springfield

Weindruch R, Gottesman SRS, Walford RL (1982) Modification of age-related immune decline in mice dietarily restricted from or after mid-adulthood. Proc Natl Acad Sci USA 79:898–902

Weindruch R, McFeeters G, Walford RL (1984) Food intake reduction and immunologic alterations in mice fed dehydroepiandrosterone. Exp Gerontol 19:297–304

Weindruch R, Walford RL, Fligiel S, Guthrie D (1986) The retardation of aging by dietary restriction: longevity, cancer, immunity and lifetime energy intake. J Nutr 116:641–654

Yu BP, Masoro EJ, Murata I, Bertrand HA, Lynd FT (1982) Life span study of SPF Fischer 344 male rats fed ad libitum or restricted diets: longevity, growth, lean body mass and disease. J Gerontol 37:130–141

CHAPTER 2

Nutritional Adequacy in Dietary Restriction

J.J. Knapka[1]

Introduction

An objective of long-term dietary restriction studies involving laboratory animal species is to decrease calorie or nutrient consumption without imposing energy or nutrient deficiencies. In nutritional terms, these animals are to be provided adequate nutrition to meet their maintenance requirements. Most frequently, this objective is achieved by offering test animals a fixed percentage of the amount of diet consumed by ad libitum-fed control animals. The use of this procedure has provided a considerable amount of data showing beneficial effects of dietary restriction (McCay et al. 1935; Masoro 1988; Weindruch and Walford 1988). However, this method of restricting energy or nutrient consumption, without consideration of the quantitative minimum nutrient requirements, may result in animals receiving less than adequate nutrition. Factors contributing to potential deficiencies when animals are fed at near maintenance levels are the inherent variations in dietary nutrient concentrations between production lots of a diet and difference in nutrient requirements due to genotype. The well-being of animals is at risk when their energy or nutrient consumption is below maintenance requirements, even for short periods; consequently, data collected during these periods may be compromised. The objectives of this presentation are to review the status of the nutrient requirements for maintenance of the laboratory animal species most frequently used in dietary restriction studies and to discuss procedures that may be used to insure that they receive adequate maintenance level nutrition.

Estimated Nutrient Requirements

Nutrient requirement data for animal species are acquired from digestibility studies; studies designed to determine specific nutrient requirements for discrete functions, i.e., reproduction, lactation, or growth; and studies which provide

---

[1]National Institutes of Health, Building 14A, Room 102, Bethesda, MD 20892, USA

nutritional information relative to performance when animals are fed experimental diets. A review of the literature indicates that there have been no studies specifically designed to determine the nutrient requirements for maintenance of laboratory animal species, nor are there published experimental results that could be used as a basis for estimating these requirements.

For the last five decades, the primary source regarding estimated nutrient requirements of farm and laboratory animal species has been the series of reports on the nutrient requirements of domestic animals issued by the National Research Council (NRC). The most recent reports on those species most frequently used in dietary restriction studies—rats and mice and nonhuman primates—were issued over a decade ago (NRC 1978a,b, respectively). Those reports provide estimates of nutrient requirements for growth and in some cases for gestation and lactation. Unfortunately, they contain only limited amounts of information regarding the requirements for maintenance. Only the requirements for crude protein, fat, digestible energy, and 10 amino acids are included for the rat. The void in the NRC reports regarding information on nutrient requirements for maintenance is caused in part by the NRC committees preparing their reports to rely only on published data as the basis for estimating nutrient requirements (Knapka 1980).

In the absence of information on the nutrient requirements for maintenance of laboratory animals, there is a tendency throughout the biomedical research community to ignore the probability that the nutrient requirements for maintenance of these species are lower than for growth or reproduction. That is, diets that have been formulated for maximum growth or reproduction are routinely fed to animals involved in long-term holding studies that extend well beyond the reproductive life span. Evidence that the nutrient requirements for maintenance are considerably lower than for growth or reproduction is provided in data associated with every species of farm animal studied as well as from work with human subjects. Data are not available to indicate that laboratory animal species are different in this regard. It should be recognized that identification of the quantitative nutrient requirements for maintenance of laboratory animals is essential for the success of studies designed to establish those mechanisms associated with the beneficial effects of dietary restriction.

Estimating Energy Requirements

Energy requirements for maintenance, or basal metabolism, can be estimated for adult animals on the basis of their metabolic body size. Based on work by Brody (1945) and Kleiber (1957), the metabolic body size of an animal is defined as weight in kilograms to the 0.75 power ($W^{0.75}$). The daily energy requirements for maintenance may be represented by the general formula kcal $= 70W^{0.75}$. This relationship is an interspecies average that is relatively constant for mature animals but varies with species, age, and sex during the growth phase. The absolute energy maintenance requirement will vary downward as the metabolic mass of an animal is reduced, particularly in very aged or diet-restricted animals. This situation imposes potentially serious nutritional consequences for animals who are at risk for anorexic behavior, have lowered gastrointestinal tract function, or diminished ability to metabolize the required amount of nutrients.

Preliminary data collected during dietary restriction studies at the National Center for Toxicological Research (NCTR) indicate it may be appropriate to increase the coefficient of 70 (in the previously mentioned equation) to as high as 110 for some strains of mice in order to insure they receive adequate amounts of energy for maintenance (S. Lewis, Personal Communication, 1989). On this basis, it appears that the most appropriate relationship between metabolic body size and maintenance energy requirements has yet to be established experimentally for specific stocks and strains of rodents involved in dietary restriction studies. However, the energy requirements based on metabolic body size can be used as a guideline to determine the minimum amount of diet that should be offered daily to animals involved in dietary restriction studies. That is, each adult animal should be fed at least the amount of diet required to provide the energy indicated by the equation $kcal = 70W^{0.75}$.

Use of Historical Data to Estimate Requirements

Feed consumption data from previously conducted studies are frequently used as the basis for estimating nutrient requirements. It may also be possible to use data from previously conducted dietary restriction studies to estimate the minimum nutrient requirements for maintenance of the species of interest. However, these estimates must be made on an nutrient consumption basis rather than on total diet consumed, particularly when natural ingredient diets are involved. There is a considerable amount of variation in nutrient concentrations among production batches of this type of diet (Rao and Knapka 1987). Therefore, minimum maintenance nutrient or energy intake could be subject to error if based upon average daily intake rather than on a specific nutrient intake basis. Ideally, only data from dietary restriction studies in which diets are routinely assayed for nutrient concentrations should be used as a basis for estimating nutrient requirements for maintenance. This would minimize the inherent errors associated with these estimates.

Assaying each production batch of diet used in dietary restriction studies for as many nutrients as is practical should be a routine procedure in order to document the amount of each nutrient consumed. This documentation may be essential to verify that the animals involved in these studies received adequate nutrition for maintenance as well as for the valid interpretation of experimental data.

Summary

It is essential that all animals involved in dietary restriction studies consume adequate amounts of nutrients and energy to meet their requirements for maintenance. This concept is important as an animal welfare issue as well as being a sound experimental practice to insure that dietary stress does not compromise experimental results. In the absence of data from controlled experiments designed to identify the nutrient and energy requirements for the maintenance of laboratory animal species, indirect methods to ascertain these requirements must be used. The most critical requirement for the future success of dietary restriction studies is the support of programs designed to obtain this basic information regarding maintenance energy and nutrient requirements.

References

Brody S (1945) Bioenergetics and growth. Reinhold, New York

Kleiber M (1947) Body size and metabolic rate. Physiol Rev 27:511–541

Knapka JJ (1980) The history and issues regarding estimated nutrient requirements for laboratory animals. Lab Anim 9:52–56

Masoro EJ (1988) Food restriction in rodents: an evaluation of its role in the study of aging. J Gerontol 42:B59–B64

McCay CM, Crowell MF, Maynard LA (1935) The effect of retarded growth upon the length of the life span and upon the ultimate body size. J Nutr 10:63

National Research Council (1978a) Committee on Animal Nutrition, Agricultural Board, Nutrient requirements of laboratory animals, 3rd rev edn, No. 10. National Academy of Sciences, Washington, DC

National Research Council (1978b) Committee on Animal Nutrition, Agricultural Board, Nutrient requirements of nonhuman primates. National Academy of Sciences, Washington, DC

Rao GN, Knapka JJ (1987) Contaminant and nutrient concentrations of natural ingredient rat and mouse diet used in chemical toxicology studies. Fundam Appl Toxicol 9:329–338

Weindruch R, Walford RL (1988) The retardation of aging and disease by dietary restriction. Thomas, Springfield

CHAPTER 3

Significance of Dietary Contaminants in Diet Restriction Studies

G.N. Rao[1]

Introduction

Diet is one of the most important factors in animal experimentation. In addition to the ingredients and nutrients, chemical contaminants and extraneous materials may markedly influence the physiological processes of animals and so their response to experimental treatment (Newberne 1975). The purpose of this report is to summarize the sources and concentrations of various dietary contaminants in open formula and closed formula nonpurified and purified rat and mouse diets (American Institute of Nutrition 1977) manufactured since 1980. A summary of biological effects of dietary contaminants and their possible influence on animal experiments, including diet restriction studies, are also discussed.

Sources of Contaminants

Diets for laboratory animals can be classified as nonpurified open formula and closed formula diets and purified diets (American Institute of Nutrition 1977). Nonpurified diets are the most commonly used diets in animal experimentation. Ingredients of these diets include, but are not limited to, ground extruded corn, ground yellow shelled corn, corn gluten meal, ground hard winter wheat, wheat middlings, wheat germ meal, ground oats, soybean meal, dried beet pulp, dry molasses, oat hulls, alfalfa meal, fish meal, dried whey, dried skim milk, blood, meat, and bone meal, vegetable oils, animal fats, dicalcium phosphate, ground limestone, rock salt, and essential mineral and vitamin supplements. Some of the contaminants in nonpurified diets are due to accumulation of contaminants from the type of soil, region of origin, use of pesticide, and season. Others are due to accidental contamination during storage and manufacture or intentional additives such as preservatives (antioxidants). Possible sources of contaminants in nonpurified diets are given in Table 3.1. Type and concentrations of contami-

[1]National Toxicology Program, National Institute of Environmental Health Sciences, Research Triangle Park, NC 27709, USA

Table 3.1. Sources of contaminants in non-purified diets.

Ingredient	Contaminants[a]
Fish meal	Pb, Cd, Hg, As, Se, organochlorine insecticides such as chlordane, DDT and its metabolites, fluoride, nitrates, nitrites, and nitrosamines
Blood, meat, and bone meal	Pb, nitrates, nitrites, nitrosamines
Corn and wheat products	Mycotoxins (aflatoxins), pesticides (especially malathion), As, Se, and fluoride
Soybean products	Estrogenic activity[b]
Alfalfa meal	Pb, nitrates, and nitrites
Vegetable oils	Mycotoxins (aflatoxins) and preservatives
Animal fats	Pesticides and preservatives
Beet pulp and dry molasses	Se and As
Dicalcium phosphate and limestone	Pb, As, Se, Cd, and fluoride

[a]NTP (unpublished data) and Zeigler Bros. Inc., Gardners, PA, (personal communication).
[b]Not due to contaminants but to naturally present phytoestrogens.

nants will vary with the type and proportion of nonpurified ingredients. Diets with high proportion of fish meal such as NIH-07 (Knapka et al. 1974) may have higher concentrations of nitrosamines, heavy metals, nitrates, and antioxidants (Rao and Knapka 1987) than diets with no fish meal or a low proportion of fish meal. Similarly, diets high in certain ingredients such as alfalfa, corn, and wheat may have high concentrations of nitrates and selenium. Since the ingredients of purified diets such as casein, soy protein, ovalbumin, sucrose, and corn starch are either purified or semipurified (American Institute of Nutrition 1977) and better defined than the nonpurified ingredients, the concentrations of contaminants in purified diets are expected to be lower than in nonpurified diets.

Concentrations of Contaminants

Common contaminants of nonpurified diets include heavy metals such as arsenic, cadmium, mercury and lead, nitrates and nitrites, volatile nitrosamines such as N-nitrosodimethylamine (NDMA), N-nitrosopyrrolidine and N-nitrosomorpholine, and pesticides such as dichlorodiphenyltrichloroethane (DDT), heptachlor, methoxychlor, chlordane, dieldrin, lindane, malathion, and ethylene dibromide (EDB). Other contaminants of nonpurified diets include the essential as well as toxic nutrients or contaminants such as fluoride and selenium. Depending upon the ingredients one or more of the preservatives or antioxidants such as butylated hydroxyanisole (BHA), butylated hydroxytoluene (BHT), and ethoxyquin may also be present.

Concentrations of some of the above contaminants in the NIH-07 diet over a 10-year period are given in Table 3.2. The concentrations in Table 3.2 include

Table 3.2. Concentrations of selected contaminants in NIH-07 diet over a 10-year period.[a]

Contaminant	Concentration	NCTR[b] (1980–1983)		NTP[c] (1980–1983)		NTP[d] (1984–1989)	
		Mean (SD)	Range	Mean (SD)	Range	Mean (SD)	Range
As	ppm	0.368(0.181)	<0.02 – 0.630	0.46(0.19)	<0.05 – 1.06	0.44(0.29)	<0.05 – 1.07
Cd	ppm	0.137(0.057)	0.07 – 0.230	0.105(0.05)	<0.05 – 0.40	0.11(0.02)	<0.10 – 0.20
Pb	ppm	1.43 (1.25)	0.48 – 4.5	0.85(0.59)	0.33 – 3.37	0.39(0.25)	<0.05 – 1.32
Se	ppm	0.268(0.061)	0.16 – 0.36	0.28(0.073)	0.10 – 0.52	0.35(0.09)	<0.05 – 0.63
Total volatile nitrosamines	ppb	12.5 (28.9)	1.1 – 99	14.23(36.47)	0.8 – 273.2	7.88(4.02)	0.08 – 20.0
Methoxychlor	ppm	NA	NA	0.037(0.022)	<0.01 – 0.13	<0.05	–
Malathion	ppm	0.064(0.028)	<0.05 – 0.120	0.116(0.146)	<0.05 – 1.02	0.16(0.18)	<0.05 – 1.29
EDB	ppb	NA	NA	2.42(2.10)	<0.05 – 6.2[e]	NA	NA
Nitrate nitrogen	ppm	NA	NA	8.13(4.48)	<0.1 – 22.0	17.60(7.75)	0.30 – 41.0
Nitrite nitrogen	ppm	NA	NA	1.79(1.51)	<0.1 – 6.9	0.23(0.39)	0.02 – 2.60
BHA	ppm	NA	NA	3.97(4.30)	<0.2 – 20.0	2.42(2.59)	<0.1 – 22.0
BHT	ppm	NA	NA	2.80(2.19)	<1.0 – 13.0	1.60(1.05)	<0.1 – 5.0
Fluoride	ppm	NA	NA	36(4)	28 – 47[e]	NA	NA

NCTR, National Center for Toxicological Research; NTP, National Toxicology Program; EDB, ethylene dibromide; BHA, butylated hydroxyanisole; BHT, butylated hydroxytoluene; NA, not done or not available.

[a] Manufactured by Zeigler Bros. Inc., Gardners, PA.

[b] 11 lots (Oller et al. 1989).

[c] 94 lots (Rao and Knapka 1987).

[d] 131 lots (NTP, unpublished data).

[e] 11 lots.

Table 3.3. Concentrations of selected contaminants in NIH-31 diet from two sources and users.

Contaminant	Concentration	NCTR[a] (1984)		NTP[b] (1985–1989)	
		Mean (SD)	Range	Mean (SD)	Range
As	ppm	0.388(0.089)	<0.02 – 0.55	0.39(0.27)	<0.05 – 0.89
Cd	ppm	0.140(0.062)	0.09 – 0.30	<0.10	–
Pb	ppm	0.291(0.125)	0.05 – 0.56	0.41(0.33)	0.10 – 1.84
Se	ppm	0.579(0.120)	0.392 – 1.00	0.31(0.08)	0.12 – 0.42
Total volatile nitrosamines	ppb	5.94(2.45)	3.74 – 8.64	8.62(4.47)	2.0 – 17.4
Malathion	ppm	0.386(0.157)	0.133 – 0.501	0.22(0.23)	<0.05 – 0.92
BHA	ppm	<1.0	–	1.96(0.76)	<1.0 – 5.0
BHT	ppm	NA	NA	1.15(0.46)	<1.0 – 3.0

NCTR, National Center for Toxicological Research; NTP, National Toxicology Program; BHA, butylated hydroxyanisole; BHT, butylated hydroxytoluene; NA, not done or not available.
[a]From 4 to 33 lots manufactured by Teklad Diets, Madison, WI (Oller et al. 1989).
[b]28 lots manufactured by Zeigler Bros., Gardner, PA (NTP, unpublished data).

results for 4 years before (Rao and Knapka 1987) and 6 years after (National Toxicology Program, NTP, unpublished data) establishing limits for contaminant concentrations. Common contaminants of NIH-07 diet are lead, arsenic, cadmium fluoride, nitrate, nitrite, BHA, BHT, methoxyclor, malathion, EDB, and NDMA. The concentrations of lead and nitrosamines decreased after establishing limits for contaminant concentrations (Rao and Knapka 1987). Concentrations of selected contaminants in NIH-07 diet for a 4-year period (1980–1983) from two users (Oller et al. 1989; Rao and Knapka 1987) are also listed in Table 3.2. Marked differences in the types and concentrations of contaminants were not observed in the NIH-07 diet manufactured for the two users.

Concentrations of detectable levels of contaminants in the NIH-31 open formula autoclavable diet (Knapka 1983) from two sources and users (Oller et al. 1989; NTP, unpublished data) for the period 1984 to 1989 are presented in Table 3.3. Type and concentrations of contaminants in NIH-31 diet were similar to the NIH-07 diet (Table 3.2).

Concentrations of selected contaminants present at detectable levels in two types of closed formula (proprietary) diets are given in Table 3.4, (Oller et al. 1989; Ralston Purina, unpublished data). Since these are nonpurified diets, the types of contaminants were similar to those present in the NIH-07 and NIH-31 open formula diets (Tables 3.2,3.3).

Selected contaminant concentrations in the AIN-76 type of purified diets from two users over a 2 to 4 year period (1980–1983) are presented in Table 3.5. Concentrations of the selected contaminants were lower and close to the minimum detectable levels. However, the arsenic, cadmium, lead, and selenium concentrations in purified diets approached the concentrations observed in some lots/batches of nonpurified diets (Tables 3.2,3.3).

Table 3.4. Concentrations of selected contaminants in two types of closed formula nonpurified diets.[a]

Contaminant	Concentration	Purina 5010[b] (1979–1984)		Purina 5002[c] (1979–1984)	
		Mean (SD)	Range	Mean	Range
As	ppm	0.351(0.173)	<0.02 – 0.80	0.31	0.1 – 0.9
Cd	ppm	0.150(0.056)	<0.05 – 0.44	0.14	<0.1 – 0.5
Hg	ppm	0.025(0.021)	<0.02 – 0.17	<0.05	–
Pb	ppm	0.712(0.21)	<0.05 – 1.52	0.32	<0.1 – 1.14
Se	ppm	0.279(0.122)	<0.05 – 0.57	0.24	0.2 – 0.6
Nitrosamines	ppb	8.0(6.2)	<1.0 – 40.0	NA	NA
Malathion	ppm	0.182(0.146)	0.037 – 0.875	NA	<0.1 – 1.18

NA, not done or not available.
[a]Manufactured by Ralston Purina, St. Louis.
[b]202 to 310 Lots (Oller et al. 1989).
[c]Purina certified Rodent Chow over 300 lots. (Ralston Purina, personal communication.)

Concentrations of contaminants in diets fed to experimental animals should be as low as possible to prevent modification of biological responses to experimental treatment. Maximum allowable concentrations or action levels of selected contaminants in nonpurified diets used by the NTP (Rao and Knapka 1987), National Center for Toxological Research (NCTR, Oller et al. 1989), and Ralston Purina (1979) or recommended by the Environmental Protection Agency (EPA; 1979), and Institute of Laboratory Animal Resources (ILAR; 1976) are listed in Table 3.6.

Biological Effects of Dietary Contaminants

The spectrum of biological effects of the common dietary contaminants may range from alterations at molecular level to carcinogenicity and gross effects such as teratogenicity and mortality. Biological effects of contaminants with the

Table 3.5. Selected contaminant concentrations in purified diets from two users.[a]

Contaminant	Concentration	NCTR[b]			NTP[c]		
		Mean (SD)		Range	Mean (SD)		Range
As	ppm	0.148(0.110)		<0.02 – 0.35	<0.05		NA
Cd	ppm	0.057(0.023)		<0.005 – 0.080	0.14	NA	<0.10 – 0.30
Pb	ppm	0.256(0.211)		<0.05 – 0.80	0.123(0.075)		<0.05 – 0.240
Se	ppm	0.205(0.074)		0.08 – 0.40	0.204(.207)		0.06 – 0.70
Nitrosamines	ppb	1.5	NA	<1.0 – 2.5	<1.0		NA
Malathion	ppm	<0.05	NA	NA	<0.05		NA

NCTR, National Center for Toxicological Research; NTP, National Toxicology Program; NA, not applicable.
[a]Manufactured by Bioserv, Inc., Frenchtown, NJ.
[b]AIN-76 (American Institute of Nutrition 1977), 22 lots manufactured from 1980 to 1983.
[c]AIN-76M (Newberne et al. 1973), 7 lots manufactured from 1980 to 1982.

Table 3.6. Limits or action levels of common dietary contaminants recommended by various organizations.[a]

Contaminant	Concentration	NTP	NCTR	ILAR	EPA	Ralston Purina
As	ppm	0.60	1.00	0.25	1.0	1.0
Cd	ppm	0.25	0.25	0.05	0.26	0.5
Hg	ppm	0.05	0.10	0.05	0.10	0.2
Pb	ppm	1.0	1.5	1.0	1.5	1.5
Se	ppm	0.50	0.65	0.5	0.6	NS
Aflatoxins	ppb	5	5	1	5	10
Nitrosamines	ppb	25	20	NS	10	NS
Lindane	ppm	0.02	0.10	0.01	0.02	0.05
Heptachlor	ppm	0.02	0.02	0.01	0.02	0.05
Methoxychlor	ppm	0.05	NS	NS	NS	NS
DDT	ppm	0.02	0.10	0.05	0.10	0.15
Dieldrin	ppm	0.02	0.01	0.01	0.02	0.05
Malathion	ppm	0.5	5.0	0.5	2.5	0.5
BHA	ppm	10	50	NS	NS	NS

NTP, National Toxicology Program; NCTR, National Center for Toxicological Research; ILAR, Institute of Laboratory Animal Resources; EPA, Environmental Protection Agency; NS, not specified
[a]Only the contaminants detected in nonpurified diets were included. For other contaminants please refer to the appropriate publications given in the text.

minimum concentrations reported to cause some biological effects are listed in Table 3.7 (Greenman et al. 1980 and the references listed therein; Anderson et al. 1979). Even at the maximum recommended levels (Table 3.6), some of the contaminants may cause biological changes. They include the aflatoxins, especially B_1 known to cause cellular changes at 1 ppb (Wogan et al. 1974), and nitrosamines, which (NDMA) may increase some spontaneous tumors at a concentration of 10 ppb (Anderson et al. 1979). Major biological effects of dietary contaminants include increases or decreases of microsomal enzyme induction, decreases in cholinesterase activity, alterations in hematopoiesis, changes in behavior and impairment of learning, increase or decrease in metabolism of xenobiotics by liver, kidney, lung, and other organs, morphologic changes in liver and kidney cells, and impairment of liver and kidney function. Other effects include changes in tumor induction, increases or decreases in spontaneous tumor rates in liver, lung, and other organs and teratologic and developmental changes in offspring, impairment of growth, reproduction, and survival.

Significance of Dietary Contaminants in Diet Restriction Studies

Lower caloric intake and low protein consumption in diet restriction studies may influence the metabolism and disposition of chemicals by the liver and other organs. Some chemicals detoxified by the enzymes of liver and other organs may

Table 3.7. Biological effects of common dietary contaminants.

Contaminant	Concentration[a] (ppm)	Biological effects[b]
As[c]	5	Growth, reproduction, and mortality. Decreased tumor incidence and may be carcinogenic
Cd[c]	0.2	Renal vasculature, hypertension, reproduction, and mortality and tumors
Hg	2.0	Renal vasculature, hypertension, behavior, learning, growth, and mortality
Pb	5	Hematopoiesis, antibody formation, growth, and mortality
Se	2	Hepatic lesions, reproduction, growth and mortality. Decreased tumor incidence
Nitrosamines[d] (NDMA)	0.01	Hepatic lesions and tumors. Increased spontaneous tumors
Aflatoxin B_1	0.001	Liver hyperplasia and tumors
Heptachlor, DDT methoxychlor, Dieldrin, lindane	1	Microsamal enzymes, liver weight, brain lesions, liver tumors, lung tumors, growth, reproduction, and mortality
Malathion	100	Liver weight and cholinesterase
BHA[e]	5	Enzyme induction, decreased incidence of tumors and may be carcinogenic

[a]Minimum concentration with some biological effect.
[b]Greenman, et al. 1980 and the references cited therein; Pal et al. 1984.
[c]IARC 1973
[d]Anderson et al. 1979
[e]Wattenberg 1972; Ito et al. 1986

be more toxic (Shakman 1974) and chemicals converted to more toxic intermediates during the metabolism and disposition may be less toxic to diet restricted animals. These changes are possibly due to the decreased induction of xenobiotic metabolizing enzymes in liver and other organs due to lowered caloric and protein consumption. Diet restriction may also decrease nuclear binding, mutagenic activity, and tumor induction by some carcinogens requiring activation by the enzyme systems of liver (Pollard and Luckert 1985; Tannenbaum 1940). Since the animals in diet restriction studies consume less diet than the ad libitum fed animals, the contaminant consumption will also be lower. In general, the common dietary contaminants at concentrations present in controlled laboratory animal diets with limits in contaminant concentrations (Table 3.6) may not have different biological effects in diet restricted animals than in ad libitum fed animals. Furthermore, the common dietary contaminants (with the possible exception of aflatoxin B_1 and NDMA) below the selected maximum concentrations (Table 3.6) may not have detectable biological effects at the cellular, organ, or tissue and whole animal level but may have some effects at the molecular level, depending upon the sensitivity of the methods. However, toxic and carcinogenic responses of chemicals in diet restricted animals could be substantially different than in ad libitum fed animals. Caloric restriction generally decreases the initiation and promotion of tumors induced by chemicals requiring metabolic activation (Tannenbaum 1940; Pollard and Luckert 1985; Pariza 1986 and the references listed therein) such as methylazoxymethanol and dimethylbenzanthracene (DMBA).

In summary, common dietary contaminant concentration in controlled diets may not have different biological effects in diet restricted animals than in ad libitum fed animals. Contaminant concentrations in diets fed to experimental animals should be as low a concentration as possible to prevent modification of biological responses in experimental treatment. This can be accomplished by establishing limits on contaminant concentrations and quality control of ingredients selected for the diets. However, if the chemical under investigation is a dietary contaminant or the dietary contaminant present in the controlled nonpurified diets modifies the biological responses of experimental treatment, it may be appropriate to use purified diets for such studies.

References

American Institute of Nutrition (AIN) (1977) Report of the AIN Ad Hoc Committee on Standards for Nutritional Studies. J Nutr 107:1340–1348

Anderson LM, Priest LJ, Budinger JM (1979) Lung tumorigenesis in mice after chronic exposure in early life to a low dose of dimethylnitrosamine. J Natl Cancer Inst 62: 1553–1555

Environmental Protection Agency (EPA) (1979) Proposed health effects test standards for toxic substances control act test rules. Good laboratory practice standards for health effects. Fed Regis 44:27334–27375

Greenman DL, Oller WL, Littlefield NA, Nelson CJ (1980) Commercial laboratory animal diets: toxicant and nutrient variability. J Toxicol Environ Health 6:235–246

International Agency for Research on Cancer IARC (1973) IARC monographs on the evaluation of the carcinogenic risk of chemicals to man. Some inorganic and organic compounds, vol 2, IARC, Lyon

ILAR Committee on Long-Term Holding of Laboratory Rodents (1976) Long-term holding of laboratory rodents. ILAR News 19:L1–L25

Ito N, Fukushima S, Tamano S, Hirose M, Hagiwara A (1986) Dose response in butylated hydroxyanisole induction of forestomach carcinogenesis in F344 rat. J Natl Cancer Inst 77:1261–1265

Knapka JJ (1983) Nutrition. In: Foster LH, Small JD, Fox JG (eds) Normative biology, immunology and husbandry. The mouse in biomedical research, vol 3. Academic, New York, pp 51–67

Knapka JJ, Smith PK, Judge FJ (1974) Effect of open and closed formula rations on the performance of three strains of laboratory mice. Lab Anim Sci 24:480–487

Newberne PM (1975) Influence of pharmacological experiments of chemicals and other factors in diets of laboratory animals. Fed Proc 34:209–218

Newberne PM, Glasser O, Friedman J, Stallings B (1973) Safety evaluation of fish protein concentrate over five generations of rats. Toxicol Appl Pharmacol 24:133–141

Oller WL, Kendall DC, Greenman DL (1989) Variability of selected nutrients and contaminants monitored in rodent diets: 6-year study. J Toxicol Environ Health 27:47–56

Pal BC, Ross RH, Milman HA (1984) Nutritional requirements and contaminant analysis of laboratory animal feeds. Report no. EPA 560/68–3–005. National Technical Information Service, Springfield, VA

Pariza MW (1986) Caloric restriction, *ad libitum* feeding and cancer. PSEBM 183: 293–298

Pollard M, Luckert EH (1985) Tumorigenic effects of direct and indirectacting chemical carcinogens in rats on a restricted diet. J Natl Cancer Inst 74:1347–1349

Purina Certified Lab Chows Specifications (1979) Certified rodent chow no. 5002. Ralston Purina, St. Louis

Rao GN, Knapka JJ (1987) Contaminant and nutrient concentrations of natural ingredient rat and mouse diet used in chemical toxicology studies. Fundam Appl Toxicol 9:329–338

Shakman RA (1974) Nutritional influences on the toxicity of environmental pollutants. Arch Environ Health 28:105–113

Tannenbaum A (1940) The initiation and growth of tumors. I. Affect of underfeeding. Am J Cancer 23:803–807

Wattenberg LW (1972) Inhibition of carcinogenic and toxic effects of polycyclic hydrocarbons by phenolic antioxidants and ethoxyquin. J Natl Cancer Inst 48:1425–1430

Wogan GN, Paglialunga S, Newberne PM (1974) Carcinogenic effects of low dietary levels of aflatoxin B_1 in rats. Food Cosmet Toxicol 12:681–685

Part II
Effects of Dietary Restriction on Toxicological Endpoints

CHAPTER 4

Dietary Restriction and Toxicological Endpoints: An Historical Overview[*]

W.T. Allaben[1], M.W. Chou[1], and R.A. Pegram[2]

Introduction

The United States and other developed countries are increasingly having to adapt to a population that has a significant percentage of individuals who are 75 years old or older (Figure 4.1). Diseases such as heart disease, cerebrovascular disease, and cancer are known to increase significantly after the age of 60 (Fig. 4.2). As the population of senior citizens increases, with a concomitant increase in age-associated disease, it can be anticipated that there will be an added burden on health care systems in the USA and other developed countries. Recently, there has been much attention paid to the modulation of the onset of disease, particularly heart disease and cancer, through diet.

   Nutritional modification of cancer has become an active area of research in the past decade due, in part, to the growing awareness that dietary excesses, deficiencies, or imbalances can play a major role in the etiology or modulation of cancer. Moreover, the linking of high fat diets to cancer, and particularly breast cancer risk, has also fostered an intense interest in the relationship of diet and cancer. Furthermore, the emphasis placed on fiber in the diet, as an intervention to prevent colon cancer, has also pointed to the importance of diet. Researchers have also "rediscovered" what is apparently the most effective preventative dietary modification: reducing caloric intake, or dietary restriction.

   Although the influence of diet on the initiation and expression of neoplastic disease has received much attention recently, nutritional effects on longevity, nonneoplastic disease, and on spontaneous and chemically induced cancer have been recognized for some time (McCay et al. 1935; Robertson et al. 1934; McCay et al 1939; Watson and Mellanby 1939; Tannenbaum 1940a; Tannenbaum and Silverstone 1953a,b). The earliest study showing the effect of underfeeding

*This work was supported in part by an appointment to the Postgraduate Research Program at the National Center for Toxicological Research administered by Oak Ridge Associated Universities through an interagency agreement between the U.S. Department of Energy and the U.S. Food and Drug Administration.
[1]National Center for Toxicological Research, Jefferson, AR 72079, USA
[2]Current address: US Environmental Protection Agency, Research Triangle Park, NC 27711

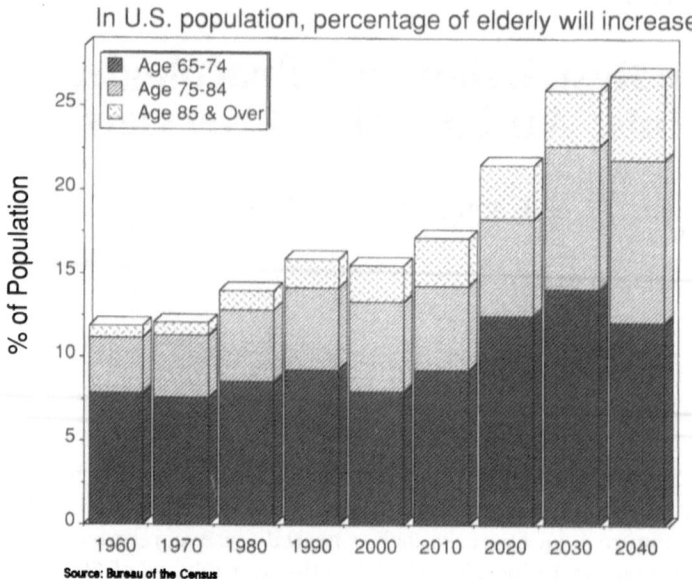

Figure 4.1. Projection of increases in the elderly population in the U.S. through the year 2040, according to the U.S. Census Bureau

on cancer was that reported by Moreschi in 1909 (Moreschi 1909). He demonstrated that transplanted tumor growth (sarcomas) was slowed by reduced feeding regimens. Rous, in 1914, underfed mice which had all but a small section of spontaneous mammary gland tumors removed and observed that the recurrence of those tumors was reduced by over 50% as compared to an ad libitum fed group (Rous 1914). Bischoff and Long (1938) also demonstrated that transplanted tumor growth is slowed by feeding diets with lowered carbohydrate or fat content. McCay et al. (1935, 1939) and Robertson et al. (1934) published results demonstrating that moderate caloric restriction retards the onset of senescence and extends longevity. While McCay's experiments did not determine the effect of caloric restriction on tumor development, they did show, for the first time, that life span could be extended (see Fig. 4.3).

In a series of experiments in the 1940's, Tannenbaum was able to demonstrate the effects of underfeeding (reduced diet) and caloric restriction (nutrient modification) on the development of spontaneous and chemically induced tumors. He found that DBA mice which were underfed by one half to one third the group given free rations had only a 7% incidence of spontaneous mammary gland tumors (compared to 30% in the free access group), and that benzo(a)pyrene-induced skin tumors in ABC mice were significantly reduced in the restricted groups (Tannenbaum 1940a,b; Table 4.1). He also demonstrated that caloric restriction significantly reduced chemically induced skin tumors as well as spontaneous mammary gland and lung tumors in four strains of mice (Tannenbaum 1942a,b). He later reported that when protein, vitamins, minerals (essentials),

Deaths per 100,000 population, 1982

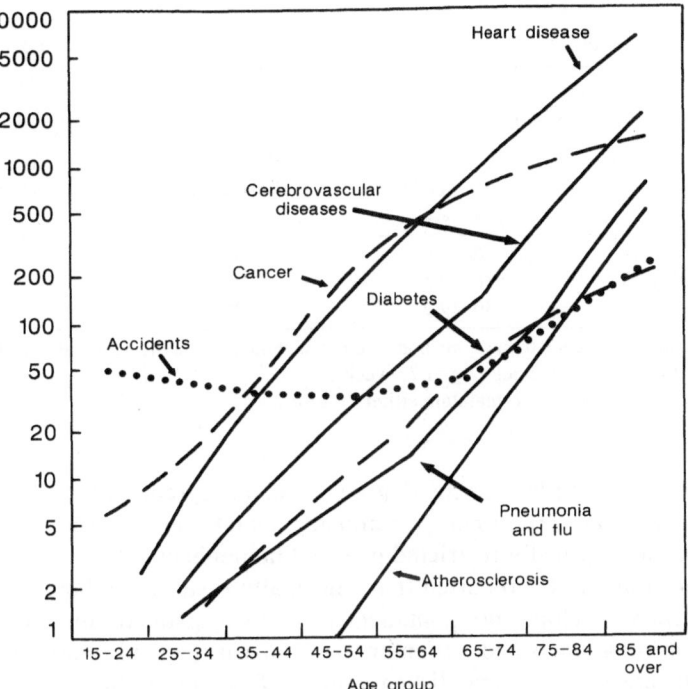

Figure 4.2. National Center for Health Statistics (per 100,000 population in 1982) shows how disease, particularly heart disease and cancer, increases with age

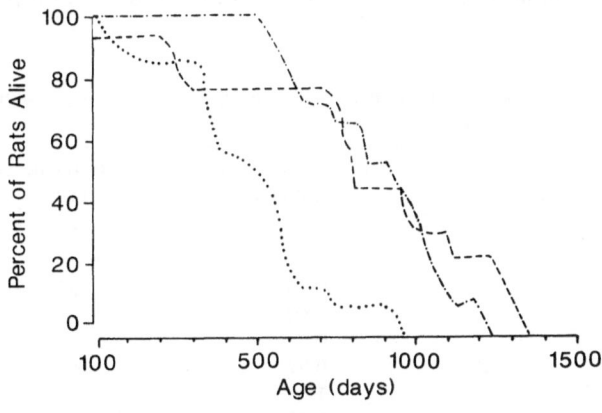

From: McCay, et al, J. Nutr. 10:63–79, 1935

Figure 4.3. Summation of data from McCay's (1935) research article. Group 1 (. . . .) animals were fed ad libitum from weaning; Group 2 (– – – –) animals were fed restricted diets from weaning; Group 3 (· – · – ·) animals were fed ad libitum for 2 weeks after weaning, then restricted diets

Table 4.1. Effects of underfeeding on initiation and growth of benzopyrene-induced skin tumors in ABC mice (from Tannenbaum 1940a).

Group A[a]		Tumor incidence at 77 weeks (%)
1. Control		44
2. Underfed (1/3–1/2 ad libitum)		18

Group B[a]	Body weight at start	Body weight after 32 weeks (average)	Tumor incidence after 32 weeks (%)
1. Control	36–39 g	42 g	14
2. Restricted (1/3)	36–39 g	25 g	4
3. Restricted (1/2)	36–39 g	20 g	0

[a]Restricted animals were fed to maintain body weight of about 20 g; control animals weighed an average of 26 g at the start and about 42 g at 77 weeks.
[b]Restricted animals were fed to maintain either 20 g or 25 g.

and fat were reduced by varying degrees, benzo(a)pyrene-induced skin tumors and, in a separate experiment, mammary gland tumors were significantly reduced in the calorically restricted groups (Tannenbaum 1945a,b; Table 4.2).

Visscher et al. (1942) reported that calorically restricted (about 67% ad libitum) C3H mice developed no spontaneous mammary gland tumors, while ad libitum fed control animals had a spontaneous mammary gland tumor incidence of 70%. White and coworkers published a series of articles in the mid 1940's which described studies that used caloric restriction (or protein deprivation) to show effects on both spontaneous and chemically induced tumors. When comparing low-cystine diets to normal rodent diets, both spontaneous and diethylstilbesterol-induced mammary gland tumors were inhibited in low-cystine (White and

Table 4.2. Diet and spontaneous mammary tumors in female C3H mice (from Tannenbaum 1945a).

Calories (kcal/day)	Fat (%)	Protein/minerals/vitamins (essentials)	Tumor incidence (%)
10	2	1.00	27
10	2	0.86	40
10	18	0.86	67
9.2	2	1.00	7
9.2	2	0.77	14
9.2	19	0.77	60
8.5	2	1.00	0
8.5	2	0.73	0
8.5	18	0.73	17
7.8	2	1.00	0
7.8	2	1.00	0
7.8	18	0.68	4

Table 4.3. Influence of fat and calories on methylchol-anthrene-induced skin tumors (from Lavik and Baumann 1943).

Diet regimen	Tumor incidence (%)
Low-fat, low-calorie	0
High-fat, low-calorie	28
Low-fat, high-calorie	54
High-fat, high-calorie	66

White 1943; White and Andervont 1943) or low-lysine (White and White 1944) fed C3H mice. Later studies, which reduced calories by about one half, found that spontaneous mammary gland tumors were reduced by 84% in the calorically restricted C3H mice (White et al. 1944). Larsen and Heston (1945) demonstrated similar effects but for spontaneous pulmonary tumors, in strain A mice. Saxton et al. (1944) reported that in a 2-year study, spontaneous leukemia lesions in AK mice was reduced by 85% in calorically restricted test groups compared with ad libitum controls. Methylchloanthrene-induced skin tumors were completely inhibited in mice fed a low-fat, low-calorie diet whereas a low-fat, high-calorie diet resulted in a 54% incidence of tumors (Lavik and Baumann 1943; Table 4.3). In an important observation Rusch et al. (1945) reported that skin tumors induced by physical agents (ultraviolet radiation) were reduced by greater than 90% in calorically restricted animals. Boutwell et al. (1948), using semipurified diets, showed that in carbohydrate-restricted animals, there was an 80% reduction in the incidence of chemically induced skin tumors as compared to a standard carbohydrate diet. Dietary restriction also had a profound effect on the growth and size of several types of transplanted mammary gland tumors in C3H mice. In all cases, the tumors were either reduced in size or regressed after 2 weeks (Tarnowski and Stock 1955).

In the 1960's Ross provided further evidence indicating that life span and tumor expression were dependent on calorie intake (and protein content) (Ross 1961, 1965). Low-protein diets retarded morbidity due to malignant lymphomas, and pancreatic and lung tumors were rarely found, whereas they routinely appeared in the high-protein diet groups (Ross 1965). Ross and Bras (1971) later published the results of a comprehensive study on caloric restriction. Rats were fed ad libitum, 60%–70% ad libitum for their life span, or 60%–70% ad libitum for 7 weeks past weaning. Both restricted groups exhibited significant reductions in tumor formation and when divided into lighter/heavier subgroups for each treatment regimen, the lighter groups had the lowest tumor incidences even within groups (Table 4.4). More recent observations of Ross et al. (1982) find animals more at risk of expressing spontaneous tumors if they have higher body weights, eat more food shortly after weaning, and consume more protein in their diets. Hence, according to Ross et al. (1982), variables which reduce the probability of developing neoplasms in rodents are:

Table 4.4. Body weight relationship and tumor incidence (from Ross and Bras 1971).

Groups	Tumor-bearing rats (%)	Mortality ratio (%)
Ad libitum (heavier)	37	
Ad libitum (lighter)	23	100
Restricted (heavier)	25	
Restricted (lighter)	12	33
Restricted/ad libitum (heavier)[a]	38	
Restricted/ad libitum (lighter)[a]	19	82

[a]Days 21–70 restricted, then ad libitum. Animal age varied according to diet regimen. On the average, ad libitum animals lived to about 650 days, restricted to about 1300 days, and restricted/ad libitum to about 750 days.

1. Low absolute protein intake shortly after weaning
2. Low degree of food utilization (efficiency) at time of puberty
3. Low level of protein intake relative to body weight during puberty and early adult life
4. Low level of food intake
5. Slow early postnatal growth rate (rate of gain)

Ross et al. (1983) also concluded from retrospective study analysis that diet and body weight-related factors during the postweaning and early adulthood periods interact to influence the outcome of spontaneous tumor formation. Feed utilization efficiency (ease of conversion) and protein intake were most often associated with body weight gain and susceptibility to neoplasia. Heavier rats consistently had a great number of tumors than lighter rats. For every 100-g increment body weight after maturation there was a corresponding 10% increase in the spontaneous tumor incidence. Haseman (1983), in examining a very large National Toxicology Program (NTP) data base, has also reported a clear association between the incidence of mammary gland fibroadenomas and body weight in females Fischer 344 rats, which strengthens the observations of Ross and coworkers.

Also in the 1960's, Berg and Simms (1961) showed that diet restriction (33% or 40%) significantly reduced the incidence of spontaneous tumors in both male and female rats. White (1961) found that feed restriction decreased tumors during induction by both chemical and physical agents and that feed restriction was effective in slowing or preventing the growth of spontaneous and transplanted tumors.

In addition to the numerous studies that have found caloric restriction to have a protective effect for neoplastic endpoints, nonneoplastic disease incidence (nephropathy, myocardial degeneration, and bile duct hyperplasia) has also been reported to be reduced or the time to onset prolonged when rats and mice were fed diets restricted in protein or that were calorically restricted (Saxton and Kimball 1941; Boutwell et al. 1948; Berg and Simms 1961; Bras and Ross 1964; Fernandes et al. 1974; Yu et al. 1982; Table 4.5).

Table 4.5. Effect of food intake on survival and incidence of glomerulonephritis, polyarteritis, and myocardial degeneration (from Berg and Simms 1961).

		Ad libitum	67% Ad libitum	54% Ad libitum
Males				
% Deaths	0–800 days	52	13	19
% Glomerulonephritis 0–800 days		97	0	7
	800 + days	100	36	0
% Polyarteritis	0–800 days	83	0	0
	800 + days	63	17	3
% Myocardial degeneration	0–800 days	69	17	0
	800 + days	96	29	3
Females				
% Deaths	0–800 days	6	4	5
% Glomerulonephritis	0–800 days	69	0	0
	800 + days			

The extent to which caloric restriction can modify tumor expression varies, however, with the experimental model. Fat enrichment of experimental diets is known to increase the incidence of cancer in most, but not all, animal model systems. The incidences of induced skin tumors and spontaneous mammary gland tumors are increased by about 25% in mice fed high-fat diets, whereas the development of spontaneous benign liver tumors was only slightly increased in animals fed the same diet (Tannenbaum 1942b; Silverstone and Tannenbaum 1951). No increase in lung adenoma or spontaneous leukemia lesions, however, was noted in mice receiving high-fat diets (Lawson and Kirschbaum 1944). Kritchevsky et al. (1984) have reported that calorically restricted Sprague-Dawley rats (40% restricted) were completely refractive to 7,12-dimethylbenz(a)anthracene (DMBA) induced carcinogenesis even though such diets had a high fat content. Kritchevsky et al. (1986) also reported that calories, rather than fat, was a more stringent determinant of DMBA-induced mammary gland tumor expression (Table 4.6). Other more recent work continues to show the impact of calories on tumor growth during the promotional phase of carcinogenesis (Kritchevsky et al. 1989). Another important finding regarding the impact of caloric restriction in rodents is the observation of Weindruch and Walford (1982) that caloric restriction started as late as 1 year of age was still impressively effective in decreasing spontaneous tumors and extending maximum survival times.

Clinton et al. (1984) have shown that ad libitum fed animals had a higher incidence of 7,12-DMBA induced tumors than those on a reduced intake diet; those animals on ad libitum diets had an increase in tumors proportional with fat content. In another study of chemically induced cancer, rats fed restricted chow diets were found to have a reduced incidence of azoserine-induced pancreatic neoplasms compared to ad libitum fed animals (Roebuck et al. 1981).

Table 4.6. Effects of fat level and caloric restriction on DMBA-induced mammary tumors in rats (from Kritchevsky et al. 1986).

Diet	Tumor incidence (%)
Ad libitum	
5% Corn oil	65
15% Corn oil	85
20% Corn oil	80
Restricted (25% ad libitum)	
20% Corn oil	60
26.7% Corn oil	30

DMBA, 7, 12-dimethylbenz(a)anthracene.

Boissonneault el al. (1986) demonstrated that animals on high-fat restricted diets (consuming about 20% fewer calories) had a much lower incidence of DMBA-induced mammary gland tumors than animals on low-fat ad libitum diets. While feed restriction reduced the incidence of methyl-azo-xymethanol-induced colon tumors, it had no effect on the incidence of colon tumors induced by methyl-nitrosourea (Pollard and Luckert 1985), i.e., an indirect-acting vs. a direct-acting carcinogen. Such results suggest an effect on metabolic enzyme systems by calorically restricted feeding regimes (Table 4.7).

Experiments designed to alter body weights independent of caloric consumption have also been performed. The feeing of dinitrophenol (a metabolic uncoupler) resulted in a 10% increase in feed intake, a decrease in body weight, and a decrease in spontaneous mammary gland and lung tumors in mice; 7,12-DMBA skin tumors were unaffected, however (Tannenbaum and Silverstone 1949). Very similar results were obtained when the investigators fed test animals thyroid extract or kept the animals at 10°C (Tannenbaum and Silverstone 1949). Exercise

Table 4.7. Effect of caloric restriction on chemically induced intestinal tumors induced by direct- and indirect-acting carcinogens (from Pollard and Luckert 1985).

Group	Carcinogen	Diet	Tumor incidence	(Total no. of tumors)
1	MAM	Ad libitum	9/10	(32)
2	MAM	Ad libitum	9/10	(28)
3	MAM	Restricted[a]	2/10	(2)
4	MNU	Ad libitum	9/9	(34)
5	MNU	Ad libitum	6/8	(15)
6	MNU	Restricted[a]	9/10	(28)
7	MAM + MNU	Ad libitum	19/19	(26)
8	MAM + MNU	Restricted[a]	15/19	(9)

[a]75% of ad libitum diet.
MAM, methylazoxymethanol; MAN.
MNU, N-methylnitrosourea.

Table 4.8. Influence of dietary fat, caloric restriction, and voluntary exercise on NMU-induced tumorigenesis (from Cohen et al. 1988).

Group/treatment	Fat (%)	Total no. of tumors
Ad libitum (sedentary)	5	38
Ad libitum (sedentary)	10	50
Ad libitum (sedentary)	20	59
Restricted (sedentary)	13	5
Ad libitum (active)	26	38

NMU, n-nitrosomethylurea

has also been shown to be an effective way to reduce body weights of laboratory rodents, thus reducing the incidence of spontaneous tumors or tumor growth (Rusch and Kline 1944; Stern 1984; Applegate et al. 1980; Cohen et al. 1988; Table 4.8).

The identification of mechanism(s) by which caloric restriction reduces tumor incidence, either spontaneous or induced, is in its hypothesis stage. Roe (1981) has proposed that caloric restriction affects the hormonal status of the animal and that ad libitum fed laboratory animals develop significant hormonal imbalances after about 6 months of feeding. The serum levels of prolactin, which acts as a classical promoter in rodent mammary tumor systems, are significantly reduced in calorically restricted animals (Roe 1981; Sarkar et al. 1982). Kritchevsky and Klurfeld (1987) have also demonstrated a reduction in plasma insulin levels with a gradation of caloric restriction (0–40%) (Table 4.9). Such observations support the changes in intermediary metabolism reported by Feuers et al. (this volume). Other peptide and steroid hormone levels may also be affected. The estrogens, in particular 17B-estradiol and estrone, are known to promote mammary gland tumorigenesis in a variety of strains of rats and mice. Gropper and Shimken (1967) found caloric restriction plus ovarectomy reduced the incidence of 3-methylcholanthrene-induced mammary gland tumors more than caloric restriction alone. Moreover, treatments that increased circulating prolactin and estrogen levels in rats promoted the development of DMBA-induced mammary tumorigenesis despite food restriction (Sylvester et al. 1981). On the other hand,

Table 4.9. Mammary gland DMBA-induced tumor incidence and plasma insulin as influenced by caloric restriction (from Kritchevsky and Klurfeld 1987).

Regimen	Tumor incidence (%)	Plasma insulin (μU/ml)
Ad libitum	60	122 ± 16
10% Restricted	60	191 ± 10
20% Restricted	60	109 ± 10
30% Restricted	40	42 ± 5
40% Restricted	5	41 ± 8

DMBA, 7, 12-dimethylbenz(a)anthracene.

increased estrogen levels can markedly reduce the detrimental effects of aflatoxin (AFB_1), including liver tumor induction in rats (Newberne and Rogers 1986; Schwartz 1974). Furthermore, caloric restriction has been shown to delay the onset of puberty, decrease serum FSH and progesterone levels, and increase serum 17B-estradiol levels in rats (Holehan and Merry 1985).

While hormonal imbalances may enhance promotion and contribute to the incidence of spontaneous tumor expression, modulation of chemically induced cancers through caloric restriction must certainly involve the metabolic capabilities of the animal and, therefore, the initiation process. As was referenced earlier, caloric restriction effectively reduced the incidence of methyl-azo-methanol-induced (indirect-acting) colon tumors, but had no effect on methylnitrosourea-induced (direct-acting) colon tumors (Pollard and Luckert 1985). The fact that inhibition of mammary tumorigenesis was noted in rats underfed only 1 week before and 1 week after DMBA administration (Sylvester et al. 1982) also supports this concept. Laboratory animal diets consist of carbohydrates, proteins, lipids, water-soluble and fat-soluble vitamins, and minerals. The interactions of any one food group are continuous and complicated. Dietary manipulation of laboratory formulations or a decrease/increase in the amount of those formulations can alter the way laboratory animals respond to xenobiotics as well as endogenous levels of circulating hormones or other important homeostatic maintenance chemicals. Such changes may account for the promotional effects of, for example, high-fat diets. Nutritional factors are known to affect the mixed-function oxidases, particularly the various P-450 isozymes that are responsible for oxidation steps in xenobiotic metabolism (Pasco et al. 1983; Hendrich and Bjeldanes 1983; Wattenberg et al. 1976; Gelboin 1980; Henning et al. 1983).

Until recently there has been a paucity of data dealing with the effects of feed restriction on xenobiotic metabolism. Sachan and Das (1982) and Sachan (1982) reported that drug-metabolizing enzyme activities were greater in 45% and 50% feed-restricted rats than in ad libitum-fed rats. A 60% decrease in hexobarbital sleeping time was also noted in rats 45% restricted for 28 days (Sachan 1982). Also, hepatic metabolism of volatile hydrocarbons was enhanced in rats following either food deprivation or restricted carbohydrate intake (Sato and Nakajima 1985). Moreover, Hashmi et al. (1986) reported that compared to ad libitum-fed control rats, hepatic cytochrome P-450 levels were greater in food-restricted male rats, while they were lower in food-restricted female rats. The activities of the drug-metabolizing enzymes aminopyrene N-demethylase and acetanilide hydroxylase were higher in food-restricted males, whereas these activities in food-restricted females were lower than in respective groups fed ad libitum. The sexual dimorphism of hepatic mixed-function oxidases, in response to caloric restriction, has been confirmed most recently by Leakey and coworkers (this volume). Koizumi et al. (1987) found that hepatic 7-ethoxycoumarin-O-deethylase activity was increased in female mice following long-term dietary restriction, while increases in aryl hydrocarbon hydroxylase and glutathione s-transferase were associated with short-term feed restriction. Moreover, enzymes which can reduce oxidative damage, such as catalase, were increased in restricted animals

(Koizumi et al. 1987). It will be important to determine how these changes in xenobiotic metabolism in calorically restricted animals can affect other specific toxicological endpoints. For example, Pegram et al. (1989) demonstrated that hepatic nuclear binding (DNA) by AFB_1 was reduced by more than half in restricted rats compared to ad libitum control rats. More recent studies have demonstrated the critical role that caloric restriction has in modifying metabolic activation/detoxication pathways affecting both initiation and promotional stages of carcinogenesis, thus reducing the risk of the expression of spontaneous or chemically induced toxic endpoints (see chapters in this volume by Chou et al., Lee et al., Feuers et al., and Leakey et al.).

While moderate caloric restriction has proven to protect against age-related and chemically induced toxic endpoints in laboratory animals, there are no well-documented records showing this effect may be extended to humans. Diet modifications which may alter human health risk has received much attention, as previously stated. In recent times, during World War II, a large population (Europeans) was undernourished, malnourished, and/or starving. During that time, there were no epidemics where there was mass undernourishment (Holland and Greece) (Keys 1948). Moreover, typhus mortality was worse among German guards than among starving Russian prisoners at Hammerstein (Markowski 1945). Evidence from World War II suggests that undernourishment did not lower resistance or heighten susceptibility to infectious diseases in general. Furthermore, mortality statistics during caloric undernutrition do not indicate any increase in cancer (from Greece and the Netherlands during World War II) (Keys 1948), and in Greece, mortality due to cancer may have decreased (Valaoris 1946). Such observations cannot, of course, be relied upon as sound scientific evidence compared to today's standards.

As reported by Kagawa (1978), there are more Okinawans aged 100 and over than in the rest of Japan. He attributed this to the diet of the Okinawan people: low-carbohydrate, high-fiber (fruit and grain), and quality protein (fish) diets. It is estimated that the total calories consumed are 15%–30% lower than on mainland Japan. While we may speculate that observations in laboratory animals (used as human surrogates by the biomedical community) may also translate to the human, that translation should follow sound scientific principles.

Conclusion

Dietary restriction has proven to be the only intervention which extends the maximum achievable life span and reduces spontaneous and chemically induced toxic endpoints. The biochemical and physiological mechanisms responsible for these life-extending and protective effects are now being identified. Through such research, interventions may be identified which can prolong life and reduce or prevent various diseases in humans.

Acknowledgments. The authors wish to thank Ms. Beverly Montgomery for her clerical assistance in the preparation of this manuscript.

References

Applegate EA, Upton DE, Stern JS (1980) Exercise and detraining: effect on food intake, adiposity, and lipogenesis in Osborne-Mendel rats made obese by a high fat diet. J Nutr 118:447–459

Berg BN, Simms HS (1961) Nutrition and longevity in the rat. III. Food restriction beyond 800 days. J Nutr 74:23–32

Bischoff F, Long ML (1928) The influence of calories per se upon the growth of sarcoma. Am J Cancer 32:418–421

Boissonneault GA, Elson CE, Pariza MW (1986) Net energy effects of dietary fat on chemically-induced mammary carcinogenesis in F344 rats. J Natl Cancer Inst 76: 335–338

Boutwell RK, Brush MK, Rusch HP (1948) Some physiological effects associated with chronic caloric restriction. Am J Physiol 154:517–524

Bras G, Ross MH (1964) Kidney disease and nutrition in the rat. Toxicol Appl Pharmacol 6:247–262

Clinton SK, Inrey PB, Alster JM, Simon J, Truex CR, Visek WJ (1984) The combined effects of dietary protein and fat on 7,12-dimethylbenz(a)-anthracene-induced breast cancer in rats. J Nutr 114:1213–1223

Cohen LA, Choi KW, Wang CX (1988) Influence of dietary fat, caloric restriction, and voluntary exercise on n-nitrosomethylurea-induced mammary tumorigenesis in rats. Cancer Res 48:4276–4283

Fernandes G, Yunis EJ, Good RA (1974) Influence of diet on survival of mice. Proc Natl Acad Sci USA 73:1279–1283

Gelboin HV (1980) Benzo(a)pyrene metabolism, activation and carcinogenesis: role and regulation of mixed-function oxidases and related enzymes. Physiol Rev 60:1107–1166

Gropper L, Shimkin MB (1967) Combination therapy of 3-methylcholanthrene-induced mammary carcinoma in rats: effect of chemotherapy, ovariectomy and food restriction. Cancer Res 27:26–32

Haseman JH(1983) Patterns of tumor incidence in two-year cancer carcinogen feeding studies in Fischer 344 rats. Fundam Appl Toxicol 3:1–9

Hashmi RS, Siddiqui AM, Kachole MS, Pawar SS (1986) Alterations in hepatic microsomal mixed function oxidase system during different levels of food restriction in adult male and female rats. J Nutr 116:682–688

Hendrich S, Bjeldanes LF (1983) Effects of dietary cabbage, Brussels sprouts, *Illicium verum*, *Schizandra chinensis*, and alfalfa on the benzo(a)pyrene metabolic system in mouse liver. Food Chem Toxicol 21:479–486

Henning EE, Demkowicz-Dobrazanski KK, Sawicki JT, Mojska H, Kujawa M (1983) Effect of dietary butylated hydroxyanisole on the mouse hepatic monooxygenase system of nuclear and microsomal fractions. Carcinogenesis 4:1243–1246

Holehan AM, Merry BJ (1985) The control of puberty in the dietary restricted female rat. Mech Ageing Dev 32:179–191

Kagawa Y (1978) Impact of westernization on the nutrition of Japanese: changes in physique, cancer, longevity and centenarians. Prev Med 7:205–217

Keys A(1948) Council on food and nutrition: caloric undernutrition and starvation, with notes on protein deficiency. JAMA 138:500–511

Koizumi A, Weindruch R, Walford RL (1987) Influence of dietary restriction and age on liver enzyme activities and lipid peroxidation in mice. J Nutr 117:361–367

Kritchevsky D, Klurfeld DM (1987) Caloric effects in experimental mammary tumorigenesis. Am J Clin Nutr 45:236–242

Kritchevsky D, Weber MM, Klurfeld DM (1984) Dietary fat versus caloric content in initiation and promotion of 7,12-dimethylbenz(a)anthracene-induced mammary tumorigenesis in rats. Cancer Res 44:3174–3177

Kritchevsky D, Weber MM, Buck CL, Klurfeld DM (1986). Calories, fat and cancer. Lipids 21:272–274

Kritchevsky D, Welch CB, Klurfeld DM (1989) Response to mammary tumors to caloric restriction for different time periods during the promotional phase. Nutr Cancer 12:259–269

Larsen CD, Heston WE (1945) Effects of cystine and calorie restriction on the incidence of spontaneous pulmonary tumors in strain A mice. J Natl Cancer Inst 6:31–40

Lavik PS, Baumann CA (1943) Further studies on tumor promoting action of fat. Cancer Res 3:749–756

Lawson FD, Kirschbaum B (1944) Dietary fat with reference to the spontaneous appearance and induction of leukemia in mice. Proc Soc Exp Biol Med 56:6–7

Markowski B (1945) Some experiences of a medical prisoner of war. Br Med J 2:361–363

McCay CM, Crowell ME, Maynard LA (1935) The effect of retarded growth upon the length of lifespan and ultimate size. J Nutr 10:63–79

McCay CM, Ellis GH, Barnes LL, Smith CAH, Sperling G (1939) Chemical and pathological changes in aging and after retarded growth. J Nutr 18:15–25

Moreschi C (1909) Beziehungen zwischen Ernahrung und Tumorwachstum. Z Immunitatsforsch 2:651–675

Newberne PM, Rogers AE (1986) The role of nutrients in cancer causation. In: Hayashi Y (ed) Diet, nutrition and cancer. Japan Scientific Press, Tokyo pp 205–222

Pasco GA, Sakai-Wong J, Soliven E, Correia MA (1983) Regulation of intestinal cytochrome P-450 and heme by dietary nutrients. Biochem Pharmacol 32:3027–3035

Pegram RA, Allaben WT, Chou MW(1989) Effect of caloric restriction on aflatoxin B_1-DNA adducts formation and associated factors in Fischer 344 rats: preliminary findings. Mech Ageing Dev 48:167–177

Pollard M, Luckert P (1985) Tumorigenesis affects of direct- and indirect-acting chemical carcinogenesis in rats on a restricted diet. J Natl Cancer Inst 74:1347–1349

Robertson TB, Marston HC, Walters JW (1934) The influence of intermittent starvation and of intermittent starvation plus nucleic acid on growth and longevity of the white mouse. Aust J Exp Biol Med Sci 12:33–39

Roe FJC (1981) Are nutritionists worried about the epidemic of tumors in laboratory animals? Proc Nutr Soc 40:57–65

Roebuck BD, Yager JD JR, Longnecker DS (1981) Dietary modulation of azoserine-induced pancreatic carcinogenesis in the rat. Cancer Res 41:888–893

Ross MN (1961) Length of life and nutrition in the rat. J Nutr 75:197–210

Ross MH (1965) Tumor incidence patterns and nutrition in the rat. J Nutr 87:245–260

Ross MH, Bras G (1971) Lasting influence of early calorie restriction on prevalence of neoplasms in the rat. J Natl Cancer Inst 47:1095–1113

Ross MN, Lustbader ED, Bras G (1982) Dietary practices of early life and spontaneous tumors of the rat. Nutr Cancer 3:150–166

Ross MH, Lustbader ED, Bras G (1983) Body weight, dietary practices, and tumor susceptibility in the rat. J Natl Cancer Inst 71:1041–1046

Rous P (1914) The influence of diet on transplanted and spontaneous mouse tumors. J Exp Med 20:433–451

Rusch HP, Kline BE (1944) The effect of exercise on the growth of a mouse tumor. Cancer Res 4:116–118

Rusch HP, Kline BE, Baumann CA (1945) The influence of caloric restriction and of dietary fat on tumor formation with ultraviolet radiation. Cancer Res 5:431–435

Sachan DS (1982) Modulation of drug metabolism by food restriction in male rats. Biochem Biophys Res Commun 104:984–989

Sachan DS, Das SK (1982) Alternations of NADPH-generating and drug-metabolizing enzymes by feed restriction in male rats. J Nutr 112:2301–2306

Sarkar NH, Fernandes G, Telang NT, Kourides IA, Good RA (1982) Low-caloried diet prevents the development of mammary tumors in C3H mice and reduces circulating prolactin level, murine mammary tumor virus expression, and proliferation of mammary alveolar cells. Proc Natl Acad Sci USA 79:7758–7762

Sato A, Nakajima T (1985) Enhanced metabolism of volatile hydrocarbons in rat liver following food deprivation, restricted carbohydrate intake, and administration of ethanol, phenobarbital, polychlorinated biphenyl and 3-methylcholanthrene: a comparative study. Xenobiotic 15:67–75

Saxton JA, Kimball GC (1941) Relation of nephrosis and other diseases of Albino rats to age and to modifications of diet. Arch Pathol 32:951–965

Saxton JA Jr, Boon MC, Furth J (1944) Observations on the inhibition of development of spontaneous leukemia in mice by underfeeding. Cancer Res 4:401–409

Schwartz AG (1974) Protective effect of benzoflavone and estrogen against 7,12-dimethyl-benz(a)anthracene and aflatoxin-induced cytotoxicity in cultured liver cells. Cancer Res 34:10–15

Silverstone H, Tannenbaum A (1951) The influence of dietary fat and riboflavin on the formation of spontaneous hepatomas in the mouse. Cancer Res 11:200–203

Stern JS (1984) Is obesity a disease of inactivity? In: Stundard AJ (ed) Eating disorders. Raven, New York

Sylvester PW, Aylsworth CF, Meites J (1981) Relationship of hormones to inhibition of mammary tumor development by underfeeding during the "critical period" after carcinogen administration. Cancer Res 41:1384–1388

Sylvester PW, Aylsworth CF, Van Vugt DS, Meites J (1982) Influence of underfeeding during the "critical period" or thereafter on carcinogen-induced mammary tumors in rats. Cancer Res 42:4943–4947

Tannenbaum A (1940a) Relationship of body weight to cancer incidence. Arch Pathol 30:509–517

Tannenbaum A (1940b) The initiation and growth of tumors. I. Effect of underfeeding. Am J Cancer 38:335–350

Tannenbaum A (1942a) The genesis and growth of tumors. II. Effects of caloric restriction per se. Cancer Res 2:460–467

Tannenbaum A (1942b) The genesis and growth of tumors. III. Effects of a high fat diet. Cancer Res 2:468–475

Tannenbaum A (1945a) The dependence of tumor formation on the degree of restriction. Cancer Res 5:609–615

Tannenbaum A (1945b) The dependence of tumor formation on the composition of the caloric-restricted diet as well as on the degree of restriction. Cancer Res 5: 616–625

Tannenbaum A, Silverstone H (1949) Effect of low environmental temperature, dinitrophenol, or sodium fluoride on the formation of mammary gland tumors in mice. Cancer Res 9:403–410

Tannenbaum A, Silverstone H (1953a) Nutrition in relation to cancer. In: Greenstein JP, Haddow A (eds) Advances in cancer research. Academic, New York, pp 451–500

Tannenbaum A, Silverstone H (1953b) Effect of limited food intake on survival of mice bearing spontaneous mammary carcinomas and on the incidence of lung metastases. Cancer Res 13:532–536

Tarnowski GS, Stock CC (1955) Selection of a transplantable mouse mammary carcinoma for cancer chemotherapy screening studies. Cancer Res 15:227–232

Valaoris VG (1946) Some effects of famine on the population of Greece. Milbank Mem Fund Q 24:215–243

Visscher MB, Ball ZB, Barnes RH, Siversten I (1942) The influence of caloric restriction upon the incidence of spontaneous mammary carcinoma in mice. Surgery 11:48–55

Watson AF, Mellanby E (1939) Tar cancer in mice. II. The condition of the skin when modified by exthermal treatment or diet as a factor in influencing the cancerous reaction. Br J Exp Pat 11:311–322

Wattenberg LW, Loeb WD, Lam LK, Speier JL (1976) Dietary constituents altering the responses to chemical carcinogens. Fed Proc 35:1327–1331

Weindruch R, Walford RL (1982) Dietary restriction in mice beginning at 1 year of age: effect on lifespan and spontaneous cancer incidence. Science 215:1415–1418

White FR, White JJ (1943) Effect of diethylstilbesterol on mammary tumor formation in strain C3H mice fed a low cystine diet. J Natl Cancer Inst 4:413–415

White JJ, Andervont HB (1943) Effect of a diet relatively low in cystine on the production of spontaneous mammary gland tumors in strain C3H female mice. J Natl Cancer Inst 3:339–451

White FR, White JJ (1944) Effect of a low lysine diet on mammary-tumor formation in strain C3H mice. J Natl Cancer Inst 5:41–42

White FR, White JJ, Mider GB, Kelly MG, Heston WE, David PW (1944) Effect of caloric restriction on mammary tumor formation in strain C3H mice and on the response of strain dba to painting with methylcholanthrene. J Natl Cancer Inst 5:43–48

White FR(1961) The relationship between underfeeding and tumor formation, transplantation and growth in rats and mice. Cancer Res 21:281–290

Yu BP, Masora EJ, Murata I, Bertrand HA, Lynd FT (1982) Life span study of SPF Fischer 344 male rats fed ad libitum or restricted diets: longevity, growth, lean body mass and disease. J Gerontol 37:130–141

CHAPTER 5

Effects of Caloric Restriction on Aflatoxin B_1 Metabolism and DNA Modification in Fischer 344 Rats*

M.W. Chou[1], R.A. Pegram[2], P. Gao[1], and W.T. Allaben[1]

Introduction

Laboratory animals maintained on a reduced calorie, but nutritionally adequate diet have extended life spans and a lowered incidence of spontaneous and chemically induced cancers compared to ad libitum-fed counterparts (Allaben et al., Chapter 4; Kritchevsky and Klurfeld 1987). Results obtained from early studies, such as the pioneering work of Tannenbaum (1942) on the effect of caloric restriction (CR) on mouse skin tumors induced by benzo(a)pyrene, and from the numerous recent studies employing a variety of restriction levels and different carcinogens to induce tumors at different tissue sites (Kritchevsky et al., 1986; Sarkar et al. 1982; Gross 1988; Lagopoulous and Stadler 1987; Newberne and Rogers 1986) demonstrate the profound impact of caloric restriction on tumorigenesis. Although mechanisms underlying the inhibitory effect of CR on carcinogenesis are still not clear, a variety of hypotheses have been proposed which generally focus on the promotion stage (Pariza and Boutwell 1987; Sarkar et al. 1982; Kritchevsky et al. 1984). Recent reports which demonstrate that CR alters xenobiotic metabolizing enzyme activities (Sachan 1982; Sachan and Das 1982; Hasmi et al 1986; Koizumi et al. 1987; Leakey et al. 1989a,b, see also this volume; Table 5.1), decreases 7,12-dimethylbenz(a)anthracene (DMBA) binding to dermal DNA in mice (Pashko and Schwartz 1983), inhibits carcinogenesis induced by indirect- but not direct-acting chemical carcinogens in rats (Pollard and Luckert 1985; Table 5.2), and reduces the binding of aflatoxin B_1 (AFB_1) to hepatic DNA in rats (Pegram et al. 1989) indicate that in addition to cancer promotion, the initiation of stage of carcinogenesis can be significantly inhibited by CR.

AFB_1 is a highly mutagenic and carcinogenic mycotoxin which has been implicated epidemiologically as a causative agent in human liver cancer (Busby and

*This research was supported in part by appointments to the ORAU Postgraduate Research Program at the National Center for Toxicological Research administered by Oak Ridge Associated Universities through an interagency agreement between the U.S. Department of Energy and the U.S. Food and Drug Administration.
[1]National Center for Toxicological Research, Jefferson, AR 72079, USA
[2]Current address: US Environmental Protection Agency, Research Triangle Park, NC 27711, USA

Table 5.1. Effects of dietary restriction on hepatic xenobiotic metabolizing enzymes in laboratory rodents.

Enzyme or activity	Species	Restriction % ad libitum	time	Effect (% change)	References
Hexobarbital sleep-ing time	Rats	55	28 days	−60	Sachan 1982
Aniline-OH-ase	Rats	55	28 days	+40	Sachan 1982
Total P-450	Rats	50	28 days	+20	Hashmi et al. 1986
Acetanilide-OH-ase	Rats	50	12 weeks	+50	Hashmi et al. 1986
UDP-GT	Rats	60	18 months	+65	Leakey et al. 1989b
GSH-S-T	Rats	60	18 months	+40	Leakey et al. 1989b
GSH-S-T	Mice	60	8 days	+60	Koizumi et al. 1987
AHH	Mice	60	8 days	+30	Koizumi et al. 1987
Catalase	Mice	60	12 months	+40	Koizumi et al. 1987

UDP-GT, uridine diphosphate-glucuronyltransferase; GSH-S-T, glutathione S-transferase; AHH, aryl hydrocarbon hydroxylase.

Wogan 1984). Metabolic activation of AFB$_1$ by microsomal xenobiotic metabolizing enzymes both in vivo and in vitro results in the formation of a reactive epoxide which binds to cellular macromolecules, such as DNA, RNA, or proteins (Essigmann et al. 1982; Swenson et al. 1974, 1977), and exerts its mutagenic and/or carcinogenic activities. The binding of AFB$_1$ to DNA in rat liver has been studied extensively, and the major AFB$_1$-DNA adduct has also been identified as 8,9-dihydro-8-(N^7-guanyl)-9-hydroxy-aflatoxin B$_1$ (AFB$_1$-N^7-Gua), which can rearrange to form AFB$_1$-N^7-formamido-pyrimidine (AFB$_1$-N^7-FAP) (Lin et al. 1977). It has been demonstrated that total covalent modification of DNA by AFB$_1$ correlates with the carcinogenic susceptibility of different species of animals that were treated with AFB$_1$ (Croy et al. 1983). AFB$_1$-induced mutagenic or carcinogenic activities were reduced by various nutritional modulations, such as changing the major components of diet, i.e., fat and protein content (Dunaif and Campbell 1987; Rogers and Newberne 1969; 1971), essential vitamins (Bhattacharya et al. 1987) and minerals (Francis et al. 1988), and restricting food consumption or caloric intake (Newberne and Rogers 1986). In the latter study (Newberne and Rogers 1986), AFB$_1$-induced liver tumors were

Table 5.2. Effect of dietary restriction on intestinal tumors induced in rats by indirect- and direct-acting carcinogens (from Pollard and Luckert 1985)

Carcinogen	Diet	Tumor incidence (%)	No. of tumors
MAM	Ad libitum	90	32
MAM	75% of Ad libitum	20	2
MNU	Ad libitum	100	34
MNU	75% of Ad libitum	90	28

MAM, methylazoxymethanol; MNU, N-methylnitrosourea

reduced by more than 50% in rats fed 75% of ad libitum (AL) consumption. In this paper, we address the results of our mechanistic studies of the effects of dietary restriction on AFB_1-induced carcinogenesis, with particular regard to (a) AFB_1 metabolism and (b) DNA modification in AFB_1-treated rats.

Materials and Methods

Animals and Diets

Male Fischer 344 rats from the National Center for Toxicological Research (NCTR) breeding colony were weaned at 21 days and fed the NIH-31 diet AL until 10 weeks of age. Half of the rats were then restricted to 60% of AL consumption using a vitamin-supplemented NIH-31 ration formulated to attain AL-equivalent vitamin consumption. The restricted group received a single daily feeding immediately prior to the dark phase of a 12-h light-dark cycle.

Both the AL and restricted rats received 1, 6, and 15 daily (Monday-Friday) oral doses of $[^3H]AFB_1$ (0.1 mg, 66.7 µCi AFB_1/kg body weight for the single dose and 0.1 mg, 25 µCi/kg for the multiple doses) dissolved in 75% $DMSO/H_2O$ at 16 weeks of age.

Isolation of Hepatic Nuclei

Rats were killed at various times as indicated in the text after receiving single or multiple doses of $[^3H]AFB_1$. The livers were rapidly excised, rinsed in cold 0.9% NaCl, frozen in liquid N_2, and stored at $-80°C$ until analyzed. Hepatic nuclei were prepared by washing with 0.5% Triton X-100 in homogenizing buffer (0.25 M sucrose, 25 mM KCl, 10 mM $MgCl_2$, 50 mM Tris-HCl, pH 7.4), according to the method of Hymer and Kuff (1964) and subsequently used for DNA isolation and $[^3H]AFB_1$-DNA binding measurements. Hepatic nuclei prepared by ultracentrifugation through hypertonic sucrose (Tata 1974) were used for the DNA alkaline unwinding assay and the in vitro nuclear binding experiment.

In Vivo and In Vitro AFB_1-DNA Binding

Hepatic nuclear DNA was purified by RNase A and proteinase K treatment followed by phenol and chloroform:isoamyl alcohol (24:1) extraction (Beland et al. 1984). DNA was quantified spectrophotometrically, and binding levels of AFB_1 to DNA were determined radiometrically using liquid scintillation counting. In the in vitro nuclear binding assay, aliquots of the nuclear suspensions from AL- and CR-fed animals containing approximately 100 µg of DNA were incubated in 2.0 nmol AFB_1 (containing 1 µCi $[^3H]AFB_1$) with equivalent aliquots from a microsomal pool for activation (Jhee et al 1988). Each microsomal aliquot contained microsomes derived from 0.1 g of liver. The DNA was isolated and analyzed as above to determine binding levels.

AFB₁-DNA Adduct Analysis

Purified DNA samples were acid hydrolyzed (1.5 N HCl incubated at 90°C for 60 min). The [³H]AFB₁-DNA adducts were separated by high-performance liquid chromatography using a reversed-phase column (DuPont Zorbax ODS, 4.6 × 250 mm) eluted with a linear gradient of 10%–18% ethanol in 20 mM potassium acetate buffer, pH 5.0. Adducts were quantified radiometrically.

Determination of Hepatic DNA Synthesis

Rats were injected i.p. with 50 μCi [methyl-³H]thymidine/100 g body weight 2 h before they were killed. Livers were removed and hepatic nuclear DNA was isolated as described above. The DNA was assayed using the diphenylamine reagent, and thymidine incorporation was determined by liquid scintillation counting.

DNA Alkaline Unwinding Assay

DNA strand breaks and abasic sites were estimated by the method of Birnboim and Jevcak (1981) using a fluorometric technique that measures the rate of unwinding of nuclear DNA upon exposure to alkaline conditions. The assay conditions were modified in this experiment since purified nuclei were used instead of cells. Briefly, 5–10 × 10⁶ rat liver nuclei were lysed in 9 M urea, containing 10 mM NaOH, 2.5 mM cyclohexylenedinitrilotetra-acetic acid and 0.1% sodium dodecyl sulfate (solution C) for 20 min at 0°C and exposed to alkaline conditions by adding 0.1 ml of 0.45 vol solution C in 0.2 N NaOH and 0.1 ml of 0.40 vol solution C in 0.2 N NaOH, and incubated at 15°C for 15 min. The solution was then neutralized and the double-stranded DNA remaining was determined fluorometrically after addition of 13.4 μg of ethidium bromide/ml of 13.3 mM NaOH. The values of double-stranded DNA remaining were proportional to the length of the 15°C incubation.

Results

In vivo AFB₁ metabolism was significantly affected by CR in rats given a single oral dose of [³H]AFB₁. The total radioactivity in plasma was 42% lower in CR rats, and plasma water-soluble radioactivity, probably the conjugated metabolites, was 2.3-fold greater than that of AL rats 3 h after dosing (Table 5.3). In addition, 9-h urinary excretion of radioactivity was significantly greater in the CR rats. These results indicate an enhancement of in vivo detoxification in CR animals.

Dietary restriction increased total hepatic microsomal cytochrome P–450 content, but reduced in vitro microsome-mediated binding of AFB₁ to calf thymus DNA (Table 5.4), indicating that epoxidation of AFB₁ is decreased by CR. These findings also suggest that CR has variable effects on P–450 activities: some isozymes may be increased while others are decreased.

Table 5.3. Plasma concentration and urinary excretion of radioactivity in ad libitum-fed and feed-restricted rats following administration of [^{3}H]Aflatoxin B$_1$.

Diet	3-h Plasma concentration of radioactivity (nCi/ml)	3-h Plasma hydrophilic radioactivity (% of protein-free fraction)	9-h Urinary excretion of radioactivity (% of dose)
Ad libitum	215±15	11.0±1.1	7.05±0.44
Restricted	125±6	25.3±3.1	9.90±0.44

[^{3}H]Aflatoxin B$_1$ was given as a single oral dose (100 µg/kg body wt; 208 mCi/mmol). Values (mean ± SEM, $n=4$) within each column differ significantly ($p \leq 0.01$).

In vivo binding of AFB$_1$ to nuclear DNA was substantially reduced by CR. Hepatic DNA-AFB$_1$ binding levels at 3, 12, 24, and 48 h after a single dose of [^{3}H]AFB$_1$ are shown in Fig. 5.1. CR reduced maximal AFB$_1$-DNA adduct formation by more than 50%. Although the CR rats received a lower AFB$_1$ dose per animal due to their smaller body size, the relative liver weights and total hepatic DNA levels of the CR animals were also significantly reduced. Therefore, compared to the AL group, the CR rats actually received a greater AFB$_1$ dose per gram liver and an equivalent dose per milligram of hepatic DNA.

In the period of 3 to 12 h after dosing, the average rates of adduct removal were 1.56 and 0.34 pmol AFB$_1$/mg DNA per hour in AL and CR rats, respectively (Fig. 5.1). The rates decreased to 0.4 and 0.28 pmol/mg per hour from 12 to 24 h after dosing and 0.25 and 0.08 pmol/mg per hour during the 24- to 48-h period. AFB$_1$-DNA adduct removal in rats that received 15 daily doses of [^{3}H]AFB$_1$ was similar to that of single-dosed rats.

The major adducts isolated from acid hydrolysates of AFB$_1$-bound DNA and separated by HPLC were tentatively identified according to Kensler et al. (1986) as AFB-N^7-Gua and AFB-N^7-FAP. The quantifications of these two major adducts from both AL and CR rats 3 h after AFB$_1$ dosing are shown in Table 5.5. Dietary restriction reduced AFB-N^7-Gau and AFB-N^7-FAP levels by 56% and 71%, respectively, compared to AL animals.

To further characterize the DNA damage that may result from AFB$_1$ adducts and subsequent depurination, DNA alkaline unwinding rates, which are corre-

Table 5.4. Effect of caloric restriction on the activities of hepatic microsomal monooxygenase enymes.

Enzyme	Ad libitum	Restricted
Cytochrome P-450 (nmol mg^{-1}) protein min^{-1}	0.50 ± 0.01	0.57 ±0.02
Microsome-mediated AFB$_1$ binding to calf thymus DNA (pmol mg^{-1} DNA mg^{-1} protein min^{-1})	36.9 ± 0.6	33.4 ± 0.4

Values (mean±SEM) within each row differ significantly ($p \leq 0.05$).

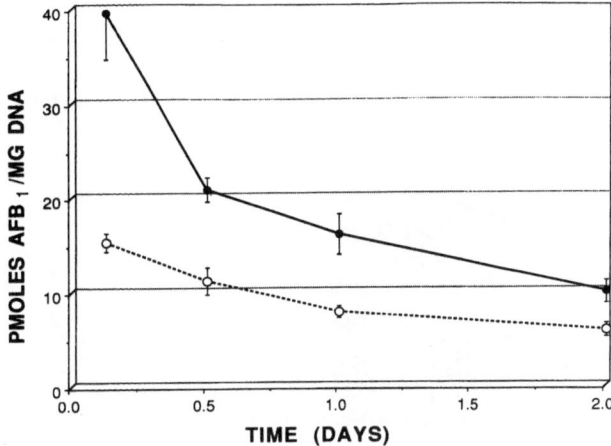

Figure 5.1. Effect of caloric restriction (CR) on hepatic nuclear DNA-AFB$_1$ binding and adduct removal after a single AFB$_1$ dose. Binding levels of the AL-fed (*solid line*) an CR (*broken line*) rats differed significantly ($p \leq 0.05$) at each *time point* ($n=4$)

lated with the quantities of DNA strand breaks and abasic sites (Birnboim 1986), were measured using hepatic nuclear DNA from AL and CR rats treated with AFB$_1$ (Fig. 5.2). The percentage of double-stranded DNA remaining after 15 min of alkaline exposure, which is inversely related to the alkaline unwinding rate, was significantly lower in the AL samples taken 3 h after AFB$_1$ dosing. The AL:CR ratio of "unwound" single-stranded DNA after alkaline treatment was 4.2, further indicating that both AFB$_1$ adduct formation and the subsequent generation of mutagenic apurinic sites and DNA strand breaks were significantly reduced in CR rats.

Table 5.5. Effect of caloric restriction on concentrations of hydrolysis products of aflatoxin B$_1$-modified hepatic nuclear DNA.

Diet	Concentration (pmol AFB$_1$ derivative/mg DNA)[a]	
	AFB$_1$-N^7-Gua[b]	AFB$_1$-N^7-FAP[c]
Ad libitum	44.2 ± 2.3	11.7 ± 2.0
Restricted	19.3 ± 1.9	3.4 ± 0.3

[a] A single oral dose of [^{3}H]AFB$_1$ (0.1 mg/kg body weight) was administered 3 h before killing. Each value is the mean ± SEM, $n=4$.
[b] 8,9-Dihydro-8-(N^7-guanyl)-9-hydroxyaflatoxin B$_1$.
[c] 8,9-Dihydro-8-(2,6-diamino-4-oxo-3,4-dihydropyrimid-5-yl formamido)-9-hydroxyaflatoxin B$_1$.

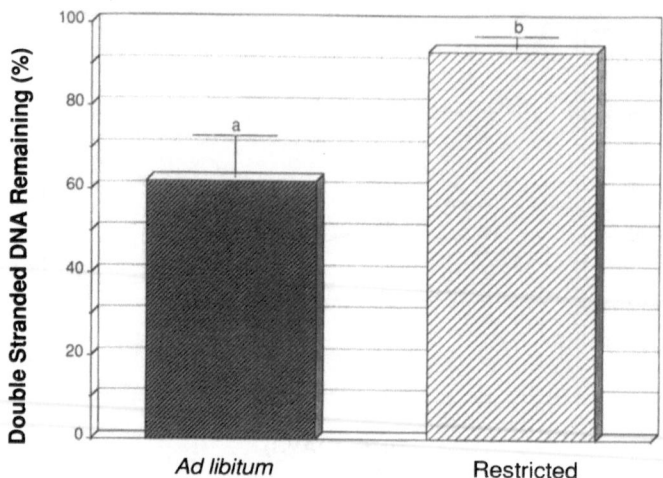

Figure 5.2. Hepatic nuclear DNA alkaline unwinding analysis after a single AFB_1 Dose in ad libitum-fed and diet-restricted rats. *Bars a* and *b* ($n=6$) differ significantly ($p \leq 0.05$)

To compare nuclear susceptibility to AFB_1, AFB_1-DNA binding was determined in purified nuclei preparations from AL and CR rats using an identical microsomal metabolic activation environment for both groups (Fig. 5.3). Although exposure to activated AFB_1 was equivalent for AL and CR nuclei, in vitro nuclear DNA binding was 37% lower in CR preparations than in the AL

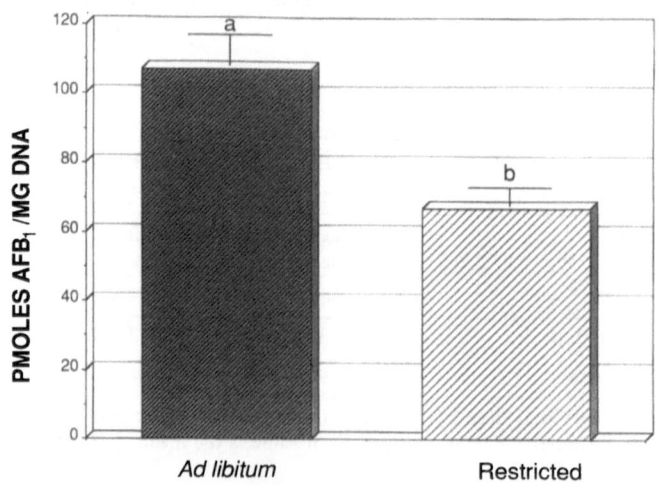

Figure 5.3. Effect of caloric restriction on in vitro binding of AFB_1 to DNA in isolated hepatic nuclei. *Bars a* and *b* ($n=6$) differ significantly ($p \leq 0.05$)

Table 5.6. Effects of caloric restriction and aflatoxin B_1 on serum alanine aminotransferase (units per liter).

Treatment	Ad libitum	Restricted
Vehicle	49.8 ± 1.3*	47.3 ± 1.2
AFB₁-10 doses	56.0 ± 2.1	48.7 ± 3.0
AFB₁-15 doses	56.2 ± 1.7	49.2 ± 2.2

*This value (mean ± SEM, $n = 6$) differs significantly ($p \leq 0.05$) from the others in the diet group.

controls. In vitro AFB_1-DNA binding in nuclei without addition of a microsomal activation system was also measured to assess the contribution of nuclear membrane-associated enzymes to AFB_1 activation. Without exogenous metabolic activation, comparatively low binding levels were noted (about 3% of the levels achieved with microsomes), but mean AFB_1-DNA binding activity was again reduced (38%) in the CR nuclei.

Serum alanine aminotransferase activity, a sensitive indicator of hepatocellular damage, was slightly but significantly elevated in AL-fed rats after 10 or 15 daily doses of AFB_1, while remaining unchanged in the CR group (Table 5.6). The protective effect of CR on hepatocellular damage was also observed by histopathological examination. After 15 AFB_1 doses, minimal individual cell necrosis was found in 83% of the AL rats compared with 50% of the CR rats. Dietary restriction also decreased the rate of hepatic DNA synthesis as measured by [³H]thymidine incorporation (Fig. 5.4). [³H]Thymidine incorporation was inhibited after

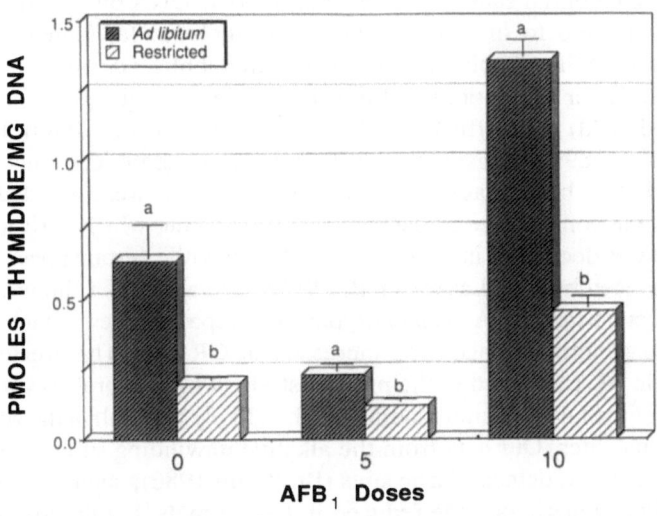

Figure 5.4. Thymidine incorporation into rat liver DNA during a multiple AFB_1 dosing regimen. *Bars a* and *b* ($p = 5$) for a given number of doses differ significantly ($p \leq 0.05$)

the first 5 AFB_1 doses, but was increased after 10 doses. These AFB_1-associated increases, which apparently reflect regenerative responses to liver damage, were much greater in the AL rats than in the CR animals.

Discussion

The metabolic activation of carcinogens by xenobiotic metabolizing enzymes to form reactive ultimate carcinogenic metabolites and the binding of these electrophilic metabolites to cellular DNA to form carcinogenic adducts is considered an important event in the initiation stage of chemical carcinogenesis. AFB_1 is metabolized to its ultimate carcinogenic derivative, AFB_1-8,9-epoxide, which can bind to DNA forming N^7-guanine adducts, bind with other macromolecules, or be conjugated with glutathione. A variety of hydroxylated derivatives of AFB_1 may be conjugated with glucuronate or sulfates. During and after the process of adduct formation, enzymatic DNA repair may also occur, resulting in the removal of carcinogen-DNA adducts. The removal of AFB_1-DNA adducts appears to occur by both enzymatic and spontaneous chemical reactions. Bennett et al. (1981) have suggested that spontaneous depurination and release of AFB_1-N^7-Gau is the major adduct response pathway operative in vivo. Others have also suggested that the spontaneous reactions seem to predominate (Croy et al. 1978; Hertzog et al. 1980; Wang and Cerutti 1979, 1980), and it now appears that the apurinic sites resulting from spontaneous depurination may be responsible for the mutagenic activity of AFB_1 (Kensler et al. 1986; Kaden et al. 1987).

The reductions of hepatic AFB_1-DNA binding and associated DNA damage in AFB_1-treated rats demonstrate that CR can inhibit the initiation stage of chemical carcinogenesis. Potential causes of lower initial AFB_1-DNA binding in the livers of CR rats compared to their AL counterparts include: (a) higher metabolic detoxification of AFB_1 (Table 5.3); (b) lower metabolic epoxidation of AFB_1 (Table 5.4); (c) lower susceptibility of the chromatin to carcinogen modification (Fig. 5.3); and/or (d) more efficient DNA repair (Lipman et al. 1989). Recently, Leakey et al. (1989a, b) have shown that hepatic xenobiotic metabolizing enzymes, including both phase I and phase II activities, were affected by CR. Interestingly, the constitutive cytochrome P450 isozyme which activates AFB_1 (P-450IIC11) was decreased in young adult CR rats, while certain phase II activities were enhanced by CR (Leakey et al. 1989b; see also this volume). Lipman et al. (1989) found that DNA excision repair, the repair process which removes AFB_1-induced genetic damage, was increased in CR rats. The greater adduct removal rate in AL rats noted in the present study, which is probably a primary consequence of the higher initial binding level, is likely to result in the generation of more apurinic sites. Our data from the alkaline unwinding strand break assay (Fig. 5.2), which also detects abasic sites (Birnboim 1986), support this contention. Not only is genetic damage reduced in CR animals, but the lower hepatic DNA synthesis rate of restricted rats after ABF_1 treatment (Fig. 5.4) indicates that the potential for the proliferation of the AFB_1-induced genetic lesions was

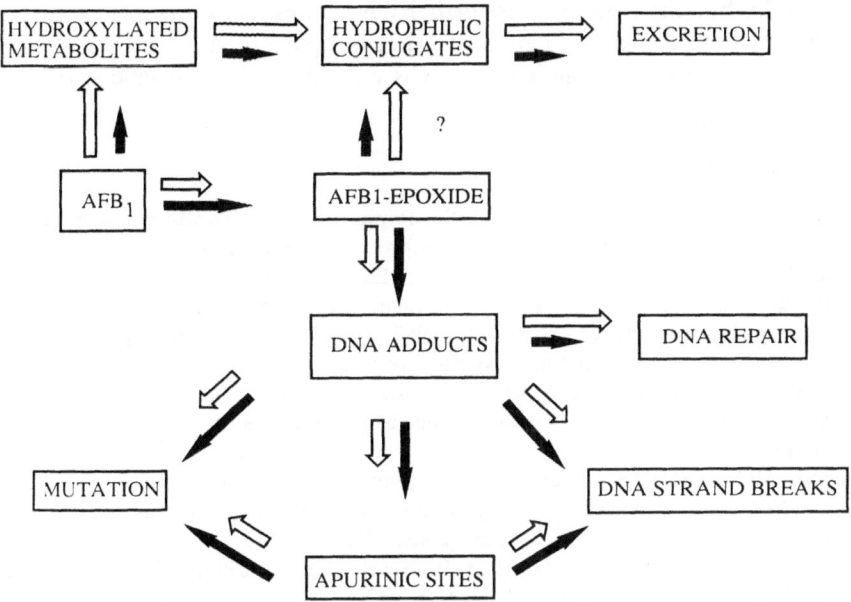

Figure 5.5. Interactive effects of caloric restriction on AFB$_1$ metabolism and DNA modification in Fischer 344 rats. The *arrows* indicate differences between dietary groups, but are not intended to represent the magnitude of the differences. *Open arrows*, restricted diet; *black arrows*, ad libitum-fed rats

lowered by CR. This may also be an important mechanistic factor in the reduction of chemical carcinogenesis by caloric restriction.

The in vitro AFB$_1$-DNA binding experiment with purified rat liver nuclear preparations provides information indicating that: (a) with identical AFB$_1$ activation systems, DNA binding in nuclei from CR rats was reduced by 37%, which suggests that there are qualitative differences in the chromatin of these two groups, leading to less carcinogen adduction in the CR rats; and (b) nuclear membrane-associated enzymes contributed somewhat to AFB$_1$ epoxidation and AFB$_1$-DNA binding, and there was an indication that nuclear activation was lower in CR rats.

A diagram which depicts possible interactive effects of CR on AFB$_1$ metabolism and DNA modification in AL and CR rats in shown in Fig. 5.5. Since total AFB$_1$-DNA binding at any given time is dependent upon the rates of AFB$_1$ activation/detoxification, AFB$_1$-DNA adduct formation, spontaneous depurination, and DNA repair, the reduction in total binding by caloric restriction could be affected at any, if not all, of these steps. However, from our data, the major effects seem to be on initial AFB$_1$-DNA adduction, perhaps resulting from altered metabolism and decreased chromatin susceptibility to binding, which may in turn be affected by the reduction in DNA synthesis. The significant reductions in initial

AFB_1-DNA binding (Fig. 5.1) as further demonstrated by the adduct analysis (Table 5.5), and the subsequent generation of apurinic sites and DNA strand breaks (Fig. 5.3) strongly indicate that CR can inhibit carcinogenic initiation.

References

Beland FA, Fullerton NF, Heflich RH (1984) Rapid isolation, hydrolysis and chromatography of formaldehyde-modified DNA. J Chromatogr 308:121–131

Bennett RA, Essigmann JM, Wogan GN (1981) Excretion of an aflatoxin-guanine adduct in the urine of aflatoxin B_1-treated rats. Cancer Res 41:650–654

Bhattacharya RK, Francis AR, Shetty TK (1987) Modifying role of dietary factors on the mutagenicity of aflatoxin B_1: in vitro effect of vitamins. Mutat Res 188:121–128

Birnboim HC, Jevcak JJ (1981) Fluorometric method for rapid detection of DNA strand breaks in human white blood cells produced by low doses of radiation. Cancer Res 41:1889–1892

Birnboim HC (1986) DNA strand breaks in human leukocytes induced by superoxide anion, hydrogen peroxide and tumor promoters are repaired slowly compared to breaks induced by ionizing radiation. Carcinogenesis 7:1511–1517

Busby WF, Wogan GN (1984) Aflatoxins. In: Searle CD (ed) Chemical carcinogen. ACS Monograph 182, Washington, D.C., pp 945–1136

Croy RG, Essigmann JM, Reinhold VN, Wogan GN (1978) Identification of the principal aflatoxin B_1-DNA formed in vivo in rat liver. Proc Natl Acad Sci USA 75:1745–1749

Croy RG, Essigmann JM, Wogan GN (1983) Aflatoxin B_1: correlations of patterns of metabolism and DNA modification with biological effects. Basic Life Sci 24:49–62

Dunaif GE, Campbell TC (1987) Relative contribution of dietary protein level and aflatoxin B_1 dose in generation of presumptive preneoplastic foci in rat liver. J Natl Cancer Inst 78:365–369

Essigmann JM, Croy RG, Bennett RA, Wogan GN (1982) Metabolic activation of aflatoxin B_1: patterns of DNA adduct formation, removal and excretion in relation to carcinogenesis. Drug Metab Rev 13:581–602

Francis AR, Shetty TK, Bhattacharya RK (1988) Modifying role of dietary factors on the mutagenicity of aflatoxin B_1: in vitro effect of trace elements. Mutat Res 199:85–93

Gross L(1988) Inhibition of the development of tumors or leukemia in mice and rats after reduction of food intake. Cancer 62:1463–1465

Hashmi RS, Siddiqui AM, Kachole MS, Pawar SS (1986) Alterations in hepatic microsomal mixed function oxidase system during different levels of food restriction in adult male and female rats. J Nutr 116:682–688

Hertzog PJ, Smith JRL, Garner RC (1980) A high pressure liquid chromatography study on the removal of DNA-bound aflatoxin B_1 in rat liver and in vitro. Carcinogenesis 1:787–793

Hymer WC, Kuff EL (1964) Isolation of nuclei from mammalian tissues through the use of Triton X-100. J Histochem Cytochem 12:359–363

Jhee EC, Ho LL, Lotlikar, PD (1988) Effect of butylated hydroxyanisole pretreatment on in vitro hepatic aflatoxin B_1-DNA binding and aflatoxin B_1-glutathione conjugation in rats. Cancer Res 48:2688–2692

Kaden DA, Call KM, Leong P-M, Komives EA, Thilly WG (1987) Killing and mutation of human lymphoblast cells by aflatoxin B_1: evidence for an inducible repair response. Cancer Res 47:1993–2001

Kensler TW, Egner PA, Davidson NE, Roebuck BD, Pikul A, Groopman JD (1986) Modulation of aflatoxin metabolism, aflatoxin-N^7-guanine formation, and hepatic tumorigenesis in rats fed ethoxyguin: role of induction of glutathione S-transferases. Cancer Res 46:3924–3931

Koizumi A, Weindruch R, Walford RL (1987) Influences of dietary restriction and age on liver enzyme activities and lipid peroxidation in mice. J Nutr 117:361–367

Kritchevsky D, Weber MM, Klurfeld DM (1984) Dietary fat versus caloric content in initiation and promotion of 7,12-dimethylbenz[a]anthracene-induced mammary tumorigenesis in rats. Cancer Res 44:3174–3177

Kritchevsky D, Weber MM, Buck CL, Klurfeld DM (1986) Calories, fat and cancer. Lipids 21:272–274

Kritchevsky D, Klurfeld DM (1987) Caloric effects in experimental mammary tumorigenesis. Am J Clin Nutr 45:236–242

Lagopoulous L, Stadler R (1987) The influence of food intake on the development of diethylnitrosamine-induced liver tumors in mice. Carcinogenesis 8:33–37

Leakey JEA, Cunny HC, Bazare J, Webb PJ, Feuers RJ, Duffy PH, Hart RW (1989a) Effects of aging and caloric restriction on hepatic drug metabolizing enzymes in the Fischer 344 rat. I. The cytochrome p–450 dependent monooxygenase system. Mech Ageing Dev 48:145–155

Leakey JEA, Cunny HC, Bazare J, Webb PJ, Lipscomb JC, Slikker W, Feuers RJ, Duffy PH, Hart RW (1989b) Effects of aging and caloric restriction on hepatic drug metabolizing enzymes in Fischer 344 rat. II. Effects on conjugating enymes. Mech Ageing Dev 48:157–166

Lipman JM, Turturro A, Hart RW (1989) The influence of dietary restriction on DNA repair in rodents: a preliminary study. Mech Ageing Dev 48:135–143

Lin JK, Miller JA, Miller EC (1977) 2,3-Dihudro-2-(guan-7-yl)-3- hydroxy aflatoxin B$_1$, a major acid hydrolysis product of aflatoxin B$_1$-DNA or -ribosomal RNA adducts formed in hepatic microsome-mediated reactions and in rat liver in vivo. Proc Natl Acad Sci USA 74:1870–1874

Newberne PM, Rogers AE (1986) The role of nutrients in cancer causation. In: Hayashi Y (eds) Diet, nutrition and cancer. Japan Scientific Press, Tokyo, pp 205–222

Pariza MW, Boutwell RK (1987) Historical perspective: calories and energy expenditure in carcinogenesis. Am J Clin Nutr 45:151–156

Pashko LL, Schwartz AG (1983) Effect of food restriction, dehydroepiandrosterone, or obesity on the binding of [^{3}H]7,12-dimethylbenz[a]anthracene to mouse skin DNA. J Gerontol 38:8–12

Pegram RA, Allaben WT, Chou MW (1989) Effect of caloric restriction on aflatoxin B$_1$-DNA adduct formation and associated factors in Fischer 344 rats: preliminary findings. Mech Ageing Dev 48:167–177

Pollard M, Luckert P (1985) Tumorigenesis effects of direct- and indirect-acting chemical carcinogens in rats on a restricted diet. J Natl Cancer Inst 74:1347–1349

Rogers AE, Newberne PM (1969) Aflatoxin B$_1$ carcinogenesis in lipotrope-deficient rats. Cancer Res 29:1965–1972

Rogers AE, Newberne PM (1971) Nutrition and aflatoxin carcinogenesis. Nature 229:62–63

Sachan DS (1982) Modulation of drug metabolism by food restriction in male rats. Biochem Biophys Res Commun 104:984–989

Sachan DS, Das SK (1982) Alterations of NADPH-generating and drug-metabolizing enzymes by feed restriction in male rats. J Nutr 112:2301–2306

Sarkar NH, Fernandes G, Telang NT, Kourides IA, Good RA (1982) Low-calorie diet prevents the development of mammary tumors in C3H mice and reduces circulating prolactin level, murine mammary tumor virus expression, and proliferation of mammary alveolar cells. Proc Natl Acad Sci 79:7758–7762

Swenson DH, Miller EC, Miller JA (1974) Aflatoxin B_1-2,3-oxide; evidence for its formation in rat liver in vivo and by human liver microsomes in vitro. Biochem Biophys Res Commun 60:1036–1043

Swenson DH, Lin JK, Miller EC, Miller JA (1977) Aflatoxin B_1-2,3-oxide as a probable intermediate in the covalent binding of aflatoxin B_1 and B_2 to rat liver DNA and ribosomal RNA in vivo. Cancer Res 37:172–181

Tannenbaum A (1942) The genesis and growth of tumors. II. Effects of caloric restriction pre se. Cancer Res 2:460–467

Tata JR (1974) Isolation of nuclei from liver and other tissues. Methods Enzymol 31:253–257

Wang TV, Cerutti PA (1979) Formation and removal of aflatoxin B_1-induced DNA lesions in epithelioid human lung cells. Cancer Res 39:5165–5170

Wang TV, Cerutti PA (1980) Effect of formation and removal of aflatoxin B_1: DNA adducts in 10T1/2 mouse embryo fibroblasts on cell viability. Cancer Res 40:2904–2909

CHAPTER 6

Dietary and Caloric Restriction: Its Effect on Cellular Proliferation in Selected Mouse Tissues

D.B. Clayson[1], F.W. Scott[2], R. Mongeau[2], E.A. Nera[1], and E. Lok[1]

Introduction

There is much experimental but very limited clinical evidence that dietary restriction has a beneficial effect in reducing the incidence of naturally occurring and induced cancer formation. Restriction has also been suggested to be advantageous in other respects (Weindruch et al. 1986), including increasing longevity and improving the effectiveness of certain aspects of the immune system. The effect of dietary and caloric restriction on experimental carcinogenesis has been recognized for many years, the sentinel observations being due to Tannenbaum (1940a, 1942; Tannenbaum and Silverstone 1957). Many others have confirmed Tannenbaum's initial observations (Andreou and Morgan 1981; White 1961; Ross and Bras 1973; Klurfeld et al. 1987). Information on the possible advantageous effect of dietary restriction in humans has proved much more elusive. Although Tannenbaum (1940b) obtained some evidence for a correlation between excessive body weight and cancer from human insurance records, further information has not been adequate to demonstrate this conclusively. In fact, there is a major controversy whether mammary and colonic tumorigenesis in humans is dependent on the high lipid content of the North American diet or on excess calories [National Academy of Sciences (USA) 1980, 1982].

There is little information in the scientific literature to indicate the manner by which dietary restriction exerts its effect on carcinogenesis. Tannenbaum (1944) examined the effect of calorie restriction on the genesis of benzo(a)pyrene-induced mouse skin tumors and concluded that restriction had its major effect on the promotional stage of carcinogenesis, although the data do not exclude an influence on the initiation stage. Others, using a variety of models, including the effect of an early partial hepatectomy on hepatocarcinogenesis, have obtained evidence for increased DNA adduct formation and increased tumor formation as

[1]Toxicology Research Division, Bureau of Chemical Safety, Food Directorate, Health Protection Branch, Health and Welfare Canada, Ottawa, Ontario, Canada KIA 0L2
[2]Nutrition Research Division, Bureau of Nutritional Sciences, Food Directorate, Health Protection Branch, Health and Welfare Canada, Ottawa, Ontario, Canada KIA 0L2

Table 6.1. Composition of the semipurified AIN-76A diet.

Constituent	Amount used (g)
Starch	65
Sucrose	–
Casein	20
Corn oil	5
AIN-76 mineral mix	3.5
AIN-76A vitamin mix	1.0
Alphacel	5.0
DL methionine	0.3
Choline bitartrate	0.2

a result of increased cellular proliferation during the initiation stage of carcinogenesis (Lawson and Dzhoiev 1970; Craddock and Frei 1974).

Carcinogenesis is a highly complicated, multistage process and it is probable that a greater understanding of whether animal results apply to humans might be obtained by focusing on one of the biological processes that contribute to the carcinogenic process. We chose to examine the effect of dietary and calorie restriction on cellular proliferation as an initial approach to the mechanism of action of restriction. We used the young adult virgin female Swiss Webster mouse as a model and examined the effect of restricted diets on several tissues.

Experimental Approach

We have conducted three experiments on the influence of dietary or calorie restriction on cellular proliferation (Lok et al. 1990). In the first experiment, we fed two groups of 10 female mice ad libitum, one group on laboratory chow (Number 5001, Ralston Purina) and one on a modified semipurified AIN-76A diet (Bieri et al. 1977; Bieri 1980). Two further groups were fed on the same diets restricted to 75% of the average amount of food consumed by the ad libitum groups during the previous week. A similar protocol was adopted in the second experiment, in which the energy provided by the diet was restricted by reducing the starch content of the AIN-76A diet, leaving the levels of the other nutrients equal to those received by the ad libitum-fed group. In the third experiment, calorie restriction levels of 0%, 10%, 20%, 30%, and 40% were maintained similarly by an overall reduction in the starch content of the AIN-76A diet (Tables 6.1, 6.2).

Mice were housed individually in plastic shoe box cages meeting the requirements of the Canadian Council for Animal Care. They received tap water ad libitum, a 12-h light/12-h dark cycle, an ambient temperature of $23 \pm 1\,^\circ\mathrm{C}$, and relative humidity was maintained at $55 \pm 5\%$. Feeding was carried out for at least 30 days. At the end of this time, or after 15 days in the third experiment, vaginal smears were examined on a daily basis. Mice were killed with excess halothane

Table 6.2. Amount of starch and calories in each diet as compared to ad libitum feeding of 100 g of diet (Lok et al., 1990).

	Ad libitum	Restriction (%)			
		10	20	30	40
Ratio[a]	–	90.4	80.8	71.1	61.5
Starch	65	55.4	45.8	36.1	26.5
kCal ingested	385	347	308	270	231

[a]Ratio is the number of grams restricted diet fed compared to 100 g ad libitum diet consumed in the previous week

2 days following the first estrus observed after the 30-day feeding, except for those mice in the 40% restricted group that did not demonstrate normal estrous cycling. These were allocated randomly to one of the sacrifice days.

Mice were injected i.p. with ^{3}H-thymidine (0.25 µCi/g body weight) 1 h prior to death. A careful necropsy was conducted, and the following tissues were processed for slide preparation: mammary gland, dermis, esophagus, urinary bladder, and three levels of the intestinal tract. Slides for radioautography and pathological examination were prepared as previously described (Nera et al. 1984).

Appropriate statistical procedures were applied to the results to test for statistical significance (Lok et al. 1990).

Results

The three experiments each indicated that reducing the overall amount of diet, or the energy content of the diet, led to a reduction in the level of cellular proliferation in each of the tissues examined. In the total dietary restriction study (Lok et al. 1988) the effect was greater in the duct cells of the mammary gland than in the other tissues which were investigated (Fig. 6.1,6.2). There are, however, a number of caveats to be considered before these results are accepted.

The first caveat concerns the possibility that dietary and calorie restriction may affect the thymidine pool size and thus the amount of tritiated thymidine that is taken up by the dividing cells. This can be investigated through the direct measurement of thymidine nucleotide levels in individual tissues, a process that is technically exacting and gives results for whole tissues rather than for individual cell types within the tissue. Instead, the ratio between the mitotic index and the labeling index was investigated (Table 6.3). In a normal cell cycle the period of DNA synthesis lasts for about 6-8 h (Fig. 6.3) while mitosis is complete in about 20 min. The constancy of the ratios between ad libitum and restricted diet in the dermis and the colorectal region suggested that pool size was not exhibiting a significant effect. In the esophagus, the overall ratio was still constant between the two dietary regimens but was much lower than would have been predicted from the normal cell cycle pattern. It is possible that this anomaly was due to the diet being presented to the restricted mice only at one time each day. In consequence,

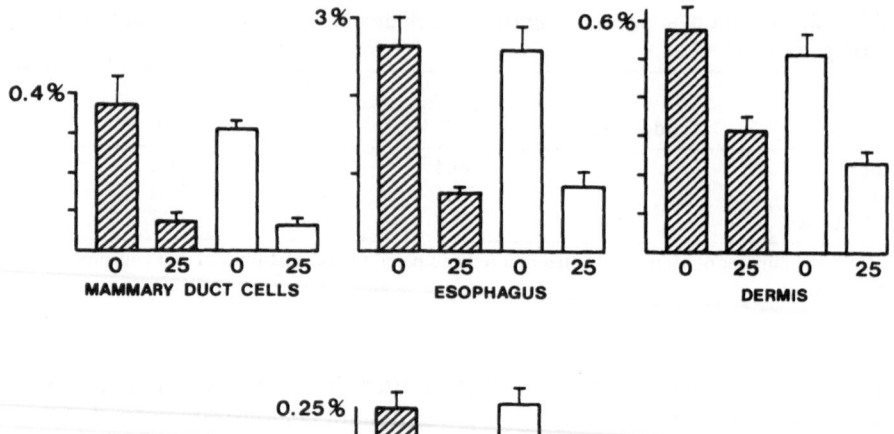

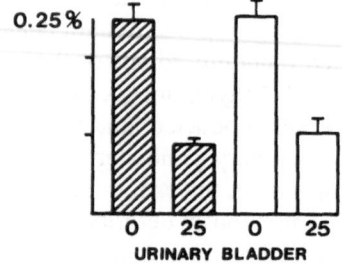

Figure 6.1. Effect of 25% dietary restriction on ³H-thymidine labeling index (*vertical axis*) of mammary duct cells, esophagus, dermis, and urinary bladder. *Shaded bars* represent laboratory chow; *open bars* AIN-76A semipurified diet. *Columns* labeled *O* represent ad libitum diet; *bars* marked *25* represent 25% dietary restriction (Lok et al. 1988)

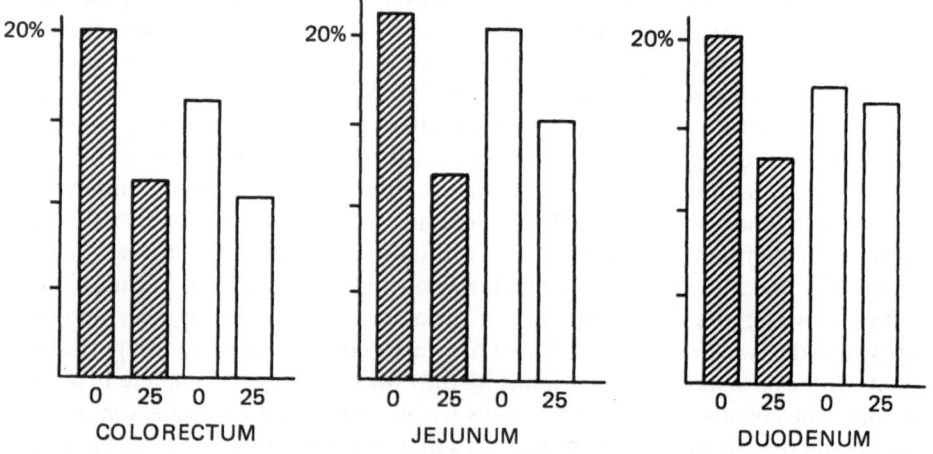

Figure 6.2. Effect of 25% dietary restriction on ³H-thymidine labeling (*vertical axis*) in the intestinal tract. *Shaded bars* represent laboratory chow; *open bars* represent AIN-76A semipurified diet. *Bars* marked *O* represent ad libitum diet: *bars* marked *25* represent 25% restricted diet (Lok et al. 1988)

Table 6.3. Effect of dietary restriction (chow diet) on ³H-thymidine labeling and mitotic indices of the dermis, esophagus, and jejunum of female Swiss Webster mice (Lok et al., 1988).

Tissue	Diet	Labeling index	Mitotic index	Ratio
Dermis	Full	0.69 ± 0.10	0.036 ± 0.007	19:2
	Restricted	0.36 ± 0.06	0.020 ± 0.006	18:0
Esophagus	Full	3.64 ± 0.47	0.89 ± 0.12	4:1
	Restricted	1.34 ± 0.12	0.50 ± 0.06	2:7
Jejunum[a]	Full	11.78 ± 0.20	0.46 ± 0.15	25:6
	Restricted	7.18 ± 0.17	0.26 ± 0.94	27:6

[a]Indices are calculated per crypt.

mice were eating the diet as a bolus which may have disrupted the normal circadian rhythm of cell proliferation and mitosis in the esophageal tissue.

The second possible anomaly concerns only the crypt cells of the intestinal tract. The cells in these crypts are divided into two compartments, an actively proliferating compartment near the base of the crypt and a nonproliferating compartment through which the differentiated cells pass to be shed from the tip of the crypt into the intestinal lumen. In consequence, a fall in the overall cell labeling of the cells in these crypts may not necessarily represent a genuine reduction in the rate of proliferation in the proliferative compartment of the crypt (Fig. 6.4). In the first two experiments, it became clear that the nature of the diet but not the degree of restriction (Fig. 6.2) affected the overall number of cells in the crypt. In the third experiment, in which the level of restriction was greater than in the first two experiments, it was decided to measure the height of the proliferative compartment from the highest labeled cell in the crypt to the base.

It appeared that both 25% restriction in the total amount of diet consumed or a 25% reduction in the amount of energy provided to the mice led to similar, but not identical, levels of reduction in cellular proliferation in the seven tissues examined (Table 6.4). It also appeared that reduction in the rate of cellular

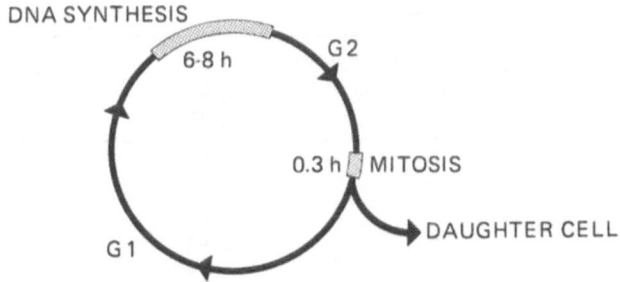

Figure 6.3. The cell cycle

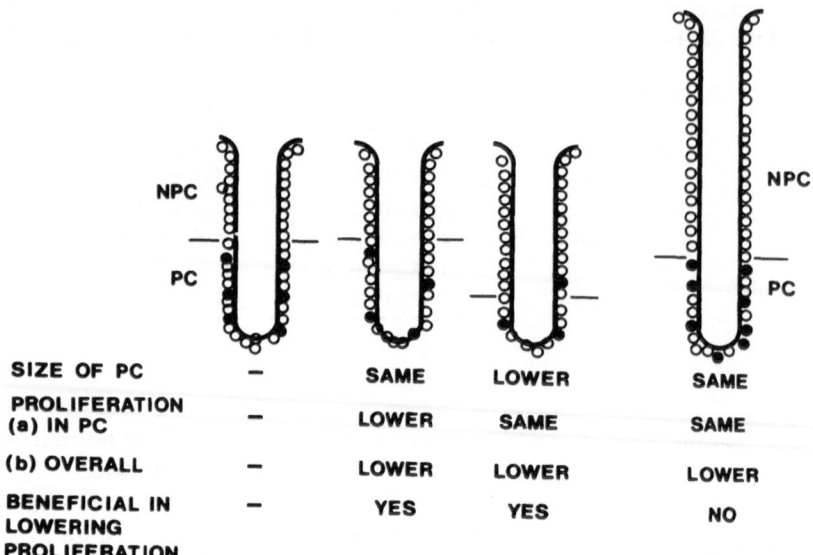

SIZE OF PC	—	SAME	LOWER	SAME
PROLIFERATION (a) IN PC	—	LOWER	SAME	SAME
(b) OVERALL	—	LOWER	LOWER	LOWER
BENEFICIAL IN LOWERING PROLIFERATION	—	YES	YES	NO

Figure 6.4. Effect of treatment on intestinal crypt cell proliferation. *PC* = proliferative compartment; *NPC*, nonproliferative compartment; *closed circles*, dividing cells; *open circles*, nondividing cells

proliferation in the duct cells of the mammary gland was consistently inhibited to a greater extent than in some of the other tissues examined. It was not possible to decide whether the differences observed were a reflection of "biological noise" or were genuine.

To obtain a better feeling for tissue differences in response to dietary or calorie restriction, a study was established to determine the effect of graded levels of calorie restriction on cellular proliferation by establishing standard groups of 10

Table 6.4. Percentage inhibition of cellular proliferation in female Swiss Webster mice fed restricted diets or restricted calories.

| | Dietary restriction | | Calorie restriction |
| | Laboratory chow | AIN-76A | AIN-76A |
Tissue	(%)	(%)	(%)
Mammary gland	73	81	72
Dermis	45	51	57
Urinary bladder	69	58	43
Esophagus	72	65	49
Duodenum	36	16	32
Jejunum	42	28	34
Colorectum	44	54	54

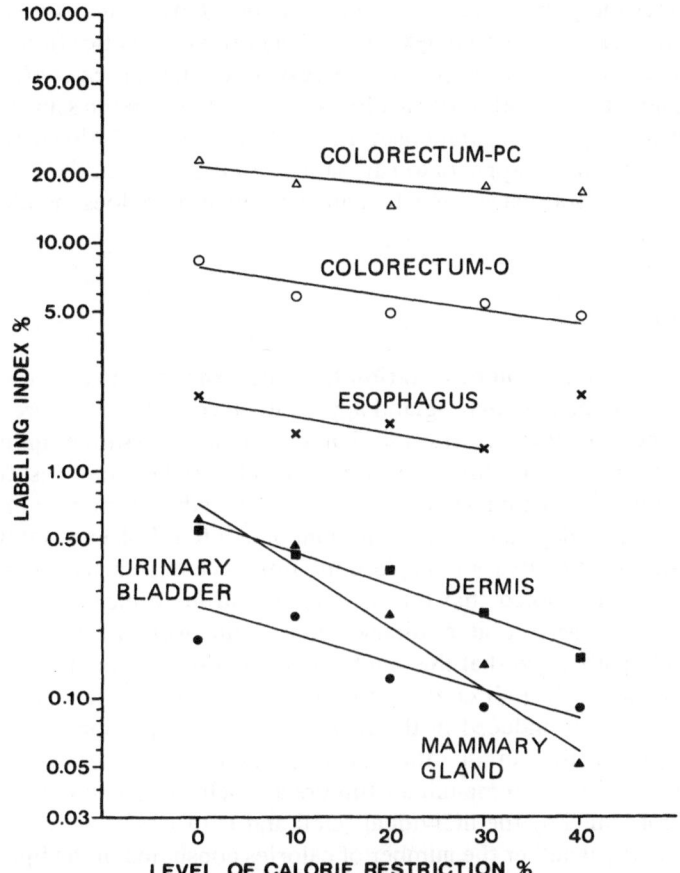

Figure 6.5. Semi-log plot of labeling indices obtained in various tissues of female Swiss Webster mice under various levels of calorie restriction. *Lines* are drawn by eye to enable reader to connect points obtained from each tissue. *O*, overall labeling of crypt cells of the colorectum; *PC*, labeling index of the proliferative compartment of the crypt cells (Lok et al., 1990)

mice on 0%, 10%, 20%, 30%, and 40% calorie-restricted diets. The labeling indices of the mammary gland, dermis, esophagus, urinary bladder, and colorectal crypts were determined.

The first observation was that vaginal smears were difficult to conduct in the 40% restricted mice because of the low level of secretion. Moreover, there were only 3 mice in this group that exhibited a single period of estrus in the 15 or more days observation period, compared to the approximately 40 that were observed in groups of 10 less severely restricted mice. The results of radioautography are summarized in Fig. 6.5. In this figure lines have been drawn by eye to help the reader connect the points referring to each tissue. Statistical analysis suggests that in specific cases more complex curves may provide a better fit to the data.

Overall, the results indicate that calorie restriction has a greater effect on the mammary gland than on other tissues for which the lines on the semi-log plot in Fig. 6.5 appear to approach parallelism. In particular, the inhibition in the overall crypt proliferative index appears to run parallel to the index of the proliferative compartment, and both have a much shallower slope than does the plot for the mammary gland.

Discussion

Perhaps the most important observation to come from this investigation is that the ductal cells of the mammary gland appear to be considerably more susceptible to the effects of calorie restriction than do the other tissues examined. This is consistent with Tannenbaum and Silverstone's (1957) data, demonstrating that naturally occurring mammary tumors, or virus-induced murine mammary tumors, were highly dependent on the amount of diet fed. Lok et al. (1988, 1990) did not obtain direct evidence to explain the influence of reduced dietary intake on this tissue. They noted that the paucity of estrous cycles in 40% calorie-restricted mice suggests that pituitary and steroid hormone levels might be affected and hypothesized that this might also affect the normal proliferation of the mammary gland (Nandi et al. 1958). It is also possible, however, that the tissue anergy directly induced in the mammary gland by calorie restriction is appreciably greater than in the other tissues examined.

Both human colonic and mammary tumors are believed to be very dependent on the diet consumed by the individual (Doll and Peto 1981). There has been a major controversy whether the number of calories consumed or the lipid composition of the diet is the major factor contributing to the induction of these important human tumors [National Academy of Sciences (USA) 1980, 1982]. If the results obtained in these studies in mice are applicable to humans, it is possible that the calorie content of the diet is of major importance in regard to mammary cancer but that other aspects of the diet are of equal or greater importance with respect to colonic cancers. Potential carcinogenic mutagens have been identified in and isolated from human feces by Bruce and his colleagues (1977) in Toronto. One of these classes of naturally occurring mutagens, the fecapentaenes (Gupta et al. 1984), do not appear to be carcinogenic on direct intrarectal administration to rodents but are generally effective on injection (Weisburger et al., 1990). The instability of the directly mutagenic fecapentaenes most probably means that they will not attain as high levels in the mammary gland or in the general circulation as they do at their site of production in the colon.

Acknowledgments. The authors thank Dr. P. Fischer, Dr. G. Sarwar, and Dr. F. Iverson for reviewing and commenting on this text.

References

Andreou KK, Morgan PR (1961) Effect of dietary restriction on induced hamster cheek pouch carcinogenesis. Arch Oral Biol 26:525–531

Bieri JG, Stoewgand GS, Briggs GM, Phillips RW, Woodward JC, Knapka JJ (1977) Report of the American Institute of Nutrition ad hoc committee on standards for nutritional studies. J Nutr 107:1340–1348

Bieri JG (1980) Second report of the ad hoc committee on standards for nutritional studies. J Nutr 110:1726

Bruce WR, Varghese AJ, Furrer R, Land PC (1977) A mutagen in the feces of normal humans. In: Hiatt HH, Watson JP, Wintsten AJ (eds) Origins of human cancer. Cold Spring Harbor Laboratory, Cold Spring Harbor, pp 1641–1646

Craddock VM, Frei JV (1974) Induction of liver cell adenomata in the rat by a single treatment with N-methyl-N-nitrosourea given at various times after partial hepatectomy. Brit J Cancer 30:503–511

Doll R, Peto R (1981) The causes of cancer: quantitative estimates of avoidable risks of cancer in the United States today. Oxford University Press, Oxford

Gupta I, Baptista J, Bruce WR, Che CT, Furrer R, Gingerich JS, Grey AA, Marai L, Yates P, Krepinsky JJ (1984) Structures of fecaspentaenes, the mutagens of bacterial origin isolated from human feces. Biochemistry 22:241–245

Klurfeld DM, Weber MM, Kritchevsky D (1987) Inhibition of chemically induced mammary and colon tumor promotion by calorie restriction in rats fed increased dietary fat. Cancer Res 47:2769–2762

Lawson TA, Dzhoiev F (1970) The binding of orthoaminoazotoluene to proliferating tissues. Chem Biol Interact 2:165–174

Lok E, Nera EA, Iverson F, Scott F, So Y, Clayson DB (1988) Dietary restriction, cell proliferation and carcinogenesis: a preliminary study. Cancer Lett 38:249–255

Lok E, Scott F, Mongeau R, Nera EA, Malcom S, Clayson DB (1990) Calorie restriction and cellular proliferation in various tissues of the female Swiss Webster mouse. Cancer Lett 51:67–73

Nandi S (1958) Endocrine control of mammary gland development and function in the C3H Crgl mouse. J Natl Cancer Inst 21:1039–1063

National Academy of Sciences (USA) (1980) Toward healthful diets: report prepared by the Food and Nutrition Board. National Academy of Sciences, Washington, DC

National Academy of Sciences (USA) (1982) Diet, nutrition and cancer. Committee on Diet, Nutrition and Cancer, National Research Council, National Academy of Sciences, Washington, DC

Nera EA, Lok E, Iverson F, Ormsby E, Karpinski KF, Clayson DB (1984) Short-term pathological and proliferative effects of butylated hydroxyanisole and other phenolic antioxidants in the forestomach of Fischer 344 rats. Toxicology 32:197–213

Ross MH, Bras G (1973) Influence of protein under- and overnutrition on spontaneous tumor prevalence in the rat. J Nutr 103:944–963

Tannenbaum A (1940a) Initiation and growth of tumors. I. Effect of underfeeding. Am J Cancer 38:335–350

Tannenbaum A (1940b) Relationship of body weight to cancer incidence. Arch Pathol 30:509–517

Tannenbaum A (1942) The genesis and growth of tumors. II. Effects of caloric restriction per se. Cancer Res 2:460–467

Tannenbaum A (1944) The importance of differential consideration of the stages of carcinogenesis in the evaluation of cocarcinogenic and anticarcinogenic effects. Cancer Res 4:678–677

Tannenbaum A, Silverstone H (1957) Nutrition and the genesis of tumors. In: Raven RW (ed) Cancer. Butterworth, London

Weindruch R, Walford RL, Fligiel S, Guthrie D (1986) The retardation of aging in mice by dietary restriction: longevity, cancer, immunity and lifetime energy intake. J Nutr 116:641–654

Weisburger JH, Jones RC, Wang C-X, Backlund J-YC, Williams GM, Kingston Dji, Van Tassel PL, Keyes RF, Wilkins TD, de Wit PP, van der Steeg M, van de Gan A (1990) Carcinogenicity tests of fecapentaene-12 in mice and rats. Cancer Lett 49:89–98

White ES (1961) The relationship between underfeeding and tumor formation, transplantation and growth in rats and mice. Cancer Treat Rev 21:281–290

CHAPTER 7

The Impact of Calorie Restriction on Mammary Cancer Development in an Experimental Model[*]

C. Ip[1]

Introduction

Epidemiological evidence from different countries worldwide has suggested a positive association between the availability of fat in the diet and variations in breast cancer mortality rate (Goodwin and Boyd 1987; Schatzkin et al. 1989; Prentice et al. 1989). In the human diet, fat and calories are so tightly linked that it is difficult to evaluate these two effects separately. This problem can only be resolved in an experimental model in which the consumption of fat and calories can be individually controlled. Previous studies by Kritchevsky and coworkers (1984) suggested that caloric intake may be a greater determinant than dietary fat in modulating the development of chemically induced mammary tumorigenesis in rats. A subsequent report from Pariza's laboratory (Boissonneault et al. 1986) indicated that tumor appearance does not depend on the amount of fat in the diet per se but rather on a complex interaction with respect to energy intake, energy retention, and body size. In a recent review article, Albanes (1987) evaluated the relation between total caloric intake, body weight, and tumorigenesis, as well as the independence of these effects from those of dietary fat, using data from 82 published experiments involving several tumor sites in mice. Multivariate regression analysis showed that, regardless of the level of dietary fat, tumor incidence increases with increasing caloric intake and body weight.

This presentation is focused on the biological effects of dietary restriction on tumor development in animals that have been treated with a carcinogen. By using specially formulated diets and adopting a stringent feeding procedure, we were able to design a series of animal experiments which were aimed at delineating the impact of fat and calories as risk factors in mammary carcinogenesis. Our model therefore provides some measure of a quantitative assessment of how much mammary cancer can be suppressed by a reduction of fat or calorie intake. All our

*This work was supported by a grant from Best Foods/CPC International.

[1]Department of Breast Surgery, Roswell Park Cancer Institute, 666 Elm Street, Buffalo, NY 14263, USA

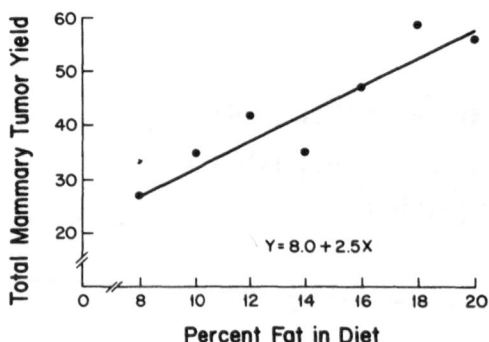

Figure 7.1. Total mammary tumor yield as a function of fat level in the diet. Each group included 30 rats

experiments described below were carried out using the dimethylbenz(a)anthracene (DMBA)-induced mammary tumor model in female Sprague-Dawley rats. The procedure of carcinogen administration and tumor palpation has been described previously (Ip 1980).

Risk Assessment in Response to Changes in Fat Intake

We reported a few years ago (Ip et al. 1985) that the level of linoleate in the diet required to elicit the maximal tumorigenic response in this mammary cancer model is around 4% by weight. A study was designed to determine the risk of mammary tumor development in rats that were given diets with increasing fat levels in the presence of an optimal supply of linoleate. Rats were fed diets containing 8%, 10%, 12%, 14%, 16%, 18%, or 20% fat. Each diet contained a constant 8% corn oil (thus providing 4.8% linoleate) with additional fat coming from hydrogenated coconut oil. The composition of these various diets was adjusted such that the intakes of protein, vitamins, minerals, and calories were similar among the different groups. Thus only the consumption of fat was varied at the expense of carbohydrate. The method of nutrient adjustment for diets containing different levels of fat to ensure a constant nutrient-to-calorie ratio has been discussed by Newberne et al. (1978).

Results of this study (Ip 1987) showed that there was a proportionate enhancement in both tumor incidence and yield with increasing levels of dietary fat. Linear regression analysis of the total tumor yield data, as illustrated in Fig. 7.1, produced an equation of $Y = 8.0 + 2.5X$ for this particular set of experimental conditions, where Y represents total tumor yield and X represents percentage of fat by weight in the diet. If the relative risk (RR) of a 20% fat diet is set as 1.0, it can be calculated using this equation that reducing the fat level to 15% of the diet will result in an RR of 0.8 when calorie intake is held constant.

Although there was no significant difference in calorie intake among all the groups in the above experiment, we observed a trend of a higher body weight gain with increasing fat level in the diet over a duration of 6 months, which was the

Figure 7.2. Average gain in body weight over a period of 6 months as a function of fat level in the diet. The rats were from the DMBA carcinogenesis experiment described in Fig. 7.1. The correlation coefficient of the regression line = 0.89

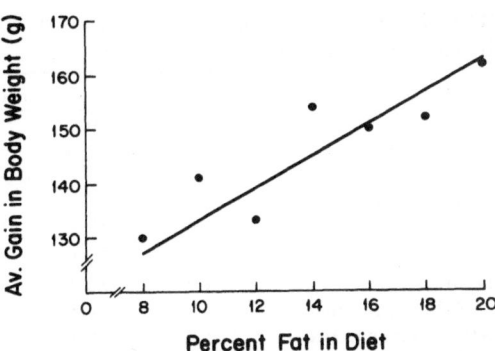

length of the tumor study (Fig. 7.2). Donato and Hegsted (1985) have proposed that a calorie from fat is not physiologically equivalent to a calorie from protein or carbohydrate due to the increase in energy utilization from fat. Compared to sucrose, fat provides about 25% more energy than that expected on the basis of the familiar Atwater values of 4 kcal/g for carbohydrate and 9 kcal/g for fat. Consequently, animals may retain more energy from a high-fat diet than those on a low-fat diet. Prompted by earlier reports that energy intake and retention may be a more important determinant than the level of dietary fat in modulating mammary carcinogenesis (Kritchevsky et al. 1984; Boissonneault et al. 1986; Albanes 1987), we proceeded to design further experiments to evaluate the impact of calorie consumption using the same method of quantitative analysis as in the fat study.

Risk Assessment in Response to Changes in Calorie Intake

In most animal nutrition studies, ad libitum feeding is still considered to be the norm, although there is some debate as to whether an ad libitum fed animal is an overly nourished animal (Berg and Simms 1960; Ross and Bras 1975; Roe 1981; Coates 1982; Conybeare 1988). For practical reasons, an experiment aimed at evaluating the impact of food or calorie intake is inherently constrained by a design which can make comparisons only between animals that eat less than what they normally will eat and the ad libitum fed animals. With this in mind, our study of risk assessment in response to changes in calorie intake was accomplished through food restriction, with ad libitum feeding used as the reference.

Rats were given free access to food or were restricted to different degrees, ranging from 10% to 30%. The composition of the diets used in the food restriction study is shown in the upper part of Table 7.1. The ad libitum fed animals (0% restriction) were given a diet containing 20% corn oil. Notice that the different diets of the food restricted animals were adjusted accordingly in relation to the control diet in order to maintain a similar intake of all nutrients (with the exception of carbohydrate), as indicated in the lower part of Table 7.1. The calorie intake data were calculated based on multiplying the measured food

Table 7.1. Composition of diets in food restriction study, with data of nutrient and calorie intake of food restricted rats.

	Food restriction regimen					
	0%	10%	15%	20%	25%	30%
	(per 100 g diet)					
Dietary components						
Corn oil	20.0	22.2	23.5	25.0	26.7	28.6
Casein	23.5	26.1	27.6	29.4	31.3	33.6
Dextrose	44.8	38.9	35.3	30.9	26.3	21.1
AIN-76 mineral mix	4.1	4.5	4.8	5.1	5.5	5.9
AIN-76 vitamin mix	1.2	1.3	1.4	1.5	1.6	1.7
Alphacel	5.9	6.5	6.9	7.4	7.9	8.4
DL-methionine	0.3	0.3	0.3	0.4	0.4	0.4
Choline bitartrate	0.2	0.2	0.2	0.3	0.3	0.3
Caloric density (cal/g)	4.53	4.60	4.63	4.66	4.71	4.76
Nutrient intake						
Food (g)	14.3	12.8	12.1	11.4	10.7	10.0
Fat (g)	2.9	2.8	2.8	2.9	2.9	2.9
Protein (g)	3.4	3.3	3.3	3.4	3.3	3.4
Minerals (g)	0.59	0.58	0.58	0.58	0.59	0.59
Vitamins (g)	0.17	0.17	0.17	0.17	0.17	0.17
Calorie (cal)	64.7	58.8	56.0	53.1	50.3	47.6
	(100%)	(90.9%)	(86.6%)	(82.1%)	(77.7%)	(73.6%)

consumption value by the caloric density of each diet. It can be seen that the reduction in calorie intake almost exactly matches the reduction in food intake. Thus through the proper formulation of the restriction diets, we were able to evaluate changes of calorie intake as the only variable in the mammary carcinogenesis experiment.

The total mammary tumor yield in each group as a function of calorie intake (expressed as a percentage of the ad libitum fed value) is shown in Fig. 7.3. Regression analysis of the plot produced an equation of $Y = -190 + 2.85X$ for this particular set of experimental conditions, where Y represents total tumor yield and X represents percentage of calorie intake of ad libitum feeding (100%). If the relative risk of ad libitum feeding is set as 1.0, it can be calculated using this equation that reducing the calorie intake by 25% (i.e., 75% of control) will result in an RR of 0.25, compared to an RR of 0.8 for a 25% reduction in dietary fat.

Distinction Between Fat and Calorie Effects

Our model predicts that calorie restriction would be more effective than fat reduction in suppressing the development of induced mammary cancer. The above calorie study was done using diets containing a high level of fat (refer to Table 7.1). The question that was of immediate interest to us was whether there

Figure 7.3. Total mammary tumor yield as a function of calorie intake (expressed as percent of ad libitum feeding). Each group included 30 rats

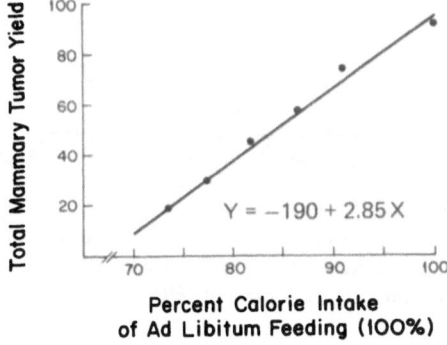

$$Y = -190 + 2.85\,X$$

**Percent Calorie Intake
of Ad Libitum Feeding (100%)**

is a threshold requirement of calorie intake in order for the fat effect to be expressed. We included three more groups of rats as a nested experiment in the calorie study. These rats were all given a normal-fat diet containing only 8% corn oil and were either allowed free access to food or were restricted by 10% or 20%. Results are shown in Fig. 7.4. Two noteworthy observations were evident from the tumor data: (a) compared to the 8% fat diet given ad libitum, the enhancing effect of a high-fat diet (20% fat) on mammary carcinogenesis could be obliterated by a 15% reduction in food consumption; and (b) the high-fat diet still produced more tumors than the low-fat diet at each of the 0%, 10%, or 20% food restriction levels, suggesting that the fat effect may be distinct from the calorie effect in modulation of mammary tumorigenesis.

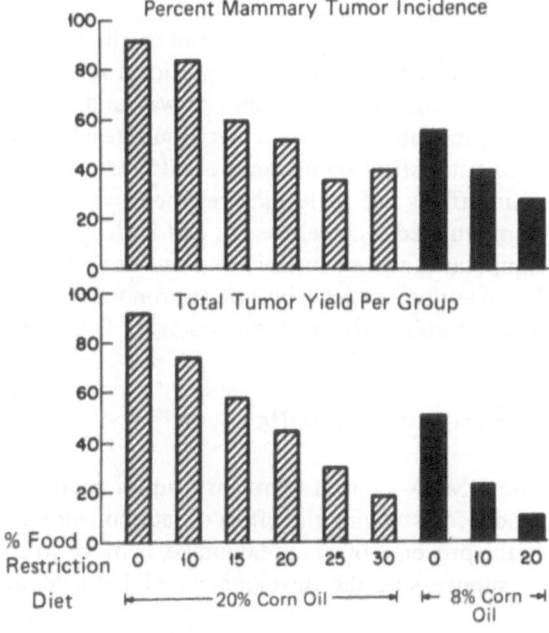

Figure 7.4. Mammary tumor incidence and total tumor yield of rats that were given 20% (*hatched bars*) or 8% fat (*black bars*) diets and were subjected to different degrees of food restriction. Each group included 30 rats. The tumor data from the 20% fat groups were derived from the experiment described under Fig. 7.3.

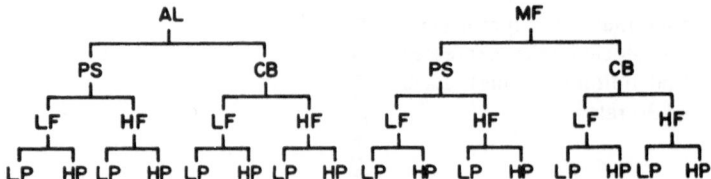

Figure 7.5. Multifactorial design for studying the interactive effects of feeding regimen and diet composition in DMBA-induced mammary carcinogenesis. *AL*, ad libitum feeding; *MF*, meal feeding; *PS*, purified synthetic diet; *CB*, cereal-based diet; *LF*, 6% fat; *HF*, 15% fat; *LP*, 12% protein; *HP*, 20% protein

Interactive Effects of Feeding Regimen and Diet Composition

In a more in-depth investigation, a multifactorial study was designed (refer to Fig. 7.5) involving four independent variables: feeding regimen (ad libitum vs. meal feeding), type of diet (cereal-based vs. semipurified), level of fat in the diet (6% vs. 15% by weight) and level of protein (12% vs. 20%). The meal feeding regimen was a time restricted regimen in which food was available to the rats for only 4 h per day during the dark cycle. Our experience indicated that a 4-h feeding schedule is equivalent to a 20% reduction in food consumption (Ip 1987). Diets were constructed so that rats in both feeding regimens under the same amount of fat and protein combination received similar quantities of all essential nutrients. This means that the rats assigned to the meal feeding regimen had a richer diet than those that were fed ad libitum. The ingredients of the cereal-based diet and the semipurified diet were formulated using the NIH-07 and the AIN-76 diets as the standard, respectively.

An analysis of variance was run on the tumor incidence rate and tumor multiplicity separately to test for significant main effects due to the four independent variables and for significant two-way and three-way interactions. For the tumor incidence rate within each diet group, results from ANOVA (multifactor program to test statistical significance of differences between groups) revealed significant main effects due to feeding regimen (ad libitum > meal feeding), type of diet (semipurified > cereal-based), and fat level (15% > 6%), but not protein level. A significant two-way interaction between feeding regimen and type of diet was also found as demonstrated by a more pronounced increase in tumor multiplicity in rats on the semipurified diet when moving from meal feeding to ad libitum feeding.

Summary of Results

The DMBA-induced mammary tumor model is a reproducible and responsive model for studying the effect of fat and calorie intake. Calorie restriction, even in the presence of a high fat intake, is more striking than a decrease in dietary fat in suppressing the development of induced mammary cancer. The fat effect appears to be distinct from the calorie effect in modulation of induced mammary carcinogenesis, although the contribution of increased energy retention due to a high-fat diet cannot be ruled out. The efficacy of calorie restriction in protection against mammary tumorigenesis appears to be modulated by other exogenous factors such as the source of dietary ingredients.

Relevancy and Extrapolation of Induced Tumor Data

What is the relevancy of studying dietary restriction on chemical-induced tumorigenesis and what can we learn from it for improvement of long-term carcinogenesis and toxicity testing? First of all, chemical-induced tumor models present an easily quantifiable endpoint within a reasonable time frame, and the results can be evaluated with the use of a relatively small group of animals. The fact that these animals are at a high risk of developing neoplastic lesions but are still sensitive to calorie restriction should be viewed as encouraging evidence that changes in dietary habit may have a significant impact on human cancer prevention. Almost a decade ago, Doll and Peto (1981) already pointed out that overnutrition should be a top priority item on a list of dietary considerations that are known to influence the incidence of cancer in humans.

There is a need to extrapolate data from induced carcinogenesis studies in rodents to spontaneous tumorigenesis studies or even the human situation. Although there are many differences (Landers et al. 1986) between the two types of rodent studies, as well as the human situation, it should be possible to control and/or assess the contribution of the most critical variables. The studies discussed here have characterized the role of several of these variables in DMBA-induced mammary carcinogenesis in rats. Similar characterization of these variables in spontaneous rodent tumor studies would improve our ability to compare results and better evaluate the utility of induced-tumor studies. Overnutrition is evident in both the spontaneous carcinogenesis studies and the human situation but the causes appear to be different; this additional variable must be considered in any attempt to extrapolate the results of rodent carcinogenesis studies to the human situation. Meal feeding or limiting access to calories is a viable way of controlling energy consumption and overnutrition in chemical-induced tumorigenesis studies. Further refinement of diet composition in conjunction with changes in feeding method could reduce the high background incidence of pathological lesions in long-term carcinogenesis and toxicity studies.

References

Albanes D (1987) Total calories, body weight, and tumor incidence in mice. Cancer Res 47:1987–1992

Berg BN, Simms HS (1960) Nutrition and longevity in the rat. II. Longevity and onset of disease with different levels of food intake. J Nutr 71:255–263

Boissonneault GA, Elson CE, Pariza MW (1986) Net energy effects of dietary fat on chemically induced mammary carcinogenesis in F344 rats. J Natl Cancer Inst 76:335–338

Coates ME (1982) Workshop on laboratory animal nutrition (letter to the editor). Food Chem Toxicol 20:149

Conybeare G (1988) Modulating factors: challenges to experimental design. *In*: Grice HC, Ciminera JL (eds) Carcinogenecity. The design, analysis, and interpretation of long-term animal studies. Springer-Verlag, Berlin Heidelberg New York, pp 149–172

Doll R, Peto R (1981) The causes of cancer: quantitative estimates of avoidable risks of cancer in the United States today. J Natl Cancer Inst 66:1191–1308

Donato K, Hegsted DM (1985) Efficiency of utilization of various sources of energy for growth. Proc Natl Acad Sci USA 82:4866–4870

Goodwin PJ, Boyd NF (1987) Critical appraisal of the evidence that dietary fat intake is related to breast cancer risk in humans. J Natl Cancer Inst 79:473–485

Ip C (1980) Ability of dietary fat to overcome the resistance of mature female rats to 7,12-dimethylbenz(α)anthracene-induced mammary tumorigenesis. Cancer Res 40:2785–2789

Ip C (1987) Fat and essential fatty acid in mammary carcinogenesis. Am J Clin Nutr 45:218–224

Ip C, Carter CA, Ip MM (1985) Requirement of essential fatty acid for mammary tumorigenesis in the rat. Cancer Res 45:1997–2001

Kritchevsky D, Weber MM, Klurfeld DM (1984) Dietary fat versus caloric content in initiation and promotion of 7,12-dimethylbenz(α)anthracene-induced mammary tumorigenesis in rats. Cancer Res 44:3174–3177

Landers RE, Norvell MJ, Bieber MA (1986) Oil gavage test-compound administration effects in NTP carcinogenesis-toxicity testing. Prog Clin Biol Res 22:357–374

Newberne PM, Bieri JG, Briggs GM, Nesheim MC (1978) Control of diets in laboratory animal experimentation. A report of the Committee on Laboratory Animal Diets, National Research Council. Inst Lab Anim Res News 21:A1–A12

Prentice RL, Pepe M, Self SG (1989) Dietary fat and breast cancer: a quantitative assessment of the epidemiological literature and a discussion of methodological issues. Cancer Res 49:3147–3177

Roe FJC (1981) Are nutritionists worried about the epidemic of tumors in laboratory animals? Proc Nutr Soc 40:57–65

Ross MH, Bras G (1975) Food preference and length of life. Science 190:165–167

Schatzkin A, Greenwald P, Byar DP, Clifford CK (1989) The dietary fat-breast cancer hypothesis is alive. JAMA 261:3284–3287

CHAPTER 8

Pathological Endpoints in Dietary Restricted Rodents—Fischer 344 Rats and B6C3F$_1$ Mice

W.M. Witt[1], W.G. Sheldon[2], and J.D. Thurman[2]

Introduction

Various types of dietary restriction imposed on weanling or middle-aged rodents have been demonstrated to increase maximum longevity (McCay et al. 1935; Ross 1961; Weindruch and Walford 1982; Yu et al. 1982), to reduce the incidence and/or delay the onset of several spontaneous and induced neoplasms (Carrol 1975; Leung et al. 1983; Maeda et al. 1985; Rao et al. 1987; Ross and Bras 1965, 1971; Sylvester et al. 1982; Tannenbaum 1942; Weindruch et al. 1986; Weindruch and Walford 1982; Yu et al. 1982), and have been established as an important factor influencing the effects of aging (Bertrand 1983; Barrows and Kokkonen 1977; Masoro 1985; Weindruch 1984). So overwhelming is the data, Pariza and Boutwell (1987) have concluded that overnutrition could be directly correlated to an increased risk of cancer, although they were unsure of the mechanism(s) responsible.

The mechanisms by which dietary restriction influences the aging process are unknown, although they are now the target of intense study. Mechanisms proposed thus far include inhibition of immunologic aging (Walford 1969; Weindruch et al. 1979), retardation of free radical-induced intracellular damage (Chipalkatti et al. 1983; Koizumi et al. 1987; Weindruch et al. 1980; Weindruch 1984), improved protein synthesis in older animals (Richardson and Cheung 1982), and influences of the endocrine and/or neural regulatory systems (Levin et al. 1981; Masoro 1988).

Dietary caloric restriction has been demonstrated to be more effective in delaying aging endpoints than is restriction of dietary protein, fat, or mineral (Iwasaki 1988). In rats, protein restriction without caloric restriction does not delay the occurrence of neoplasia, but does retard chronic nephropathy and cardiomyopathy, although much less effectively than caloric restriction (Maeda

[1]Division of Veterinary Services, HFT-240, National Center for Toxicological Research, Jefferson, AR 72079, USA
[2]Pathology Services Contract, MC–923, Pathology Associates, Inc., Suite 1, 15 Worman's Mill Court, Frederick, MD 21701, USA

et al. 1985). Under those defined conditions, restriction of protein alone caused a small increase (15%) in longevity which may have, in fact, been a benefit of the retardation of chronic nephropathy.

Most dietary restriction studies to date have utilized some type of uniquely formulated diet. In 1985, the National Center for Toxicological Research (NCTR) and the National Institute on Aging (NIA) began a joint project to establish colonies of variably aged ad libitum and restriction-fed rodents to be used in gerontologic and toxicity/carcinogenicity studies. An open formula diet was selected for use in seven of the nine colonies. Two of the rodent species/strains included in the project were the Fischer 344 (F344) rat and the B6C3F$_1$ hybrid mouse. These animals were included in the project because of the extensive historical pathological data base available for comparison (Haseman et al. 1984).

This report presents data which demonstrate the effects of a calorically restricted open formula diet in male and female F344 rats and B6C3F$_1$ mice examined at 12, 18, 24, and 30 months of age. The effort was not to relate findings to the aging process per se, but rather to present data that may be beneficial to investigators using the NCTR/NIA-produced animals in their research.

Materials and Methods

Animal Maintenance and Dietary Procedures

Male and female F344 rats and B6C3F$_1$ mice were obtained from NCTR stock, bred and maintained under specific-pathogen-free (SPF) conditions. Both the rat and mouse stocks were originally obtained from the National Institutes of Health colonies. Female C57BL/6N and male C3H/HeN MTV(-) mice were crossed to produce the F1 hybrid. The animals were maintained under SPF conditions throughout their lives in the NCTR barrier facility as part of the Project on Caloric Restriction (PCR) as described previously (Witt et al. 1989). The animals were weaned at 3 weeks of age and allocated to study at 4 weeks of age. Once on study the animals were singly housed in modified polycarbonate mouse cages (Witt et al. 1989). Through 13 weeks of age, all animals were fed ad libitum. To establish the restriction feeding regimen, ad libitum-fed animals were loaded on study first. Their total food consumption was measured for 2 weeks prior to the loading of their restriction-fed cohorts, and this data was used to calculate the restricted feeding regimen. At 14 weeks of age, a ramping procedure was initiated for the restricted animals to obtain the desired level of restriction (40%) over a 2-week period. During the 14th week of age, those animals were placed on a 10% feed restriction, i.e., they received 90% of the food consumed by the ad libitum-fed group. During the 15th week, the restriction was increased to 25% and finally, at 16 weeks, the rats were restricted to the targeted 40% restriction level. Hand feeding of the restriction-fed animals was initiated to ensure accuracy in delivering the proper amount of food to each.

Table 8.1. NIH-31 Open formula diet.[a]

Ingredients	Kilograms per ton
Fish meal (60% protein)	81.6
Soybean meal (48.59% protein)	45.4
Alfalfa meal (17% protein)	18.1
Corn gluten meal (60% protein)	18.1
Ground whole hard wheat	322.1
Ground #2 yellow shelled corn	190.5
Ground whole oats	90.7
Wheat middlings	90.7
Brewer dried yeast	9.1
Soy oil	13.6
Salt	4.5
Decalcium phosphate	13.6
Ground limestone	4.5
Premixes[b]	4.5
TOTAL	907.00

[a]The vitamin-fortified version of the diet fed to "restricted" animals contained approximately 1.67x the vitamin premixes as the standard diet.
[b]Provides 22000000 IU vitamin A; 3800000 IU vitamin D_3; 20 g vitamin K; 15 g dialpha-tocopheryl acetate; 700 g choline; 1 g folic acid; 20 g niacin; 25 g D-panothenic acid; 5 g riboflavin; 65 g thiamine; 14000 meg vitamin B_{12}; 2 g pyridoxine; 120 mg D-biotin; 400 mg colbalt carbonate; 5 g copper sulfate; 60 g iron sulfate; 400 g magnesium oxide; 100 g manganese oxide; 10 g zinc oxide; 1500 mg calcium iodate per ton of finished product

The NIH-31 open formula diet (Purina Mills, Inc., Richmond, Indiana) was fed to all four groups of animals in this study. Refer to Table 8.1 for composition of the diet. The restricted diet was fortified with approximately 1.67 times the amount of vitamins in the ad libitum diet. The restriction-fed animals therefore had access to the same level of vitamins as the ad libitum-fed animals.

Procedure for Animal Study

At 12, 18, 24, and 30 months of age, 12 rats and 15 mice from the ad libitum-fed and 12 rats and 15 mice from the restriction-fed groups were randomly selected, weighed, anesthetized by CO_2 inhalation for blood sample collection, and killed by CO_2 asphyxiation. The animals were examined for gross pathological lesions. The adrenal glands, sternum, brain, esophagus, eyes, Harderian glands, pituitary gland, heart, gastrointestinal tract, kidneys, salivary glands, liver, lung, lymph nodes, mammary gland, turbinates, pancreas, parathyroid glands, spleen, urinary bladder, trachea, thyroid glands, skeletal muscle, abdominal skin, spinal cord, thymus, tongue, and Zymbal's glands were examined. Additionally, in

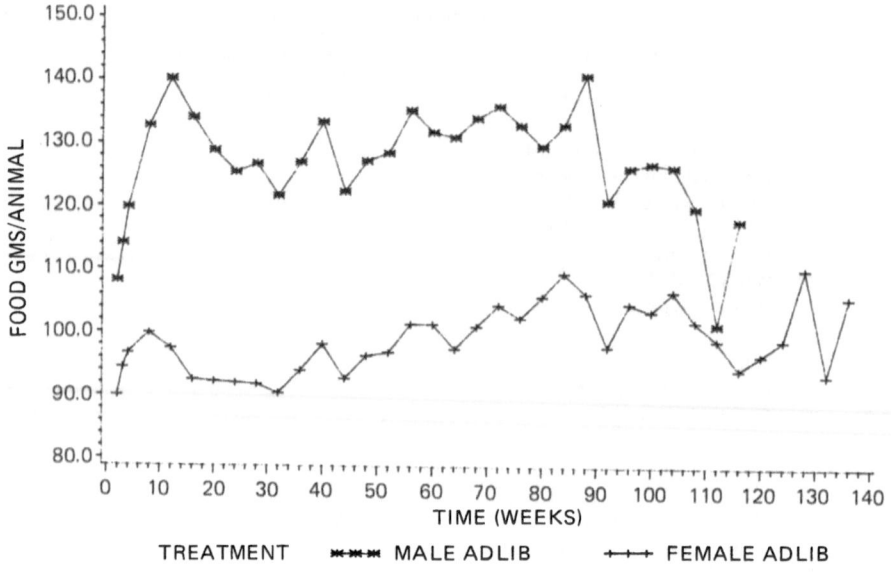

Figure 8.1. Mean food consumption of F344 rats (grams per animal)

males the following organs were examined: testes, epididymis, prostate, seminal vesicles, penis, and preputial glands. In females, vagina, clitoral glands, uterus, and ovaries were examined. The brain, thymus, kidneys, liver, and spleen were weighed and fixed immediately in 10% neutral buffered formalin. Other tissues and organs were not weighed, but were fixed immediately. Any other tissue or organ in which lesions were observed by gross inspection was excised and fixed. Fixed tissues were embedded in paraffin, sectioned at 4 μm and stained with hematoxylin-eosin for histopathological review.

Grading of Lesions

When present in the F344 rats, nephropathy and testicular interstitial cell hyperplasia were graded in a manner similar to reported methods (Coleman et al. 1977). Both types of lesions were graded on a scale from 0 to 4, 0 representing normal, 1 minimal, 2 mild, 3 moderate, and 4 severe.

Results

Food Intake

Mean food consumption for F344 rats and B6C3F$_1$ mice expressed as grams per animal is presented in Figs. 8.1 and 8.2, respectively. Food intake was measured from an ad libitum-fed cohort of animals. The food consumption data were

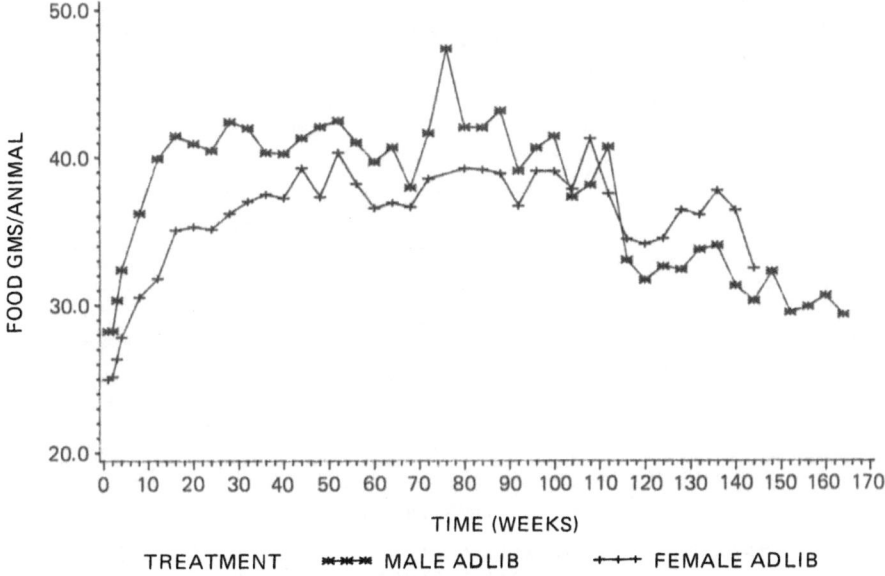

Figure 8.2. Mean food consumption of B6C3F₁ mice (in grams per animal)

used to calculate the restriction feeding regimen, i.e., the restriction-fed animals received daily 60% of the amount of food consumed by their ad libitum-fed cohorts.

Body Weights

Mean body weights for F344 rats and B6C3F₁ mice are presented in Figs. 8.3 and 8.4, respectively. In both species, the mean body weights of the ad libitum-fed animals steadily progressed to a peak (approximately 95 weeks for male rats, 110 weeks for female rats, and 105 weeks for male and female mice), then declined. Those of the restriction-fed animals remained relatively constant until approximately 125 weeks of age in the rats and 140 weeks of age in the mice, where gradual declines in mean body weights were observed in all but the female mice.

Longevity

The survival curves for F344 rats and B6C3F₁ mice are shown in Figs. 8.5 and 8.6, respectively. Special cohorts of rats and mice, separate from those used to collect pathological data, were used to establish the colony-specific longevity data. In the F344 rats, females lived longer than males in the respective ad libitum- and restriction-fed cohorts. The median length of life was significantly greater for the restriction-fed rats (126 and 133 weeks for males and females,

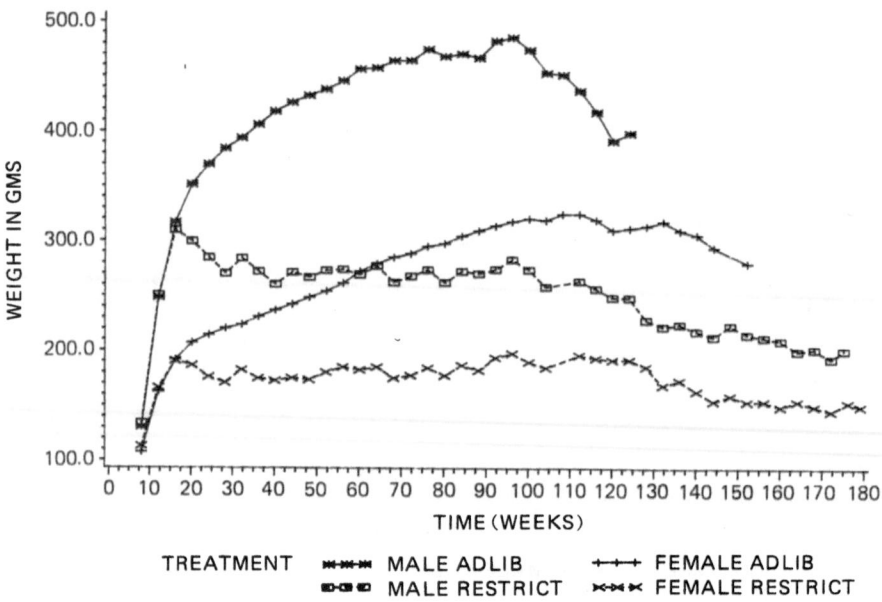

Figure 8.3. Mean body weights of F344 rats (in grams)

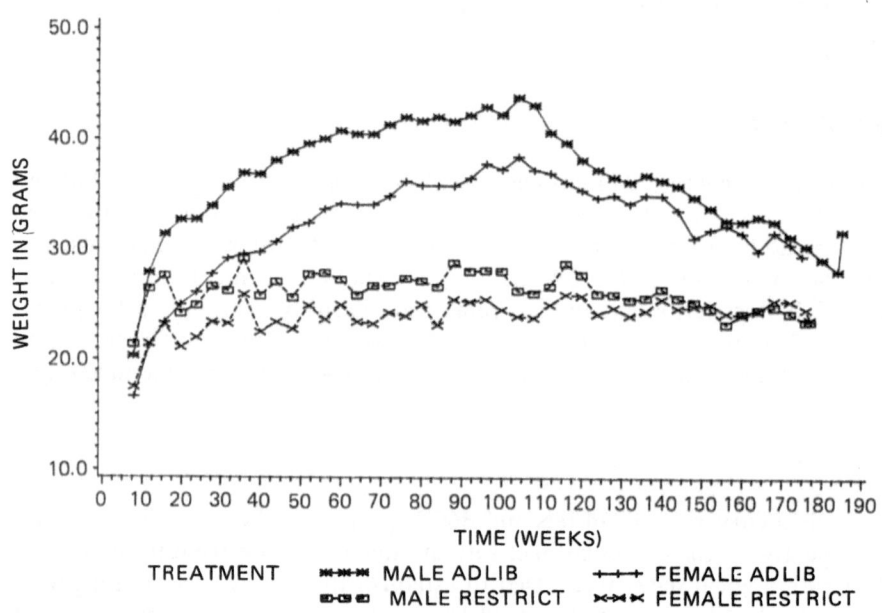

Figure 8.4. Mean body weights of B6C3F₁ mice (in grams)

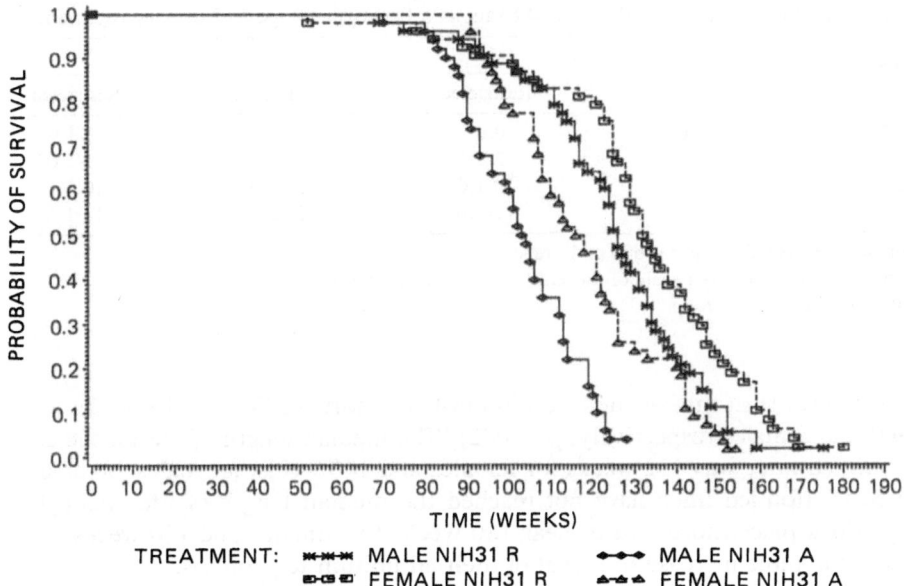

Figure 8.5. Survival curves for F344 rats

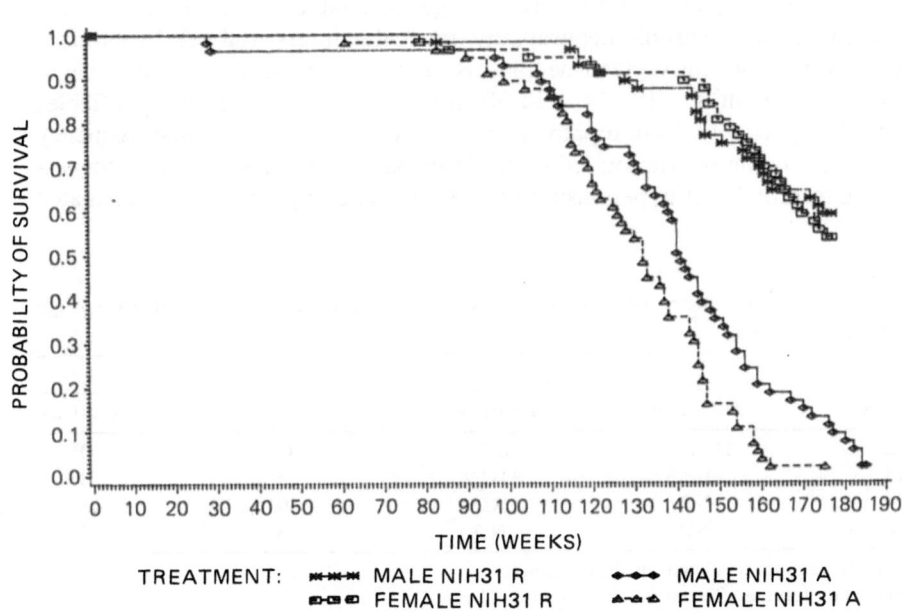

Figure 8.6. Survival curves for B6C3F₁ mice

Table 8.2. Chronic nephropathy in F344 rats as a function of age and dietary regimen.

Age (months)	Females		Males	
	Ad libitum	Restricted	Ad libitum	Restricted
12	0[a]/0[b]	0/0	7/1.0	1/1.0
18	11/1.1	2/1.0	12/1.9	9/1.0
24	12/2.0	10/1.0	12/2.6	10/1.1
30	12/2.4	11/1.0	9[c]/3.4	12/1.7

[a]Number of affected rats per group of 12 rats.
[b]Average severity on a 1–4 scale of those rats having nephropathy.
[c]Only nine rats in this group.

respectively) than that of their ad libitum-fed cohorts (103 and 116 weeks for males and females respectively; $p < .002$). The median length of life for the ad libitum-fed mice was 132 weeks for females and 140 weeks for males. To date, the restriction-fed mice have not reached the median length of life although projections place those figures near 180 weeks for females and 190 weeks for males, a notable increase over that of their ad libitum-fed cohorts.

Lesions

Both male and female rats and mice on the restricted regimen, with rare exceptions, had less frequency and severity of age-associated lesions at each period examined. Data on chronic nephropathy, testicular interstitial cell hyperplasia and adenomas, pituitary gland neoplasms, and mononuclear cell leukemias of F344 rats killed at 12, 18, 24, and 30 months of age are reported in Tables 8.2–8.5, respectively. Nephropathy had a later onset, with decreased frequency and severity in the restriction-fed group. This pattern was also apparent for testicular interstitial cell hyperplasia in male rats. The hyperplasia was tabulated

Table 8.3. Testicular interstitial cell hyperplasia/adenoma in F344 rats as a function of age and dietary regimen.

Age (months)	Hyperplasia		Adenoma	
	Ad libitum	Restricted	Ad libitum	Restricted
12	11[a]/2.2[b]	0/0	0[a]	0[a]
18	11/3.3	11/2.1	5	0
24	ND	ND	10	3
30	ND	ND	9[c]	2

ND hyperplasia not tabulated for these animals.
[a]Numbers on left are affected rats per group of 12 rats.
[b]Numbers on right represent average severity on a 1–4 scale of those rats having hyperplasia.
[c]Only nine rats in this group.

Table 8.4. Pituitary gland neoplasms in F344 rats as a function of age and dietary regimen.

Age (months)	Females		Males	
	Ad libitum	Restricted	Ad libitum	Restricted
12	1[a]/11[b]	0/12	3/12	0/12
18	3/10	0/10	6/10	2/12
24	11/12	4/12	9/11	4/12
30	11/12	4/12	9/9	7/11

[a]Numbers on left are affected rats.
[b]Numbers on right are animals examined per group.

only for the 12- and 18-month periods because interstitial cell adenomas frequently obliterated much of the testicular parenchyma in the ad libitum-fed group at the later periods.

The number of pituitary neoplasms was greatly reduced in both sexes of the restriction-fed group. Additionally, the ad libitum-fed group had two pituitary carcinomas (females, 30 months of age) while the restriction-fed group had only adenomas. The incidence of mononuclear cell leukemia showed no clear difference between the dietary regimens. The restriction-fed group had substantially fewer miscellaneous neoplasms than their ad libitum-fed cohorts, as presented in Table 8.6.

The number of neoplasms in the B6C3F$_1$ mice, especially those of lung, liver, and lymphoreticular tissues, were either nonexistent or were substantially less in restricted animals than in their ad libitum-fed cohorts (Table 8.7). This pattern was most dramatic in the 18- to 30-month-old males and the 30-month-old females. Uterine cystic hyperplasia, a common finding in the ad libitum-fed females of all age groups, was seldom observed in restriction-fed animals. A decrease in intrahepatocyte fat droplets was observed in both sexes of each age group in the restriction-fed mice.

Table 8.5. Mononuclear cell leukemia in F344 rats as a function of age and dietary regimen.

Age (months)	Females		Males	
	Ad libitum	Restricted	Ad libitum	Restricted
12	1[a]/12[b]	0/12	0/12	0/12
18	0/10	0/12	0/10	0/12
24	1/12	4/12	6/12	4/12
30	7/12	3/12	5/9	6/12

[a]Numbers on left are affected rats.
[b]Numbers on right are animals examined per group.

Table 8.6. Neoplasms[a,b] in F344 rats as a function of age and dietary regimen.

Neoplasm	12 Months				18 Months				24 Months				30 Months			
	Female		Male		Female		Male		Female		Male		Female		Male	
	A	R	A	R	A	R	A	R	A	R	A	R	A	R	A	R
Keratoacanthoma	0	0	0	0	0	0	1	0	0	0	0	0	0	0	0	0
Fibroma—skin	0	0	0	0	0	0	0	0	0	0	0	0	0	1	0	0
Fibroma—ear	0	0	0	0	0	0	0	0	0	0	1	0	0	0	0	0
Fibroadenoma (mammary)	0	0	0	0	1	0	0	0	5	0	0	0	6	1	2[c]	0
Mammary gland adenoma	0	0	0	0	0	0	0	0	2	0	0	0	0	0	0	0
Mammary gland adenocarcinoma	0	0	0	0	0	0	0	0	1	0	0	0	0	0	0	0
Hepatic adenoma	0	0	0	0	0	0	1	0	2	0	0	0	0	0	0	0
Brain neoplasm	0	0	0	0	0	0	0	0	0	0	1	0	0	0	0	0
Pheochromocytoma	0	0	0	0	0	1	0	0	0	0	0	2	3	0	3[c]	2
Pheochromocytoma (malignant)	0	0	0	0	0	0	0	0	0	0	0	0	0	0	1[c]	0
Thyroid C-cell adenoma	1	0	0	0	0	0	0	0	1	1	0	0	1	0	2[c]	0
Thyroid follicular adenocarcinoma	1	0	0	0	0	0	0	0	0	0	0	0	0	0	0	0
Islet cell adenoma	0	0	0	0	0	0	0	0	0	0	0	1	1	0	4[c]	2
Islet cell carcinoma	0	0	0	0	0	0	0	0	0	0	0	0	1	0	1[c]	0
Parathyroid adenoma	0	0	0	0	0	0	0	0	0	0	0	0	0	0	0	1
Renal adenoma	0	0	0	0	0	0	0	0	0	0	0	1	0	0	0	0
Clitoral gland adenoma	0	0	•	•	0	0	•	•	4	0	•	•	2	0	•	•
Leimyosarcoma (cervix)	0	0	•	•	0	0	•	•	0	0	•	•	1	0	•	•
Uterine adenocarcinoma	0	0	•	•	0	0	•	•	0	0	•	•	0	1	•	•
Uterine stromal polyp	0	0	•	•	0	0	•	•	0	0	•	•	0	1	•	•
Luteoma	0	0	•	•	0	0	•	•	2	0	•	•	1	0	•	•
Chondroma	0	0	0	0	0	0	0	0	0	0	1	0	0	0	0	1
Adrenal gland tumor (mixed)	0	0	0	0	0	0	0	0	0	0	0	0	0	0	0	0
Intestinal carcinoma	0	0	0	0	0	0	0	0	0	0	0	0	0	1	0	0
Retinal schwannoma	0	1	0	0	0	0	0	0	0	0	0	0	0	0	0	0
Malignant neoplasm (eyelid)	0	0	0	0	0	0	0	0	0	0	0	0	0	1	0	0
Mesothelioma	0	0	0	0	0	1	0	0	0	0	0	0	0	0	1[c]	0
Preputial adenoma	•	•	0	0	•	•	0	0	•	•	1	0	•	•	0	0

A, ad libitum; R, restricted

[a]Neoplasms other than testicular interstitial adenomas, pituitary neoplasms, and mononuclear cell leukemias.

[b]Numbers are affected rats per group of 12.

[c]Only nine rats in this group.

Table 8.7. Neoplasms in B6C3F$_1$ mice[a] as a function of age and dietary regimens.

Neoplasm	12 Months Female A	R	Male A	R	18 Months Female A	R	Male A	R	24 Months Female A	R	Male A	R	30 Months Female A	R	Male A	R
Hepatic neoplasms	0	0	0	0	1[b]	0[b]	4[b]	0[b]	1[c]	0[b]	4	0[b]	2	0	5	3
Hemangiosarcomas— hepatic	0	0	0	0	0	0	0	0	0	0	1	0[b]	0	0	1	0
Hemangiosarcoma— spleenic	0	0	0	0	0	0	0	0	0	0	0	0	0	0	1	0
Teratoma (ovary)	0	0	•	•	0[b]	1[b]	•	•	0	0	•	•	0	0	•	•
Cystic papillary adenoma (ovary)	0	0	•	•	1[b]	0[b]	•	•	0	0	•	•	0	0	•	•
Granulosa cell tumor (ovary)	0	0	•	•	0	0	•	•	0	0	•	•	1	0	•	•
Uterine polyp	1	0	•	•	0	0	•	•	1[c]	0[b]	•	•	2	0	•	•
Mammary adenocarcinoma	0	0	0	0	0	0	0	0	0	0	0	0	1	0	0	0
Lung neoplasms	0	0	0	0	1[b]	0[b]	1[b]	0[b]	0[c]	1[b]	6	1[b]	1	0	6	0
Thyroid follicle adenoma	0	0	0	0	0	0	0	0	0	0	0	0	4	1	0	0
Pituitary adenoma	0	0	0	0	0	0	0	0	0	0	0	0	4[c]	0	0	0
Adrenal cortical adenoma	0	0	0	0	0	0	0	0	0	0	0	0	0	0	1	0
Harderian gland adenoma	0	0	0	0	0	0	0	0	0	0	0	0	0	0	1	0
Lymphoma	0	0	0	0	0	0	0	0	1[c]	1[b]	0	1[b]	4	0	3	0
Sarcoma (skin)	0	0	0	0	0	0	0	0	1[c]	0[b]	0	0	0	0	0	0
Gastric papilloma	0	0	0	0	0	0	0	0	1[c]	0[b]	0	0	0	0	0	0
Gastric squamous cell carcinoma	0	0	0	0	0	0	0	0	0	0	0	0	1	0	0	0

A, ad libitum; R, restricted
[a]Numbers are affected mice per group of 15.
[b]Number of affected mice per group of 14.
[c]Number of affected mice per group of 13.

Discussion

The present findings are in accord with previous reports demonstrating that dietary restriction can increase mean and maximum life span (McCay et al. 1935; Ross 1961; Weindruch and Walford 1982; Yu et al. 1982) and have a positive influence on the occurrence and progression of many late-life diseases, both neoplastic and nonneoplastic (Maeda et al. 1985; Ross and Bras 1965; Tannenbaum 1942; Weindruch and Walford 1982; Yu et al. 1982). The most striking finding of this study is that these observations were found in F344 rats and B6C3F$_1$ mice fed reduced amounts (40% restriction) of an "over the counter" open formula rodent chow.

Many studies have demonstrated that dietary restriction is capable of tremendous influence on the aging process (Cutler 1981; Ross 1978; Sacher 1977; Walford 1969). Because most of those studies achieved similar results, i.e., some expression of life span extension, using a wide range of food sources (Weindruch 1985), researchers are now beginning to focus more on the role of caloric restriction as opposed to restriction of a specific dietary ingredient such as protein, fat, or minerals. Support for this line of reasoning was recently offered by Maeda et al. (1985), who demonstrated that protein restriction without caloric restriction had little effect on longevity, except for the retardation of kidney disease, or on the physiological and pathological processes of aging. Similarly, Iwasaki et al. (1988) showed that dietary restriction of fat or minerals in the same way had no influence on longevity.

The F344 rat, especially the male, has been a major animal model for aging and toxicological research, as has been the B6C3F$_1$ mouse for toxicological research. Present results indicate that use of proper caloric restriction feeding regimens can prevent or at least delay the onset of several pathological endpoints (chronic nephritis, pituitary adenomas, adrenal pheochromocytomas, and testicular interstitial cell tumors in the rats and benign and malignant liver neoplasms and pituitary adenomas in the mice), which may limit the use of these two rodent strains in gerontologic research. This is important in that current gerontologic literature is filled with reports dealing with the effects of nutrition on aging although most are based on many different rodent strains and just as many different diets and dietary regimens. For this reason, there is a great need for standard or base rodent models (to include diet and dietary regimen) with well-documented pathological profiles to be available for researchers studying the aging process.

A specific goal of the joint NCTR/NIA Project on Caloric Restriction is to achieve positive modulation of the aging process in specific rodent strains by dietary caloric restriction, thus providing gerontologists with a powerful experimental model. Present findings indicate that this goal is being accomplished. As data becomes available, future reports depicting the longitudinal and cross-sectional pathology of each of the Project on Caloric Restriction rodent strains are anticipated.

References

Barrows CH, Kokkonen GC (1977) Relationship between nutrition and aging. Adv Nutr Res 1:253–298

Bertrand HA (1983) Nutrition-aging interactions: life-prolonging action of food restriction. In: Rothstein M (ed) Review of biological research in aging, vol 1. Liss, New York, pp 359–378

Carrol KK (1975) Experimental evidence of dietary factors and hormone-dependent cancers. Cancer Res 35:3374–3384

Chipalkatti S, De AK, Aiyar AN (1983) Effect of diet restriction on some biochemical parameters related to aging in mice. J Nutr 113:944–950

Coleman GL, Barthold SW, Osbaldiston GW, Foster SJ, Jonas AM (1977) Pathological changes during aging in barrier-reared Fischer 344 male rats. J Gerontol 32:258–278

Cutler RG (1981) Life span extension. In: McGaugh JL, Kiesler SB (ed) Aging: biology and behavior. Academic, New York, pp 31–76

Haseman JK, Huff J, Boorman GA (1984) Use of historical control data in carcinogenicity studies in rodents. Toxicol Pathol 12:126–135

Iwasaki K, Gleiser CA, Masoro EJ, McMahan CA, Seo E, Yu BP (1988) The influence of dietary protein source on longevity and age-related disease processes of Fischer rats. J Gerontol (Biol Sci) 43:B5–B12

Koizumi A, Weindruch R, Walford RL (1987) Influences of dietary restriction and age on liver enzyme activities and lipid peroxidation in mice. J Nutr 117:361–367

Leung FC, Aylsworth CF, Meiters J (1983) Counteraction of underfeeding-induced inhibition of mammary tumor growth in rats by prolactin and estrogen administration. Proc Soc Exp Biol Med 173:159–163

Levin P, Janda JK, Joseph JA, Ingram DK, Roth GS (1981) Dietary restriction retards the age-associated loss of striatal dopaminergic receptors. Science 214:561–562

Maeda H, Gleiser CA, Masoro EJ, Murata I, McMahan CA, Yu BP (1985) Nutritional influences on aging of Fischer 344 rats. II. Pathology. J Gerontol 40:671–688

Masoro EJ (1985) Nutrition and aging—a current assessment. J Nutr 115:842–848

Masoro EJ (1988) Minireview: food restriction in rodents: an evaluation of its role in the study of aging. J Gerontol 43:B59–B64

McCay CM, Crowell MF, Maynard LF (1935) The effect of retarded growth upon the length of life span and upon the ultimate body size. J Nutr 10:63–79

Pariza MW, Boutwell RK (1987) Historical perspective: calories and energy expenditure in carcinogenesis. Am J Clin Nutr 54:151–156

Rao GN, Piegorsch WW, Haseman JK (1987) Influence of body weight on the incidence of spontaneous tumors in rats and mice of long-term studies. Am J Clin Nutr 45:252–260

Richardson A, Cheung HF (1982) The relationship between age-related changes in gene expression, protein turnover, and the responsiveness of an organism to stimuli. Life Sci 31:605–613

Ross MH (1961) Length of life and nutrition in the rat. J Nutr 75:197–210

Ross MH (1978) Nutritional regulation of aging. In: Behnke JA, Finch CE, Monent GB (ed) The biology of aging. Plenum, New York, pp 173–189

Ross MH, Bras G (1965) Tumor incidence patterns and nutrition in the rat. J Nutr 87:245–260

Ross MH, Bras G (1971) Lasting influence of early caloric restriction on prevalence of neoplasms in the rat. J Natl Cancer Inst 47:1095–1113

Sacher GA (1977) Life table modification and life prolongation. In: Finch CE, Hayflick L (ed) Handbook of the biology and aging. Van Nostrand Reinhold, New York, pp 582–638

Sylvester PW, Aylsworth CF, Van Vugt DA, Meiter J (1982) Influence of underfeeding during the "critical period" or thereafter on carcinogen-induced mammary tumors in rats. Cancer Res 42:4943–4947

Tannenbaum A (1942) The genesis and growth of tumors. II. Effects of caloric restriction per se. Cancer Res 2:460–467

Walford RL (1969) The immunologic theory of aging. Munksgaard, Copenhagen, pp 104–111

Weindruch R (1984) Dietary restriction and the aging process. In: Armstrong D, Sohal R, Culter R, Slater TF (ed) Free radicals in molecular biology, aging, and disease. Raven, New York, pp182–202

Weindruch R (1985) Aging in rodents fed restricted diets. J Am Geriatr Soc 33:125–132

Weindruch R, Walford RL (1982) dietary restriction in mice beginning at 1 year of age: effects on life span and spontaneous cancer incidence. Science 215:1415–1418

Weindruch RH, Cheung MK, Verity MA, Walford RL (1980) Modification of mitochondrial respiration by aging and dietary restriction. Mech Age Dev 12:375–392

Weindruch RH, Kristie JA, Cheney KE, Walford RL (1979) Influence of controlled dietary restriction on immunologic function and ageing. Fed Proc 38:2007–2016

Weindruch R, Walford RL, Fligel S, Guthrie D (1986) The retardation of aging in mice by dietary restriction: longevity, cancer, immunity and lifetime energy intake. J Nutr 116:641–654

Witt WM, Brand CD, Attwood VG, Soave OA (1989) A nationally supported study on caloric restriction of rodents. Lab Anim 18:37–43

Yu BP, Masoro EJ, Murata I, Bertrand HA, Lynd FT (1982) Life span study of SPF Fischer 344 male rats fed ad libitum or restricted diets: longevity, growth, lean body mass and disease. J Gerontol 37:130–141

CHAPTER 9

Effects of Dietary Restriction on Age-Associated Pathological Changes in Fischer 344 Rats

K. Imai[1], S. Yoshimura[1], K. Hashimoto[1], and G.A. Boorman[2]

Introduction

Since McCay et al. reported beneficial effects of caloric restriction on longevity in rats (McCay et al. 1935), the influence of dietary factors on longevity and spontaneously occurring pathological lesions have been studied in experimental animals for several decades by many authors (Berg and Simms 1961; Maeda et al. 1985; Tannenbaum 1945; Tucker 1979; Yu et al. 1985). Food or caloric restriction inhibits the development of some tumors or age-related pathological lesions such as chronic nephropathy and myocardiopathy. However, there are few available data explaining a rational relationship between food restriction and hematological, biochemical, or pathological findings in Fischer rats.

This report describes the effects of restricted food intake on hematological and biochemical effects and age-related pathological lesions in Fischer rats.

Materials and Methods

Experimental Design

A total of 380 Fischer (F344/Du; Crj) rats of both sexes, 4 weeks old, were purchased from Charles River Japan (Kanagawa, Japan). They were housed in barrier-sustained animal quarters at 24°C ± 1°C, and 55% ± 5% humidity. Animals were kept individually in a wire bottom metal cage (243 X 300 X 190 mm). Following quarantine for 2 weeks, the rats that appeared clinically normal were divided into three groups, each containing 60 females and 60 males.

Group 1 was fed a standard commercial diet CRF-1 (Oriental Yeast Co., Tokyo, Japan) ad libitum. Group 2 was fed daily 8 g and 12 g of diet for females and males, respectively (approximately 67% of ad libitum intake). Group 3 was

---

[1]Food and Drug Safety Center, Hadano, Kanagawa 257, Japan
[2]National Institute of Environmental Health Science, Research Triangle Park, NC 27709, USA

fed twice a week a total of 67% of ad libitum intake (28 g for females and 42 g for males each time). Tap water was available ad libitum.

Ten animals from each group were killed after 12 months. The remaining 50 animals of each group were maintained for 2 years. Clinical observations and mortality were recorded every day. Body weights in all groups and food consumptions in group 1 were recorded once a week during the first 3 months and every month thereafter.

Blood Chemistry

A blood sample was taken from the inferior caval vein of rats at scheduled autopsy, using a heparinized syringe under pentobarbital anesthesia and the plasma separated by centrifugation. The plasma concentrations of glucose, total protein, albumin, total cholesterol, urea nitrogen, creatinine, lactic dehydrogenase (LDH), glutamate oxalate transaminase (GOT), glutamate pyruvate transaminase (GPT), alkaline phosphatase (Alp), calcium, sodium, potassium, chloride, and lipid peroxide were measured.

Hematology

Blood samples for hematological examinations were obtained from the tail vein of all survivors about 20 h before scheduled autopsy. Total red blood cell counts, hemoglobin, hematocrit, total white blood cell counts, and platelets were evaluated using an electric cell counter. Blood smears were stained with Wright-Giemsa and differential cell count of white blood cells was calculated with light microscopy. Reticulocytes were scored by the Brecher's method, using blood samples collected from the inferior caval vein of survivors of each group at scheduled autopsy.

Pathological Examinations

All animals that died or became moribund during the experimental period were autopsied by standard procedures. At the end of the experiment, rats were exsanguinated from the inferior caval vein under pentobarbital anesthesia, and the blood samples were utilized for the blood examinations mentioned above. Autopsy was performed, and organs and tissues were fixed in 10% neutral formalin. The sections processed by standard technique were stained with hematoxylin and eosin. Histological examinations were primarily completed in the Food and Drug Safety Center, and the duplicate slides were reevaluated by members of the Pathological Working Group in the National Institute of Environmental Health Sciences (NIEHS) under a joint program.

Results and Discussion

Clinical Observations

Food consumption of male rats in group 1 was almost constant throughout the experimental period, while for female rats food consumption increased with age. Average food consumption during the experimental period was 12.06 g/day per rat and 17.65 g/day per rat for females and males, respectively.

The movement of rats of both sexes in groups 2 and 3 was more active than those in group 1. Animals in groups 2 and 3 appeared to be more sensitive and aggressive than those in group 1. Body weights of rats in groups 2 and 3 declined slightly during the first 2 weeks and gradually increased thereafter. At the end of the experiment, weight gains were suppressed at 52.5% in female and 63.2% in male rats of group 2, and 55% in female and 60.5% in male rats of group 3, as compared with those of group 1.

Twelve females and 15 males in group 1, 7 females and 10 males in group 2, and 10 females and 6 males in group 3 died during the experimental period. A statistically significant mortality difference was noted between the males of group 1 and group 3 ($p < 0.01$).

Blood Chemistry

Groups Treated for 12 Months

Decreased plasma levels of total protein, albumin, total cholesterol, and lipid peroxide were noted in both sexes in groups 2 and groups 3 at 12 months (Fig. 9.1).

Groups Treated for 24 Months

The plasma concentrations of total protein, urea nitrogen, total cholesterol, lipid peroxide, and calcium were significantly lower in both sexes in group 2 and group 3 than those of group 1. Alp and LDH activity was also decreased in the former groups at 24 months (Fig. 9.2), while a slight increase of GPT activity was also noted in male rats of groups 2 and 3 (Fig. 9.2).

Hematological Examinations

Groups Treated for 12 Months

A slight increase in red blood cell count and marked increases in hemoglobin and hematocrit were observed in both sexes in groups 2 and 3. A marked decrease of total white blood cell count was noted in the males in group 2 and group 3, while the count only tended to decrease in the females in groups 2 and 3. No significant differences in differential white blood cell count were noted between groups as shown in Fig. 9.3.

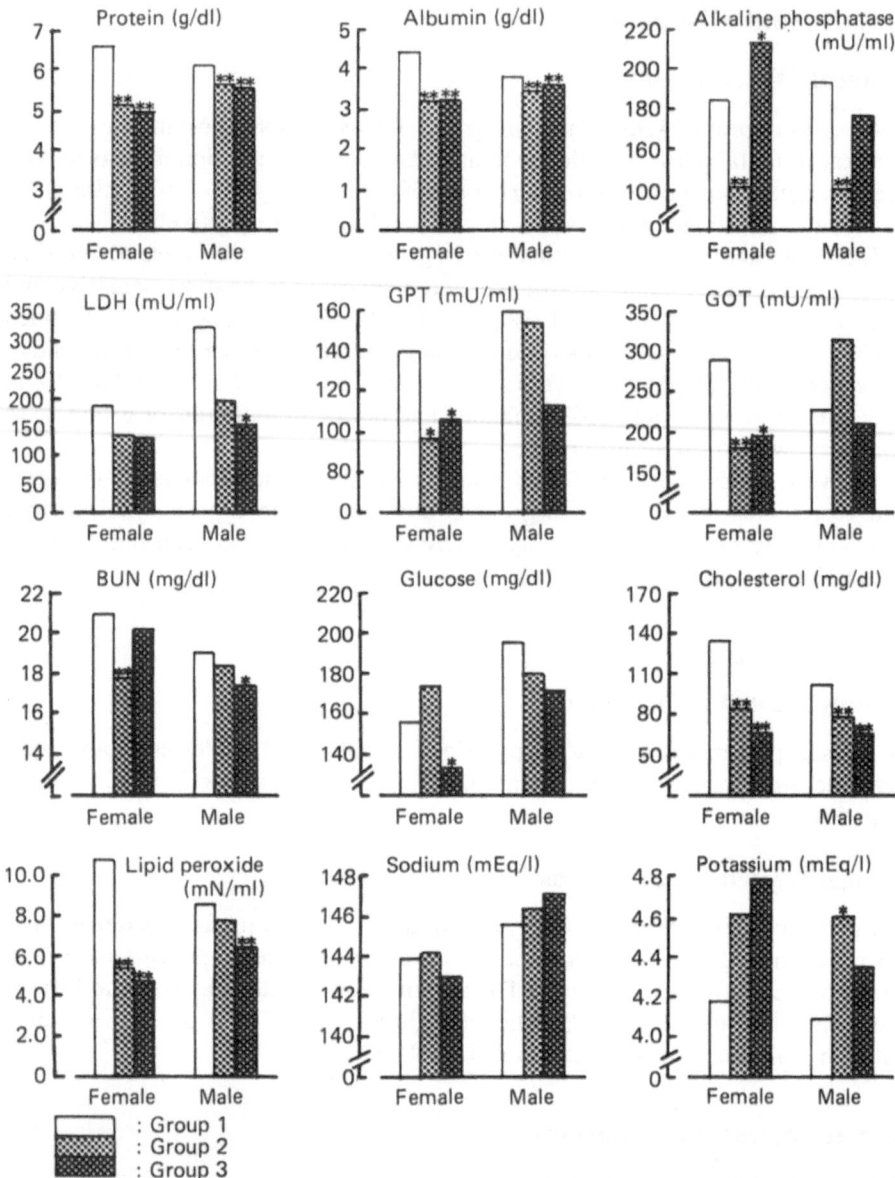

Figure 9.1. Biochemical findings of blood plasma in F344 rats fed restricted amounts of diet for 12 months. *LDH*, lactic dehydrogenase; *GPT*, glutamate pyruvate transaminase; *GOT*, glutamate oxalate transaminase; *BUN*, blood urea nitrogen. Group 1: 10 females, 10 males; group 2: 10 females, 10 males; group 3: 10 females, 10 males. *, significantly different from group 1, $p < 0.05$; **, significantly different from group 1, $p < 0.01$

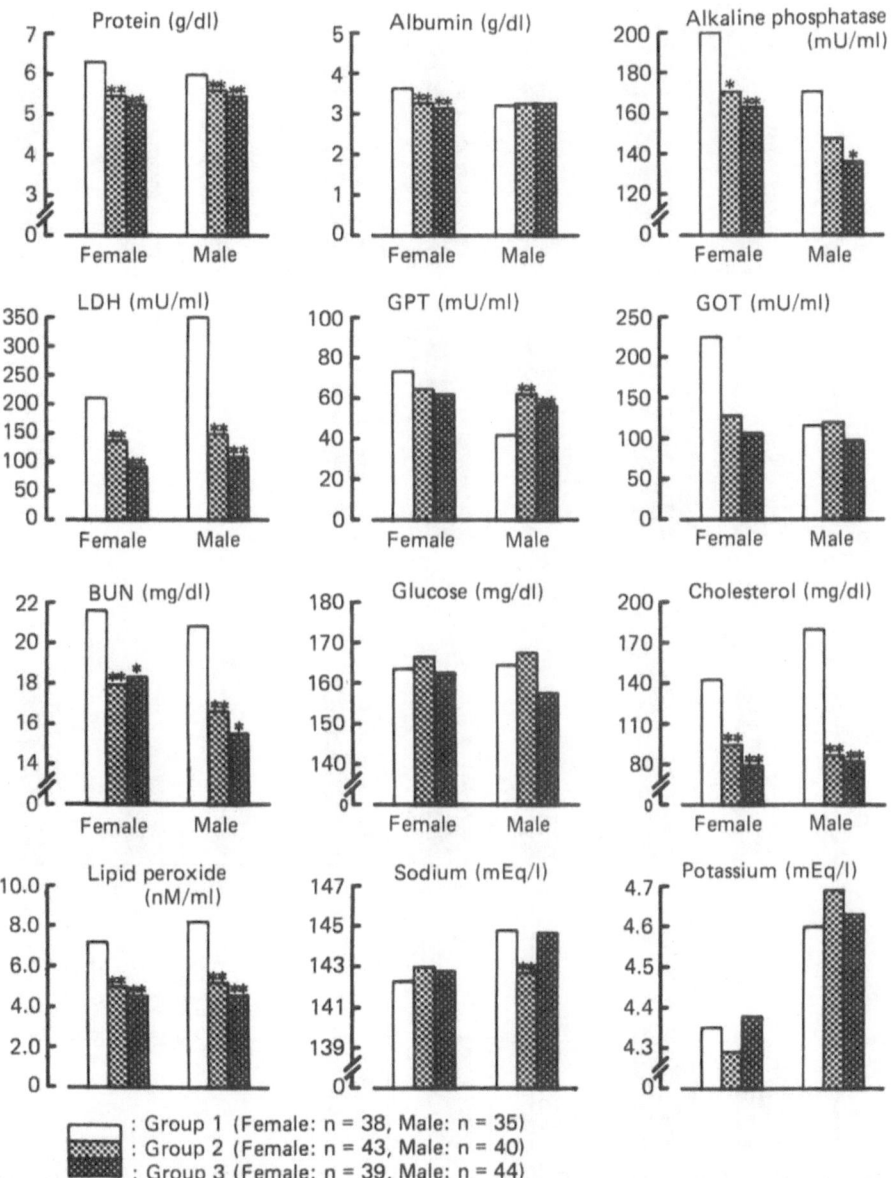

: Group 1 (Female: n = 38, Male: n = 35)
: Group 2 (Female: n = 43, Male: n = 40)
: Group 3 (Female: n = 39, Male: n = 44)

Figure 9.2. Biochemical findings of blood plasma in F344 rats fed restricted amounts of diet for 24 months. Group 1: 38 females, 35 males; group 2: 43 females, 40 males; group 3: 39 females, 44 males

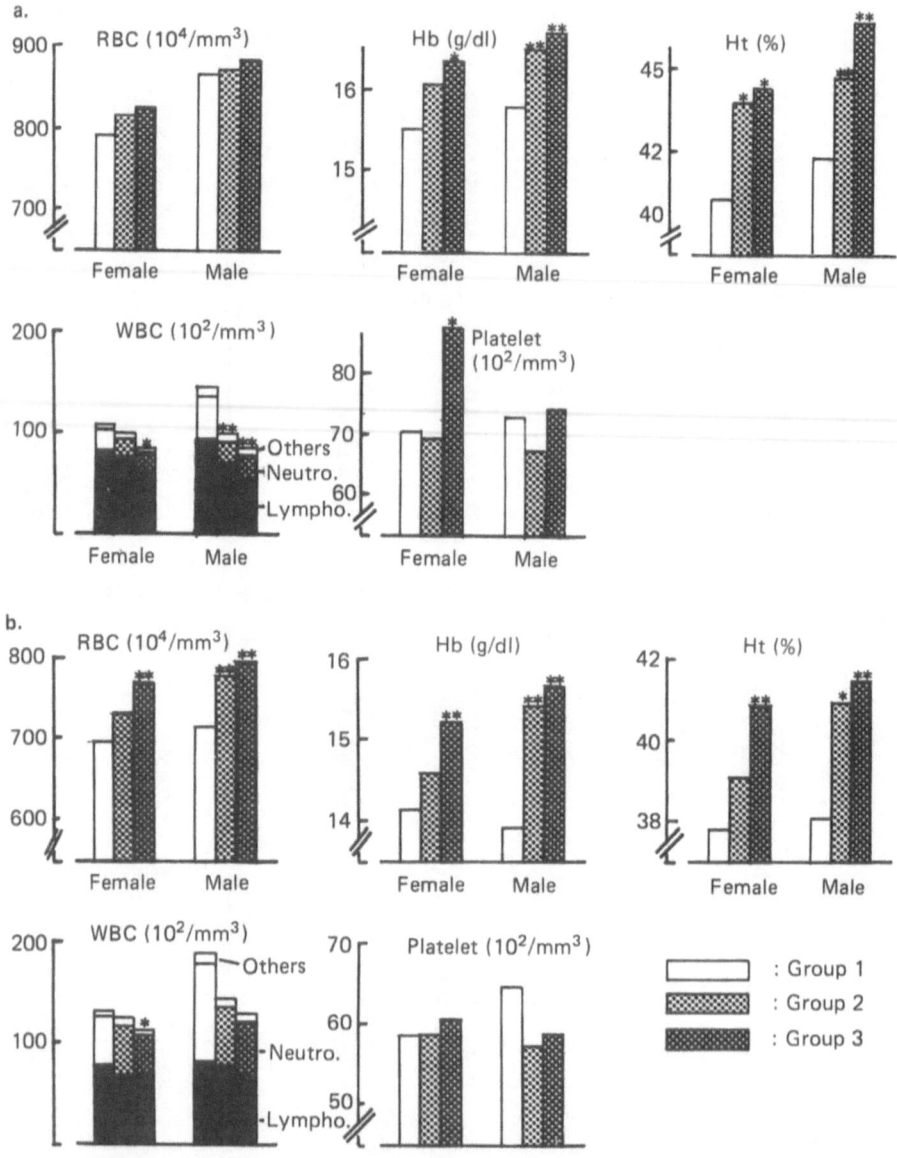

Figure 9.3a,b. Hematological findings of peripheral blood in F344 rats fed restricted amounts of diet for 12 months (**a**) and 24 months (**b**). *RBC*, red blood cell count; *Hb*, hemoglobin; *Ht*, hematocrit; *WBC*, white blood cell count. Groups and statistical data is the same as in Fig. 9.1 for (a) and for (b) is the same as in Fig. 9.2

Groups Treated for 24 Months

Significant increases in red blood cell count, hemoglobin, and hematocrit were noted in the female rats in group 3 and in the male rats in groups 2 and 3 (Fig. 9.3). The number of neutrophilic leukocytes decreased in the female rats in group 3 and in the male rats in groups 2 and 3.

Pathological Examinations

Groups Treated for 12 Months

Several preneoplastic lesions, including focal hyperplasia of the pituitary glandular cells, focal hyperplasia of testicular interstitial cells, C-cell hyperplasia in the thyroid, or altered cell foci in the liver, were noted in animals in all groups. The incidence, however, of focal hyperplasia of testicular interstitial cells was significantly lower in groups 2 and 3 than in group 1. As for overt neoplasia, two pituitary adenomas and a bronchiolar/alveolar cell adenoma were observed only in group 2.

Groups Treated for 24 Months

In general, the number of tumor-bearing animals was significantly lower in groups 2 and group 3 in both sexes (Table 9.1). Especially, the incidence of mononuclear cell leukemia, pituitary adenoma, testicular interstitial cell tumors, adrenal pheochromocytoma, and bronchiolar/alveolar cell tumor was far lower in male rats of both food-restricted groups. Significantly fewer pituitary adenoma and endometrial stromal polyps were noted in female rats in groups 2 and 3, and no breast tumors were observed in these groups.

The incidence of preneoplastic lesions and focal hyperplasia of testicular interstitial cells, however, was higher in groups 2 and 3, and a high incidence of focal hyperplasia of the pituitary glandular cells was also noted in group 3 in both sexes. On the other hand, the number of rats having altered cell foci of the liver was significantly lower in the females in group 3 and in the males in groups 2 and 3.

The nonneoplastic lesions most frequently observed in the ad libitum feeding group (group 1) in both sexes were chronic nephropathy, chronic myocardiopathy, and simple bile duct hyperplasia. The kidneys were more severely affected in males than in females, and severity and incidence of nephropathy were far lower in groups 2 and 3 in both sexes (Table 9.2).

Significant differences in the severity and incidence of chronic myocardiopathy were noted between group 1 and group 2 or group 3 in both sexes (Table 9.3). The incidence of hyperplasia of the bile ducts was also significantly lower in male rats in group 3 (Table 9.4).

In this experiment, a restriction of food intake by approximately 30% induced a marked reduction in the incidence of some tumors, including mononuclear cell leukemia, pituitary adenoma, testicular interstitial cell tumor, adrenal

Table 9.1. Incidence of preneoplastic and neoplastic lesions of F344 rats fed restricted amount of diet for 24 months.

Organ and histological diagnosis	Female			Male		
	1	2	3	1	2	3
No. of rats used	50	50	50	50	50	50
No. of tumor-bearing rats	35	13[a]	13[a]	50	44[b]	34[a,d]
Mononuclear cell leukemia	3	3	2	11	4[b]	0[a]
Pituitary						
Focal hyperplasia(A)	5	3	12[d]	5	9	14[b]
Adenoma(B)	15	7[a]	3[a]	21	6[a]	11[b]
(A+B)	20	10[b]	15	26	15[b]	25
Breast						
Adenoma/fibroadenoma(A)	5	0[b]	0[b]	0	0	0
Adenocarcinoma(B)	1	0	0	1	0	0
(A+B)	6	0[b]	0[b]	1	0	0
Fibroma	0	0	0	1	2	0
Uterus						
Endometrial stromal polyp	10	3[b]	1[a]	–	–	–
Adenocarcinoma	1	0	0	–	–	–
Leiomyosarcoma	1	0	0	–	–	–
Maligmant schwannoma	1	0	0	–	–	–
Testis						
Focal hyperplasia of interstitial cells(A)	–	–	–	1	9[a]	12[a]
Interstitial cell tumor(B)	–	–	–	45	34[a]	25[a]
(A+B)	–	–	–	46	43	37[b]
Adrenal						
Focal hyperplasia of pheochromocyte(A)	0	0	0	2	1	0
Pheochromocytoma(B)	3	0	1	11	3[b]	2[b]
(A+B)	3	0	1	13	4[b]	2[b]
Focal hyperplasia of cortical cells(A)	1	0	1	0	0	0
Cortical cell adenoma(B)	1	0	0	0	1	0
(A+B)	2	0	1	0	1	0
Thyroid						
C-cell hyperplasia(A)	1	3	0	8	2[b]	2[b]
C-cell tumor(B)	1	1	0	2	1	5
(A+B)	2	4	0	10	3[b]	7
Lung						
Focal hyperplasia of alveolar epithelium(A)	0	0	0	2	1	0
Bronchiolar/alveolar cell tumor(B)	1	0	0	5	0[b]	1
(A+B)	1	0	0	7	1[b]	1[a]
Ovary						
Theca cell tumor	1	0	0	–	–	–
Brain						
Astrocytoma	0	1	1	0	1	0
Malignant reticulosis	0	1	0	0	0	0
Ear						
Neurofibroma	1	0	0	0	0	0
Ear Duct						
Squamous cell carcinoma	1	0	0	1	0	0

Table 9.1. (*Continued*).

Organ and histological diagnosis	Female			Male		
	1	2	3	1	2	3
Skin						
Papilloma	0	0	0	2	0	0
Keratocanthoma	0	0	0	1	0	0
Basal cell carcinoma	0	0	0	0	1	0
Subcutaneous						
Rhabdomyosarcoma	0	0	0	0	1	0
Malignant schwannoma	0	0	0	0	0	1
Oral cavity						
Papilloma	0	0	1	0	0	0
Stomach						
Papilloma	0	0	0	0	1	0
Undifferentiated cell sarcoma	0	0	1	0	0	0
Small intestine						
Adenoma	0	0	0	0	1	0
Adenocarcinoma	1	0	0	0	0	0
Leiomyosarcoma	1	0	0	0	0	0
Liver						
Neoplastic nodule	0	0	0	1	0	1
Histiocytic sarcoma	0	0	0	1	0	0
Pancreas						
Islet cell tumor	1	0	1	4	4	2
Kidney						
Nephroblastoma	0	0	0	1	0	0
Transitional cell carcinoma	0	1	0	0	0	0
Urinary bladder						
Transitional cell papilloma	0	1	0	0	0	0
Thymus						
Thymoma	0	1	0	0	0	0
Spleen						
Fibrosarcoma	0	0	0	1	0	0
Abdominal cavity						
Mesothelioma	0	0	0	1	2	1
Paraganglioma	0	0	0	1	0	1
Malignant fibrous histiocytoma	0	0	0	1	0	0
Vertebra						
Chordoma	0	0	0	1	0	0
Bone						
Osteosarcoma	0	0	0	1	1	0

1, Ad libitum feeding group; 2, daily restricted group; 3, intermittent feeding group
[a]Statistically significant difference compared with groups 1 and 2 or 3 ($p < 0.01$).
[b]Statistically significant difference compared with groups 1 and 2 or 3 ($p < 0.05$).
[c]Statistically significant difference compared with groups 2 and 3 ($p < 0.01$).
[d]Statistically significant difference compared with groups 2 and 3 ($p < 0.05$).

Table 9.2. Effect of dietary restriction on chronic nephropathy in F344 rats.

Sex	Group	n^a	Grades of nephropathy					Analysis[b]
			0	1	2	3	4	
	Ad libitum	38	3	21	12	2	0	–
Female	Daily restricted	43	39	4	0	0	0	0.01[c]
	Intermittent feeding	40	34	6	0	0	0	0.01
	Ad libitum	35	0	0	10	22	3	–
Male	Daily restricted	40	19	21	0	0	0	0.01
	Intermittent feeding	44	32	12	0	0	0	0.01

[a]Number of rats examined.
[b]Mann-Whitney's U test was used to compare the incidence and severity of chronic nephropathy.
[c]$p < 0.01$ as compared to ad libitum feeding group.

pheochromocytoma, bronchiolar/alveolar cell tumor, breast tumor, and endometrial stromal polyps. In contrast to the development of neoplastic lesions, the incidence of preneoplastic lesions, including focal hyperplasia of testicular interstitial cells and pituitary glandular cells, was significantly greater in food restriction groups, presumably reflecting that those lesions had not progressed to tumors. In addition, food restriction also reduced the incidence of some non-neoplastic diseases, chronic nephropathy, and myocardiopathy, and showed markedly lower plasma lipid peroxide, protein, and cholesterol concentrations.

Walford et al. (1987) recently reviewed the possible mechanism for retardation of aging by dietary restriction, including effect on the immune system, on basal state and proliferation potential, metabolic rate, DNA repair, and levels of free radical scavengers. Although the mechanism concerning the inhibitory effect of dietary restriction on tumorigenesis is not clarified as yet, the following possibilities are being considered: (a) Dietary restriction may inhibit tumor development through a retardation of the decline in DNA repair capacity that occurs with aging. (b) Food restriction may suppress tumor growth by affecting generation or persistence of free radicals. (c) Dietary restriction has been shown to cause

Table 9.3. Effect of dietary restriction on myocardiopathy in F344 rats.

Sex	Group	n^a	Grades of fibrosis					Analysis[b]
			0	1	2	3	4	
	Ad libitum	38	2	12	22	2	0	–
Female	Daily restricted	43	14	20	9	0	0	0.01[c]
	Intermittent feeding	40	10	23	7	0	0	0.01
	Ad libitum	35	0	0	19	15	1	–
Male	Daily restricted	40	0	5	31	4	0	0.01
	Intermittent feeding	44	0	1	41	2	0	0.01

[a]Number of rats examined.
[b]Mann-Whitney's U test was used to compare the incidence and severity of myocardiopathy.
[c]$p < 0.01$ as compared to ad libitum feeding groups.

Table 9.4. Incidence of altered cell foci and simple hyperplasia of bile duct in F344 rats fed restricted amount of diet for 12 or 24 months.

Age (Months)	Group	Altered cell foci Female	Altered cell foci Male	Simple hyperplasia of bile duct Female	Simple hyperplasia of bile duct Male
12	1	0/10	0/10	0/10	10/10
	2	0/10	0/10	0/10	10/10
	3	0/10	0/10	0/10	3/10[a,b]
24	1	28/50	28/50	13/50	45/50
	2	22/50	12/50[a]	7/50	48/50
	3	6/50[a,b]	5/50[a]	2/50[a]	29/50[a,b]

Group 1, ad libitum feeding group; group 2, daily restricted group; group 3, intermittent feeding group

[a]Statistically significant difference compared with group 1 and groups 2 or 3 ($p < 0.01$).
[b]Statistically significant difference compared with groups 2 and 3 ($p < 0.01$).

decrease of some tropic hormones, including LHRH and TRH in the hypothalamus, and dysfunction or atrophy of these endocrine organs was also noted (Badger et al. 1985; Connors et al. 1985). Therefore, it is suggested that dietary restriction may be effective in suppressing cell growth in the endocrine organs, thus resulting in varied effects on different tumor types.

The present study showed significantly lower weight gains in association with activated spontaneous movement and increased sensitivity or aggression in the food-restricted rats. We also suggest that lipid peroxide in the blood plasma may be correlated with the development of these age-related diseases, including not only neoplasms but also nephropathy and myocardiopathy. Finally, there was no essential difference between daily and intermittent restriction in these effects.

References

Badger TM, Lynch EA, Fox PH (1985) Effects of fasting on luteinizing hormone dynamics in the male rats. J Nutr 115:788–797

Berg BN, Simms S (1961) Nutrition and longevity in the rat. J Nutr 74:23–32.

Connors JN, Devito WJ, Hedge GA (1985) Effects of food deprivation on the feedback regulation of the hypothalamic-pituitary-thyroid axis of the rat. Endocrinology 117: 900–906

Maeda H, Gleiser CA, Masoro FJ, Murata I, Mc Mahn CA, Yu BH (1985) Nutrition influences on aging of Fischer 344 rats: II. Pathology. J Gerontol 40:671–688

McCay CM, Crowell MF, Maynard LA (1935) The effect of retarded growth upon the length of life span and upon the ultimate body size. J Nutr 10:63–79

Tannenbaum A (1945) The dependence of tumor formation on the composition of the calorie-restricted diet as well as on the degree of restriction. Cancer Res 5:616–625

Tucker MJ (1979) The effect of long-term food restriction on tumours in rodents. Int J Cancer 23:803–807

Yu BP, Masaro EJ, Mc Mahn CA (1985) Nutritional influences on aging of Fischer 344 rats: I. Physical metabolic and longivity characteristics. J Gerontol 40:657–670

Walford RL, Harris SB, Weindruch R (1987) Dietary restriction and aging: historical phases, mechanisms and current directions. J Nutr 117:1650–1654

CHAPTER 10

Influence of Diet Restriction on Toxicity and Hormonal Status in Chemically Treated Mice

K.M. Abdo[1], R. Irwin[1], and J. Johnson[2]

Introduction

The influence of quality and quantity of diets on longevity and susceptibility of animals to cancer and other age-related degenerative diseases has long been recognized (McCay et al. 1935; Tannenbaum 1942, 1945). Chronic dietary restriction in male Osborne-Mendel, Sprague-Dawley, or SPF Fischer 344 rats was associated with a decrease in severity and a delay in the onset of age-related diseases such as nephropathy, interstitial cell hyperplasia of the testis, myocardial degeneration, and bile duct hyperplasia (Saxton and Kimball 1941; Berg and Simms 1960; Bras and Ross 1964; Yu et al. 1982). In female CFY Sprague-Dawley rats, dietary restriction leading to 50% depression in body weight caused a delay in the onset of puberty and prolonged the duration of fertility (Merry and Holehan 1979). Food restriction to amounts ranging from 10% to 75% caused decreases in the absolute weight of liver, spleen, heart, lung, kidney, testes, ventral prostate and seminal vesicles, epididymal fat pads, bones (tibia, femur, and humerus), and six different muscles (Hegarty et al. 1977; Myers et al. 1984; Oishi et al. 1979). Of these organs, only the relative weight of the brain and the testes were increased while that of the liver was decreased (Oishi et al. 1979). Short-term (4-5 weeks) food restriction (10%–35%) was associated with decreased leukocyte count and increased erythrocyte count, hemoglobin concentration, and hematocrit (Oishi et al. 1979; Pickering and Pickering 1984; Scharer 1977).

Restriction of energy intake may lead to change in hormonal balance described as "pseudohypophysectomy" (Pariza and Boutwell 1987; Pariza 1987). Rats which were restricted feed every other day from 1.5 through 28 months of age had a significant decrease in the serum levels of T_3 but not T_4 (Haley-Zitlin et al. 1989). Thyroid hormone (T_3, T_4, TSH) levels in Lobund-Wistar rats were not

[1]National Institute of Environmental Health Sciences, Research Triangle Park, NC 27709, USA
[2]Battelle Columbus Laboratories, Columbus, OH 43201, USA

affected but testosterone levels were increased by mild dietary restriction (≤ 12 g feed/day) (Snyder et al. 1988). Dietary restriction greater than 30% was shown to cause a decline in thyroid hormones in rats (Campbell et al. 1977; Merry and Holehan 1981; Ortiz-Caro et al. 1984).

In toxicological studies, ad libitum feeding is the common practice and quite often body weights of treated groups are depressed relative to the controls. A survey of National Cancer Institute/National Toxicology Program (NCI/NTP) toxicity and carcinogenicity studies (140 rat and 120 mouse studies) conducted prior to July 1985 showed that the body weights of rats and mice treated with chemicals average 10%–15% below the control and in some cases reductions of up to 40% were observed. In the light of the above discussion, it is clear that interpretation of toxicity studies becomes difficult when body weight depression occurs in chemically treated groups, particularly when reduced food consumption may be a factor.

The purpose of this study is to investigate the influence of dietary restriction/body weight reduction on some toxicological parameters in chemically treated mice and to distinguish changes due to test chemical from changes due to reduction in food intake/body weight.

Scopolamine hydrobromide, an antimuscarinic compound, is used as a preanesthetic agent to provide an antisilagogue effect and sedative and amnesic effects and to prevent reflux bradycardia (Mirakhur 1979). Toxic effects observed previously with this compound at doses of 25 mg/kg or more included a decrease in body weight and liver and kidney weights in ad libitum-fed mice. The present study was designed to determine whether the changes in organ weights were solely due to body weight depressions or whether reduced feed consumption was a contributing factor. Additionally, we investigated the influence of food restriction/body weight reduction, with or without scopolamine hydrobromide on some toxicological endpoints such as hematology parameters and T_3, T_4, and testosterone levels. This objective was achieved by pair-feeding groups of mice to the same growth rate and body weight as that of mice receiving 25 mg/kg scopolamine hydrobromide and fed ad libitum and by including additional groups of mice with or without treatment with the compound in which body weight was reduced by restricting food consumption to 70% that of the ad libitum-fed controls.

Materials and Methods

Chemicals

Scopolamine hydrobromide ($\geq 89\%$ pure) was obtained from Henley & Co., New York, NY. The purity of the chemical was determined by high-performance liquid chromatography. The impurities were not identified. Results of elemental analyses of the bulk chemicals were in agreement with theoretical values. Infrared, ultraviolet, visible, and nuclear magnetic resonance spectra were consistent with the structure of the respective chemical. Scopolamine hydrobromide was mixed in corn oil to form the dosing solution. This solution was stable for at least 2 weeks when stored at room temperature. Scopolamine hydrobromide was one of four chemicals selected for studying the influence of food restriction/reduced body weight on toxicity and carcinogenicity of test chemicals. The chemical met one or more of the following criteria: (1) nonmutagen, (2) mutagen, (3) structurally related to a known carcinogen, (4) available information from the literature suggests that it may not be a carcinogen. Scopolamine hydrobromide is not mutagenic.

Animals

Male and female $B6C3F_1$ mice (4-5 weeks old) were obtained from Simonson Laboratories, Gilroy, CA. They were individually housed. Each treatment group consisted of 60 animals of each sex. All mice were quarantined prior to initiation of studies. Room conditions were maintained at $72 \pm 3\,°F$ and $50 \pm 15\%$ relative humidity, light cycle 12 h on and 12 h off, and 15 air changes per hour. Food consumption was determined once weekly for the first 13 weeks and once monthly thereafter.

Treatments

Dietary restriction was investigated in mice by adding three special dietary restriction groups to the standard carcinogenicity study design of three dose groups and a control. The study design for scopolamine hydrobromide consisted of seven treatment groups. These are: ad libitum-fed control (AL), low-dose sopolamine (LD), middle-dose scopolamine (MD), high-dose scopolamine (HD), pair-fed control (PC; pair-fed to result in the same body weight as the HD group by controlling the amount of food offered), restricted controls (RC; daily food allotment restricted to 70% of that consumed by the AL group), restricted high dose (RHD; daily food allotment restricted to 70% of that consumed by the AL group). The first four groups represent the standard carcinogenicity study groups and the remaining three groups represent food-restricted groups. An interim evaluation of toxicity and hormonal status as influenced by dietary restriction was made on the AL, PC, HD, RC, and RHD groups. Unless specified otherwise, this evaluation was made on 10 male and 10 female mice at week 65 of the study. The 65-week interim evaluation results will be the subject of this report. However, the studies for this chemical will be continued for up to 3 years to investigate the influence of dietary restriction on tumor incidence in chemically treated mice. The high dose groups of mice in the scopolamine hydrobromide study received 25 mg/kg in corn oil by gavage 5 days per week. Control groups received only corn oil at the same dosing schedule.

Hematology and Clinical Chemistry

Blood samples were collected from the vena cava of anesthetized (70% CO_2/ 30% air anesthesia) mice. Hematology determinations for hematocrit (HCT), hemoglobin (HGB), red blood cell (RBC), white blood cell (WBC), and platelet

(PLT) count were determined using an Ortho ELT-8 hemocytometer. Total serum triiodothyronine (T_3) and thyroxine (T_4) were determined according to the I-125 radioimmunoassay method described by Chopra (1977). Total serum testosterone was determined according to the I-125 radioimmunoassay described by Abraham (1977).

Clinical and Pathological Examination

Body weights were recorded at the start of the study, weekly for 13 weeks, and every 4 weeks thereafter. A complete necropsy was conducted on all male and female mice in the studies. Liver, brain, lung, heart, thymus, right kidney, and testis were removed, trimmed, and dissected from all animals at necropsy. Analyses of absolute organ weights as well as organ to body weight ratios were made. A total of 42 tissues per animal were collected and fixed in neutral buffered formalin. After sectioning and staining with hematoxylin and eosin a complete histopathological examination was performed.

Statistical Evaluation

The data were evaluated by analysis of variance (Snedecor and Cochran 1967) and Duncan's new multiple range test (Kramer 1957). The difference between a treatment and control was considered significant when the calculated p value was ≥ 0.01.

Results

Food Consumption, Body Weight, Organ and Relative Organ Weights

Growth curves for male and female mice are shown in Fig. 10.1. The growth rate was proportional to the amount of food consumed. The highest growth rate was achieved by the ad libitum-fed controls and the lowest was achieved by the RHD groups. Final mean body weight and organ weights of the various treatment groups are shown in Table 10.1. Final mean body weights of male and female mice in the PC, HD, RC, or RHD were lower ($p \leq 0.01$) than those of the AL group. The differences in final mean body weights between males in the PC, HD, RC, or RHD groups were not significant. Differences in the final mean body weight between females in the PC or HD groups and between females in RC or RHD groups were also not significant. The final mean body weights of restricted groups (70% restriction), with or without the chemical, were the lowest compared to the remaining groups. Liver weights of males in the PC group and females in the PC, RC, and RHD groups were lower ($p \leq 0.01$) than those of the AL groups. Liver weights were lower for males and females in the PC group than

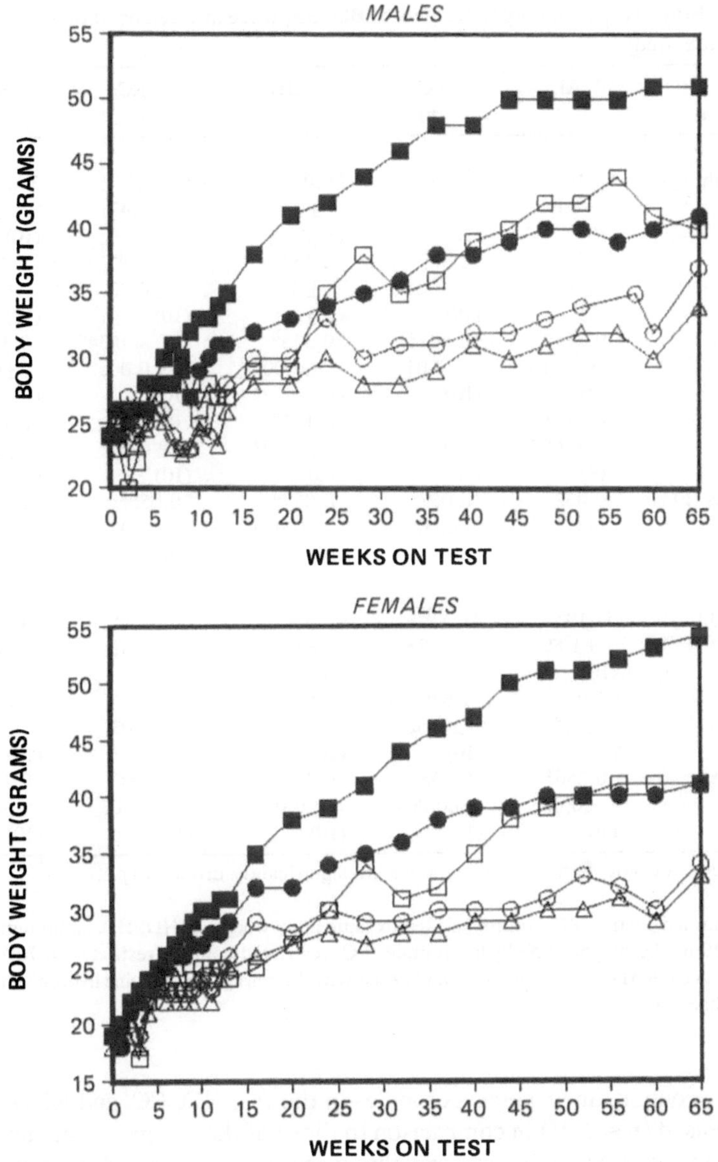

MALES

BODY WEIGHT (GRAMS)

WEEKS ON TEST

FEMALES

BODY WEIGHT (GRAMS)

WEEKS ON TEST

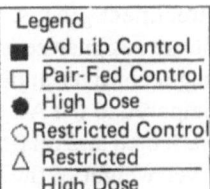

Legend
■ Ad Lib Control
□ Pair-Fed Control
● High Dose
○ Restricted Control
△ Restricted
  High Dose

Figure 10.1a,b. Growth curves of male (*a*) and female (*b*) B6C3F$_1$ mice in the scopolamine hydrobromide study. *Solid squares*, ad libitum controls; *open squares*, pair-fed controls; *solid circles*, high dose; *open circles*, restricted controls; *open triangles*, restricted high dose

Table 10.1. Body weight and organ weights of B6C3F$_1$ mice at week 66 of the scopolamine hydrobromide study.

Treatment	AL (*n*)	PC (*n*)	HD (*n*)	RC (*n*)	RHD (*n*)
Males					
Body weight (g)	50.16[a]	39.34[b]	39.02[b]	38.22[b]	36.57[b]
	±0.64	±1.33	±0.64	±0.47	±0.61
	(9)	(10)	(10)	(10)	(10)
Liver (g)	2.34[a]	1.30[b]	2.15[a]	1.77[a,b]	1.71[a,b]
	±0.12	±0.06	±0.35	±0.02	±0.02
	(9)	(10)	(10)	(10)	(10)
Kidney (g)	0.375[a]	0.303[c]	0.325[b,c]	0.34[b]	0.314[b,c]
	±0.015	±0.01	±0.006	±0.006	±0.004
	(9)	(10)	(10)	(10)	(10)
Testis (g)	0.119[a]	0.159[a]	0.117[a]	0.12[a]	0.117[a]
	±0.003	±0.004	±0.003	±0.002	±0.003
	(9)	(10)	(10)	(10)	(10)
Epididymis (g)	0.062[a]	0.057[a]	0.057[a]	0.055[a]	0.053[a]
	±0.003	±0.003	±0.003	±0.002	±0.003
	(9)	(10)	(10)	(10)	(10)
Females					
Body weight (g)	53.19[a]	41.37[b]	40.66[b,c]	35.82[c,d]	34.41[d]
	±1.89	±1.23	±1.72	±0.81	±0.56
	(10)	(10)	(10)	(10)	(10)
Liver (g)	1.92[a]	1.61[b]	1.82[a]	1.63[b]	1.62[b]
	±0.06	±0.02	±0.05	±0.03	±0.03
	(10)	(10)	(10)	(10)	(10)
Kidney (g)	0.256[a]	0.233[a]	0.237[a]	0.238[a]	0.243[a]
	±0.005	±0.004	±0.008	±0.003	±0.007
	(10)	(10)	(10)	(10)	(10)

Values within a row not sharing the same letter are significantly different ($p \leq 0.01$); ± standard error

AL, ad libitum fed control; PC, control pair-fed to same body weight as HD; HD, ad libitum fed and dosed with 25 mg/kg scopolamine hydrobromide; RC, restricted control (restricted to 70% of food consumed by AL); RHD=restricted as RC and dosed with 25 mg/kg scopolamine hydrobromide; (*n*), number of observations

in the HD group. Kidney weights of males in the PC, HD, RC, and RHD groups were decreased ($p \leq 0.01$) in comparison to those of the AL group, but no differences in the testis or epididymis weights were observed. No differences were observed in the kidney weights of females among the various treatment groups.

The relative organ weights of male and female mice at week 66 are summarized in Table 10.2. The effect of food restriction on the relative weights of liver and kidney in males and females in the PC groups was variable. Compared to the AL group, the relative kidney weight of the PC group males was unaffected and relative liver weight was decreased, whereas in females relative liver weight of the PC groups was unaffected and relative kidney weight was increased. The relative

Table 10.2. Relative organ weights (% body weight) of B6C3F$_1$ mice at week 66 of the scopolamine hydrobromide study.

Treatment	AL (n)	PC (n)	HD (n)	RC (n)	RHD (n)
Males					
Liver	4.65[a]	3.31[b]	4.60[a]	6.43[a]	4.69[a]
	±0.23	±0.09	±0.08	±0.05	±0.08
	(9)	(10)	(10)	(10)	(10)
Kidney	0.750[c]	0.773[b,c]	0.832[a,b]	0.860[a]	0.891[a]
	±0.023	±0.022	±0.009	±0.018	±0.009
	(9)	(10)	(10)	(10)	(10)
Testis	0.237[b]	0.299[a]	0.299[a]	0.313[a]	0.319[a]
±0.005	±0.016	±0.004	±0.006	±0.006	
	(9)	(10)	(10)	(10)	(10)
Epididymis	0.123[b]	0.147[a]	0.145[a]	0.143[a]	0.145[a]
	±0.005	±0.008	±0.006	±0.004	±0.007
	(9)	(10)	(10)	(10)	(10)
Females					
Liver	3.61[b]	3.93[b]	4.52[a]	4.56[a]	4.71[a]
	±0.06	±0.12	±0.15	±0.011	±0.09
	(10)	(10)	(10)	(10)	(10)
Kidney	0.483[c]	0.568[b]	0.587[b]	0.668[a]	0.706[a]
	±0.010	±0.021	±0.016	±0.015	±0.023
	(10)	(10)	(10)	(10)	(10)

Values within a row sharing the same letter are not significantly different ($p \leq 0.01$); ± standard error. Abbreviations as in Table 10.1.

weights of testis and epididymis in all treatment groups were lower than those of the AL group; this change is reflective of the difference in body weights. The relative kidney weights of the restricted groups (RC and RHD) of males and the relative weights of liver and kidney of restricted groups of females were significantly increased. The relative weights of kidney in males and liver and kidney of females in the HD groups were significantly higher than those of the AL groups. The increases in relative weight of the various organs appear to be inversely related to body weight.

Hematology Profile

Hematological profiles are shown in Table 10.3 for males and Table 10.4 for females. In males, a significant decrease in HCT, mean corpuscular volume (MCV), and platelet count and a significant increase in mean corpuscular hemoglobin concentration (MCHC) were observed in the RC and RHD groups relative to the AL group. The same results were observed in females, with the exception of PLT; this was increased in the RHD group.

Table 10.3. Hematological profile of male B6C3F$_1$ mice at week 66 of the scopolamine hydrobromide study.

Treatment	AL	PC	HD	RC	RHD
WBC	7.69[a]	6.36[a,b]	5.88[b]	7.40[a]	7.43[a]
(thousand/µl)	±0.70	±0.44	±0.52	±0.31	±0.25
PLT	1314.7[a,b]	1185.8[b]	1357.1[a]	1164.0[b]	1156.1[b]
(thousand/µl)	±40.7	±36.8	±60.8	±34.8	±13.3
RBC	9.82[a]	10.06[a]	9.66[a]	9.99[a]	9.82[a]
(million/µl)	±0.17	±0.11	±0.16	±0.05	±0.09
HCT	46.81[a]	47.67[a]	45.67[a,b]	44.23[b]	44.07[b]
(%)	±0.87	±0.57	±0.75	±0.41	±0.46
HGB	14.66[a,b]	14.98[a]	14.39[b]	15.09[a]	14.82[a,b]
(g/dl)	±0.20	±0.11	±0.17	±0.07	±0.13
MCH	14.94[a]	14.92[a]	14.92[a]	15.09[a]	15.09[a]
(pg)	±0.11	±0.10	±0.18	±0.08	±0.07
MCHC	31.36[a]	31.44[a]	31.53[a]	34.12[b]	33.66[b]
(g/dl)	±0.25	±0.17	±0.25	±0.22	±0.32
MCV	47.60[a]	47.50[a]	47.30[a]	44.20[b]	44.80[b]
(µm³)	±0.52	±0.27	±0.56	±0.33	±0.39

Values within a row not sharing the same letter are significantly different ($p \le 0.01$); ± standard error for 10 observations. Treatment abbreviations as in Table 10.1. WBC, white blood cell count; PLT, platelet count; RBC, red blood cell count; HCT, hematocrit; HGB, hemoglobin; MCH, mean corpuscular hemoglobin; MCHC, mean corpuscular hemoglobin concentration; MCV, mean corpuscular volume

Table 10.4. Hematological profile of female B6C3F$_1$ mice at week 66 of the scopolamine hydrobromide study.

Treatment	AL	PC	HD	RC	RHD
WBC	4.84[a]	5.26[a]	5.10[a]	6.66[a]	4.86[a]
(thds/µl)	±0.50	±0.45	±0.46	±0.65	±0.54
PLT	928.4[b]	936.2[b]	1018.7[a,b]	1001.1[a,b]	1073.4[a]
(thds/µl)	±20.4	±13.4	±28.8	±39.6	±26.5
RBC	9.96[a,b]	10.15[a,b]	9.57[b]	10.03[a]	9.86[a,b]
(mil/µl)	±0.08	±0.12	±0.07	±0.10	±0.13
HCT	48.39[a]	49.25[a]	46.54[b]	44.44[c]	43.72[c]
(%)	±0.87	±0.57	±0.75	±0.41	±0.46
HGB	14.98[a,b]	15.31[a]	14.52[b]	15.08[a]	14.99[a,b]
(g/dl)	±0.13	±0.12	±0.10	±0.12	±0.13
MCH	15.03[a]	15.09[a]	15.16[a]	15.03[a]	15.21[a]
(pg)	±0.08	±0.10	±0.06	±0.08	±0.10
MCHC	30.96[b]	31.1[b]	31.23[b]	33.96[a]	34.32[a]
(g/dl)	±0.19	±0.25	±0.24	±0.19	±0.24
MCV	48.60[a]	48.70[a]	48.50[a]	44.20[b]	44.33[b]
(µm³)	±0.34	±0.42	±0.37	±0.25	±0.24

Values within a row not sharing the same letter are significantly different ($p \le 0.01$); ± standard error for 10 observations. Treatment abbreviations as in Table 10.1; parameters as in Table 10.3

Table 10.5. T_3, T_4, and testosterone levels in B6C3F$_1$ mice at week 66 of the sopolamine hydrobromide study.

Treatment	AL (n)	PC (n)	HD (n)	RC (n)	RHD (n)
T_3 (ng/l)	230.3[b2]	174.1[b,c]	458.9[a]	66.2[d]	81.0[c,d]
	±42.3[3]	±24.0	±38.2	±5.9	±9.4
	(9)[4]	(10)	(8)	(10)	(10)
T_4 (μg/dl)	3.92[a]	4.43[a]	4.02[a]	3.77[a]	3.61[a]
	±0.33	±0.12	±0.20	±0.24	±0.30
	(9)	(10)	(10)	(9)	(10)
Testosterone	2.39[a]	2.15[a]	2.74[a]	2.26[a]	2.19[a]
	±0.76	±0.92	±0.87	±0.64	±0.84
	(5)	(9)	(10)	(8)	(10)
Females					
T_3 (ng/l)	87.86[a,b]	–	77.41[a,b]	50.94[b]	152.61[a]
	±15.39	–	±20.01	±3.64	±29.54
	(6)		(8)	(7)	(10)
T_4 (μg/dl)	3.51[b]	4.56[a,b]	3.43[b]	5.40[a]	3.82[b]
	±0.18	±0.47	±0.28	±0.66	±0.27
	(10)	(9)	(10)	(7)	(10)

Values within a row not sharing the same letter are significantly different ($p \leq 0.01$); ± standard error. Treatment abbreviations as in Table 10.1.

T_3, T_4, and Testosterone Levels

Serum T_3, T_4, and testosterone levels in male and female mice are summarized in Table 10.5. In males, T_3 levels were depressed in the RC and RHD groups and increased in the HD groups relative to those of the AL group. T_4 and testosterone levels did not differ among the various groups. In females, the T_4 levels in the PC, HD, and RHD groups were comparable to those of the AL group. T_3 level in the RC group was significantly lower than that of the RHD but was not different from that of the HD group. T_4 levels in the AL, HD, and RHD groups were lower than those of the RC group.

Pathology Findings

Significant lesions observed in mice that were killed at week 66 of the study and in mice that died prior to this time period are shown in Table 10.6. A decrease in the number of mice with primary tumors was observed in diet-restricted groups (RC and RHD) and the PC compared to the AL group. This effect is more obvious in males than females.

A decrease in the incidence of nephropathy and hyperplasia of the uterus was observed in all treatment groups compared to the AL group.

Table 10.6. Incidence of primary tumors, nephropathy, and uterine hyperplasia in mice at week 66 of the scopolamine hydrobromide study.

Treatment	AL (%)	PC (%)	HD (%)	RC (%)	RHD (%)
Males					
Primary tumors	4/12	1/16	3/14	0/10	1/11
	(33.3)	(6.3)	(21.4)	(0)	(9.1)
Nephropathy	11/12	7/16	6/14	8/10	6/12
	(91.7)	(43.7)	(42.9)	(80)	(50)
Females					
Primary tumors	1/12	0/11	1/13	0/10	0/12
	(8.3)	(0)	(7.7)	(0)	(0)
Nephropathy	4/12	2/11	1/13	0/10	0/12
	(33.3)	(18.2)	(7.7)	(0)	(0)
Uterus, hyperplasia	6/12	1/11	4/13	0/10	2/12
	(50)	(9.1)	(30.8)	(0)	(16.7)

Lesion incidence = number of animals with the lesion/number of animals examined. The denominator = 10 animals that were killed at week 66 of study + those that died prior to this time.

Discussion

The decreases in liver and kidney weights and the increase in relative weights of liver, kidney, testis, and epididymis observed in mice given scopolamine hydrobromide, irrespective of the method of feeding (ad libitum vs. restricted), appear to be related to the reduction in body weight. This is supported by our finding that the same effects were observed in PC mice of the same body weight as that of mice receiving this compound and fed ad libitum. The influence of body weight on organ weight reduction in animals on a food-restricted diet is well documented. A decrease in rat body weight and liver and kidney weights and an increase in their relative weights as a result of food restriction have been reported (Scharer 1977; Gold and Costello 1975; Pickering and Pickering 1984). In our study, we found that the absolute weights of the testis and epididymis were unaffected, but their relative weights were increased due to reduction in body weight gain induced by chemical treatment and/or food restriction. Similar effects were observed in food-restricted intact rats and swine (Scharer 1977; Calloway et al. 1962).

Food restriction with or without treatment with scopolamine hydrobromide caused a decrease in HCT and MCV and an increase in MCHC but did not affect the parameters WBC, RBC, or HGB. As in the case of the organ weight effects, the changes observed in the above parameters appear to be associated with reduced body weights. Our hematology findings contrast with those reported in food-restricted rats (Oishi et al. 1979; Pickering and Pickering 1984). This contrast could be due to differences in the strain of animals used and/or the degree of food restriction imposed. These researchers reported a decrease in leukocyte count and an increased erythrocyte count and HCT values in rats restricted to 25%–50% of ad libitum consumption. The food restriction imposed by these authors was severe compared to that imposed in our study. This severe restriction may have resulted in hemoconcentration and a decrease in the number of lymphocytes.

The influence of treatment with scopolamine hydrobromide and/or food restriction on serum levels of thyroid hormones is difficult to explain. T_3 levels were decreased in food-restricted male mice with or without scopolamine treatment but were increased in ad libitum-fed scopolamine-treated male mice as compared to ad libitum-fed controls. Serum levels of T_4 and testosterone in males and females and T_3 in females were unaffected by food restriction, body weight reduction, and/or chemical treatment. It is possible that the degree of restriction and/or body weight reduction was too small to be effective in causing a significant change in the level of these hormones. In the majority of the studies where changes in the levels of these hormones were observed, starvation (feeding every other day) or severe food restriction ($<70\%$ of ad libitum consumption) were imposed (Campbell et al. 1977; Merry and Holehan 1981; Ortiz-Caro et al. 1984). Food restriction to no more than 70% of ad libitum in Lobund-Wistar rats had no influence on serum levels of TSH, T_3, and T_4 (Snyder et al. 1988).

In the present study we found that food restriction/reduced body weight resulted in a decrease in the number of mice bearing tumors as well as a decrease in the incidence of nephropathy and hyperplasia of the uterus. This decrease relative to AL occurred in all treatment groups (PC, HD, RC, RHD). The finding that food restriction/body weight reduction resulted in a decrease in the number of tumor-bearing animals and the incidence of nonneoplastic lesions is similar to previous findings obtained by other researchers (McCay et al. 1935; Tannenbaum 1942, 1945; Bras and Ross 1964). It is too early in the present study to make a definite conclusion as to the impact of reduced body weight on the outcome of the carcinogenicity study of scopolamine hydrobromide. However, there is an indication that mice with or without scopolamine treatment that have a reduced body weight have a lower incidence of tumors than heavier ad libitum fed mice. It is quite likely that one may arrive at a different conclusion if comparisons for tumor incidence are made between ad libitum fed controls and chemically treated groups with reduced body weight rather than between chemically treated groups and controls pair-fed to similar body weight.

It is apparent from the results of this study that body weight plays an important role in the outcome of toxicology studies. Some toxic effects ascribed to a chemical which causes a significant depression in body weight in a conventional toxicity study could be due to differences in body weights between the control and treated animals. Toxic endpoints affected by severe body weight depressions include organ weights, hematology, hormone levels, and pathology findings (tumor and nontumor). To circumvent the confounding effect of body weight, it is suggested that experimental designs include controls pair-fed to the same body weight as chemically treated animals. This is particularly important for chemicals in which the major toxic effect is body weight reduction, as in the case of scopolamine hydrobromide.

References

Abraham GE (1977) Radioimmunoassay of steroids In: Abraham G (ed) Handbook of radioimmunoassay. Dekker, New York, NY, p 591

Berg BN, Simms HS (1960) Nutrition and longevity in rats. II. Longevity and onset of diseases with different levels of food intake. J Nutr 71:255–266

Bras G, Ross MH (1964) Kidney disease and nutrition in the rat. Toxicol Appl Pharmacol 6:247–262

Calloway DH, Hilf R, Huson AH (1962) Effect of chronic food restriction in swine. J Nutr 76:365

Campbell GA, Kurcz M, Marshall S, Meites J (1977) Effect of starvation in rats on serum levels of follicle stimulating hormone, luteinizing hormone, thyrotropin, growth hormone, and prolactin; response to LH-releasing hormone and thyrotropin-releasing hormone. Endocrinology 100:580–587

Chopra IJ (1977) Radioimmunoassay of iodothyronines. In: Abraham G (ed) Handbook of radioimmunoassay. Dekker, New York, NY, p 679

Gold AJ, Costello LC (1975) Effect of semistarvation on rat liver, kidney, and heart mitochondrial fraction. J Nutr 105:208

Haley-Zitlin V, Dellow M, Wright DL, Beauchene RE (1989) Effect of diet restriction and exercise on adrenal and thyroid hormones and liver and muscle enzymes in aging male Wistar rats. Fed Am Soc Exp Biol Med 3(3):A462

Hegarty IVJ, Kim S, Ahn P (1977) Effect of dietary restriction and source of rats on growth rate and efficiency of conversion of food into muscles, bones, and organs. Growth 41:221–234

Kramer CY (1957) Extension of multiple range tests to group means with unequal numbers of replication. Biometrics 12:307

McCay CM, Cromwell ME, Maynard LA (1935) The effect of retarded growth upon the length of life span and ultimate size. J Nutr 10:63–79

Merry BJ, Holehan AM (1979) Onset of puberty and duration of fertility in rats fed a restricted diet. J Reprod Fertil 57:253–259

Merry BJ, Holehan AM (1981) Serum profiles of LH, FSH, testosterone and 5a-DHT from 21 to 1000 days of age in ad libitum fed and dietary restricted rats. Exp Gerontol 16:431–444

Mirakhur RK (1979) Anticholinergic drugs. Br J Anaesth 51:671

Myers B, Dubick M, Reiser K, Gerriets J, Last J, Rucker R (1984) Ozone exposure food restriction and protein deficiency: changes in collagen and elastin in rodent lung. Toxicol Lett 23:43–49

Oishi O, Oishi H, Hiraga K (1979) The effect of food restriction of 4 weeks on common toxicity parameters in male rats. Toxicol Appl Pharmacol 47:15–22

Ortiz-Caro J, Gonzalez C, Jolin T (1984) Diurnal variations of plasma growth hormone, thyrotropin, thyroxine, and triiodothyronine in streptozotocin-diabetic and food-restricted rats. Endocrinology 115:2227–2231

Pariza MW (1987) Fat, calories and mammary carcinogenesis: net energy effects. Am J Clin Nutr 45:261–263

Pariza MW, Boutwell RK (1987) Historical perspective: calories and energy expenditure in carcinogenesis. Am J Clin Nutr 45:151–156

Pickering RG, Pickering CE (1984) The effects of reduced dietary intake upon the body and organ weights and clinical chemistry and haematological variates of the young Wistar rat. Toxicol Lett 21:271–277

Saxton JA Jr, Kimball GC (1941) Relation of nephrosis and other diseases of albino rats to age and to modification of diet. Arch Pathol 32:951–965

Scharer K (1977) The effect of chronic underfeeding on organ weights of rats. Toxicology 7:45–56

Snedecor GW, Cochran WG (1967) Statistical methods, 6th ed. Iowa State University Press, Ames

Snyder DL, Wostmann BS, Pollard M (1988) Serum hormones in diet restricted and conventional Lobund-Wistar rats. J Gerontol 43(6):B168–173

Tannenbaum A (1942) The genesis and growth of tumors. Cancer Res(2) 468–475

Tannenbaum A (1945) The dependence of tumor formation on the degree of caloric restriction. Cancer Res 5:609–615

Yu BP, Masoro EJ, Murata I, Bertrand HA, Lynd FT (1982) Life span study of SPF Fischer 344 male rats fed ad libitum or restricted diets: longevity, lean body mass and disease. J Gerontol 37:130–141

Part III
Effects and Mechanisms
of Dietary Restriction:
Physiological Consequences

CHAPTER 11

Physiological Consequences of Dietary Restriction: An Overview

E.J. Masoro[1]

Age Changes in Physiological Systems

The early studies on food restriction focused primarily on its ability to extend the life span and to retard age-associated disease processes (Weindruch 1985). Not until the 1970s was there a major emphasis on the effects of food restriction on physiological processes and on age changes in these processes.

Our reason, and probably that of others, for focusing on the effects of food restriction on age changes in the physiological processes was the belief that such information might shed light on the mechanisms by which the life span is extended. Moreover, we felt that knowledge of these mechanisms would provide insights on the nature of the primary aging processes.

Scope of Effects on Physiological Processes

Our studies and those of other groups have shown that food restriction influences a broad array of age changes in the physiological processes of rats and mice (Masoro 1989). These actions include delaying or blunting or preventing the following: age-related increases in the concentrations of plasma lipids, age-related increases in the concentrations of serum parathyroid hormone and calcitonin, age-related losses in the response of fat cells to hormones, the loss of neurotransmitter receptors with advancing age, the age-associated loss of γ-crystallins from the lens of the eye, female reproductive senescence, the decrease in spontaneous locomotor activity with age, age-associated changes in immune function, and many others – including senile bone loss. However, not all age changes in the physiological systems are influenced by food restriction. Although the number not influenced appears to be small, this assessment may be an underestimation because investigators often do not attempt to publish negative findings nor do journals encourage their publication.

[1]Department of Physiology, The University of Texas Health Science Center, 7703 Floyd Curl Drive, San Antonio, TX 78284-7756, USA

Role of Disease

The question arises as to whether food restriction retards age changes in the physiological processes because of its ability to prevent, delay, or slow the progression of age-associated diseases. Although most age-associated diseases are retarded by food restriction (Maeda et al. 1985), data obtained in our and other laboratories indicate that most of the effects of food restriction on physiological processes are not secondary to the retardation of these diseases. Specifically, food restriction influences age changes in most physiological processes starting at young ages, long before the occurrence of serious age-associated diseases (Maeda et al. 1985).

However, a few of the actions of food restriction on physiological processes are at least in part secondary to protecting the animal from an age-associated disease. An example is the blunting by food restriction of the marked rise in serum parathyroid hormone concentration occurring late in the life of male Fischer 344 rats. This increase in serum parathyroid hormone levels in the ad libitum-fed rats appears to be primarily due to severe nephropathy, a disease process almost totally prevented by food restriction (Kalu et al. 1988).

Significance of the Findings on the Physiological Processes

The breadth of the effects on the physiological processes is strong evidence that food restriction is acting at a basic biological level, probably retarding one or more of the primary aging processes. That is, by modulating the primary aging processes, food restriction slows, delays, or prevents a broad range of age changes secondary, tertiary, and further removed from these primary processes. Unfortunately, the very breadth of the physiological processes influenced makes it difficult to use these effects for pinpointing either the mechanisms by which food restriction retards the primary aging processes or the basic nature of these processes.

Tool for Evaluating the Theories of Aging

There are many theories of aging. One reason for such a large number is the difficulty in designing experiments which test the validity of a theory. Because of its marked ability to retard the aging processes, food restriction appears to provide a powerful tool for testing theories. Moreover, the influence of food restriction on specific physiological processes provides a means of doing so. We have used this approach to test three theories of aging.

Metabolic Rate Theory

Shortly after the turn of this century, Rubner (1908) reported an inverse correlation between the life span of domestic animal species and the metabolic rate per unit of body mass of the species. On the basis of this and other evidence Pearl

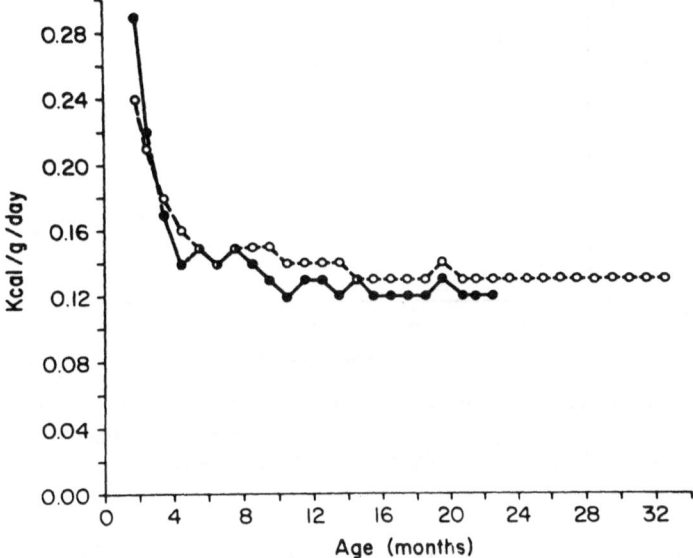

Figure 11.1. Energy intake per unit of body mass in ad libitum-fed (Group A; *solid circles*) and food-restricted (Group R; *open circles*) male Fischer 344 rats. Food restriction was initiated at 6 weeks of age (from Masoro 1985)

(1928) proposed the rate of living theory of aging, which states that the duration of life is determined by the exhaustion of a vital substance consumed at a rate proportional to the metabolic rate.

Sacher (1977) concluded that the decrease in the rate of aging caused by food restriction is a consequence of a decrease in the rate of energy metabolism. If this conclusion is correct, not only does it point to the mechanism by which food restriction retards the aging processes but it also provides support for the metabolic rate theory or rate of living theory of aging.

However, research carried out in our laboratory does not support Sacher's conclusion. Measurement of food intake (Masoro et al. 1982) revealed that within 6 weeks of the initiation of food restriction, the kcal of food ingested per unit of body mass of male Fisher 344 rats is not lower but rather higher in food-restricted rats than ad libitum-fed rats and remains so for the rest of the life span (Fig. 11.1). The reason for this unexpected finding is that the lean body mass adjusts to the reduced food intake so that caloric intake per unit of lean body mass becomes the same in food-restricted rats as in ad libitum-fed rats. McCarter et al. (1985) and McCarter and McGee (1989) found that the daily metabolic rate per unit of lean body mass of the food-restricted rats becomes the same as that of the ad libitum-fed by 6 weeks of food restriction and remains that way and the basal metabolic rate by 18 weeks (Fig. 11.2).

These data show that food restriction does not retard the aging processes by slowing the metabolic rate. Moreover, they make evident that metabolic rate cannot be the sole process responsible for aging. Of course, our data do not rule out

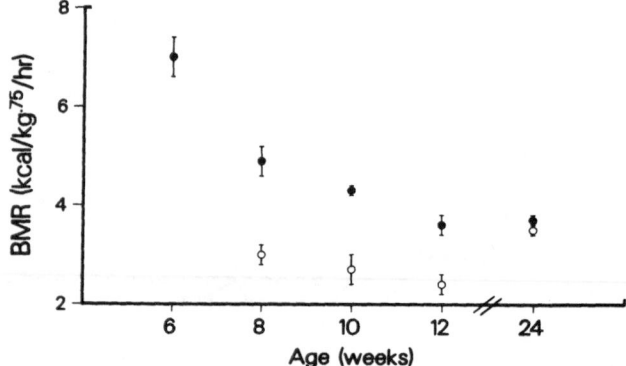

Figure 11.2. Basal metabolic rate per unit of lean body mass of ad libitum-fed (*solid circles*) and food-restricted (*open circles*) male Fischer 344 rats. Food restriction was initiated at 6 weeks of age (from McCarter and McGee 1989)

the metabolic rate theory as describing a component of the aging processes but indicate that, at most, it defines only a minor aspect.

Nevertheless, our explorations (Masoro 1988) indicate that it is the reduction in the intake of energy and not of another specific nutrient that is responsible for the action of food restriction on the aging processes. However, it is not reduction in energy intake per unit of "metabolic mass" (i.e., a reduction of metabolic rate), but rather per rodent that is responsible. The challenge is to learn how the reduction per rodent rather than per unit of "metabolic mass" is coupled to the aging processes.

Glucocorticoid Cascade Hypothesis of Aging

The hypothesis proposed by Sapolsky et al. (1986) is based on the view that stress-induced increases in the concentration of plasma glucocorticoids downregulate glucocorticoid receptors in neurons of the hypothalamus and that these neurons are involved in the negative feedback regulation of glucocorticoid secretion. The result is periods of hypersecretion of glucocorticoids. It is proposed that at young ages, removal of the stress is followed by a self-correction of the receptor loss but there is some cumulative effect. At some point in life, prolonged hypersecretion of glucocorticoids coupled with an insult such as ischemia results in a permanent loss of hippocampal neurons. Ultimately, it is envisioned that a feed-forward cascade occurs involving sustained hyperadrenocorticism, the pathophysiological consequences of which are similar to what is seen at advanced ages (e.g., immunosuppression, osteoporosis, cognitive impairment, etc.).

This hypothesis is primarily based on research with rats in which the recovery following stress of plasma corticosterone levels was found to be delayed at advanced ages and an age-associated loss of hippocampal neurons and glucocorticoid receptors occurred (Sapolsky et al. 1986). We felt that the food-restricted rat model would be useful for testing this hypothesis.

Basal Plasma Corticosterone Concentrations

A detailed longitudinal study of the circadian pattern of plasma corticosterone concentrations in 21 ad libitum-fed and 21 food-restricted male Fischer 344 rats was undertaken (Sabatino et al. 1990). The food-restricted rats were given their daily food allotment of 60% of the mean intake of the ad libitum-fed rats at 1500 hours and all rats were on a 12-h light–12-h dark cycle, lights on at 0400 hours and off at 1600 hours. Both groups of rats had similar circadian patterns of plasma corticosterone concentrations. Peak concentrations occurred between 1500 and 1700 hours and minimal concentrations at 0400 hours. The characteristics of this circadian pattern changed little with age through 25 months of age, the oldest rats studied to date.

Mean 24-h plasma corticosterone concentrations were calculated. Through 13 months of age both the ad libitum-fed rats and food-restricted rats had mean 24-h plasma corticosterone concentrations of about 100 ng/ml. The food-restricted rats continued to have such levels through 25 months of age. By 15 months of age the ad libitum-fed rats had mean 24-h plasma corticosterone levels of about 130 ng/ml which remained at that level through 25 months of age. Plasma corticosterone binding globulin (CGB) concentrations progressively decreased with advancing age in food-restricted but not ad libitum-fed rats through 25 months of age. Mean 24-h plasma free corticosterone concentrations, calculated on the basis of the CGB concentration, were higher at most ages in food-restricted rats than in ad libitum-fed rats. Indeed, at advanced ages (21–25 months of age), the food-restricted rats had mean 24-h plasma free corticosterone concentration about twice that of the ad libitum-fed rats of the same age.

Restraint Stress

We also measured the response of the plasma corticosterone concentrations to restraint stress in ad libitum-fed and food-restricted male Fischer 344 rats. At 5–6 months of age the increase in plasma corticosterone levels in response to restraint was similar for ad libitum-fed and food-restricted rats. In the ad libitum-fed rats of 18–19 months of age the increase in plasma corticosterone levels during restraint was somewhat greater than the 5–6 month old rats but returned to that level by 23–24 months of age. In the food-restricted rats, the increase in plasma corticosterone levels during restraint tended to marginally decrease with advancing age, but falling CBG levels indicate that this is not the case for plasma free corticosterone levels.

Return of the plasma corticosterone concentration to basal levels following the restraint had the same time course in ad libitum-fed and food-restricted rats. The recovery was somewhat slower at 18–19 months of age than at 5–6 months of age but no further slowing was observed at 23–24 months of age.

Conclusions

Our data with the ad libitum-fed and food-restricted rats provide strong evidence against the validity of the glucocorticoid cascade hypothesis of aging for the following reasons:

1. The 24-h mean plasma free corticosterone levels were markedly higher at advanced ages in food-restricted rats than in ad libitum-fed rats.
2. The recovery following restraint of plasma corticosterone concentrations had the same time course in ad libitum-fed and food-restricted rats of the same age even though food-restricted rats are biologically younger than ad libitum-fed rats.

Moreover, our findings clearly indicate that food restriction does not retard the aging processes by preventing the occurrence of hyperadrenocorticism.

Glycation Hypothesis of Aging

Cerami (1985) proposed that glucose may serve as a mediator of aging by non-enzymatically glycating proteins. A series of reactions are involved between the aldehydic groups of glucose and the amino groups of proteins to ultimately form advanced glycation end products (Pongor et al. 1984). Excessive glycation has detrimental effects on the functioning of proteins such as inactivation of enzymatic activity, inhibition of the binding of regulatory molecules, and alterations in other biological activities (Vlassara et al. 1986). DNA can similarly be influenced by glycation resulting in altered genomic function (Bucala et al. 1984). A major factor determining the extent of glycation of macromolecules is the level of the sustained glucose concentration in their environment (Cerami 1985).

Although this hypothesis of aging is intriguing, there is little hard evidence in its support. We therefore decided to use the food restriction rat model to further explore it.

Basal Plasma Glucose Concentrations

We executed a detailed longitudinal study of the circadian pattern of plasma glucose concentrations in 21 ad libitum-fed and 21 food-restricted male Fischer 344 rats. Our findings with rats in the age range of 4–6 months are shown in Fig. 11.3 (Masoro et al. 1989). Plasma glucose concentrations of the food-restricted rats were considerably below those of the ad libitum-fed rats for most of the day through 25 months of age (the oldest age studied to date).

Mean 24-h plasma glucose concentrations were calculated and found to increase modestly with age in both the ad libitum-fed and food-restricted rats. At all age ranges studied, food-restricted rats maintained plasma glucose concentrations about 15 mg/dl below those of the ad libitum-fed rats.

Conclusions

A food restriction regimen which retards the aging processes also sustains lower plasma glucose concentrations through most, if not all, of the life span. These findings are consistent with the glycation hypothesis of aging but, of course, do not provide evidence for a causal relationship between a lower sustained plasma

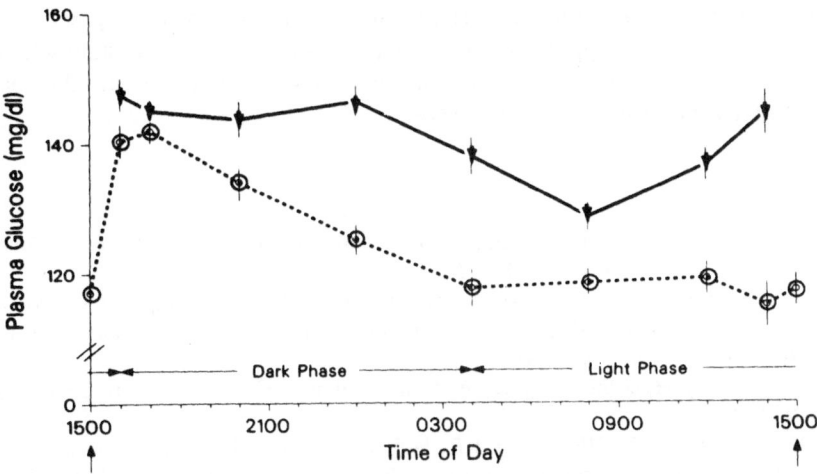

Figure 11.3. Diurnal pattern of plasma glucose concentrations in ad libitum-fed (*arrow-heads*) and food-restricted (*circles*) male Fischer 344 rats in the 4- to 6-month-old age range. Food-restricted rats were fed at 1500 hours as indicated by *vertical arrow* (from Masoro et al. 1989)

glucose level and the retardation of the aging processes. Research is in progress to further evaluate the possibility of a causal relationship.

It is evident that the sustained reduction in plasma glucose concentration in food-restricted rats is not due to a decreased intake of carbohydrate per unit of body mass (Masoro et al. 1982). Moreover, body composition and metabolic rate measurements indicate that the flux of glucose through the metabolic system per unit of lean body mass is not lower in food-restricted than in ad libitum-fed rats (McCarter and McGee 1989). Thus, the sustained reduction in the plasma glucose concentration in food-restricted rats must be due to changes in the characteristics of the metabolic system. Either an increase in "insulin sensitivity" or in "glucose effectiveness" as defined by Bergman (1989) or in both must be involved. Further research is needed to choose between these possibilities.

Summary

Food restriction delays, blunts, or prevents a broad spectrum of age changes in the physiological systems. Indeed, the number of physiological processes influenced is so great that these findings do not provide strong guidance for the exploration of the basic biological processes involved in aging or for learning the mechanisms by which food restriction influences these processes.

However, food restriction has proven to be a useful tool for testing the validity of theories of aging. In our laboratory, the effects of food restriction on specific physiological processes have been utilized for such studies. We have tested three major theories of aging by this means: (1) the metabolic rate theory, (2) the glucocorticoid cascade hypothesis, and (3) the glycation hypothesis. The results of our studies provide little support for the metabolic rate theory or for the glucocorticoid cascade hypothesis. However, our findings are consistent with the glycation hypothesis but further research is needed to evaluate the causal involvement of glycation in the aging processes.

References

Bergman RH (1989) Toward physiological understanding of glucose tolerance. Minimal model approach. Diabetes 38:1512–1527

Bucala R, Model P, Cerami A (1984) Modification of DNA by reducing sugars: a possible mechanism for nucleic acid aging and age-related dysfunction in gene expression. Proc Natl Acad Sci USA 81:105–109

Cerami A (1985) Hypothesis: glucose as a mediator of aging. J Am Geriatr Soc 33:626–634

Kalu DN, Masoro EJ, Yu BP, Hardin RR, Hollis BW (1988) Modulation of age-related hyperparathyroidism and senile bone loss in Fischer rats by soy protein and food restriction. Endrocinology 122:1847–1854

Maeda H, Gleiser CA, Masoro EJ, Murata I, McMahan CA, Yu BP (1985) Nutritional influences on aging of Fischer 344 rats. II. Pathology. J Gerontol 40:671–688

Masoro EJ (1985) Metabolism. In: Finch C, Schneider EL (eds) Handbook of the biology of aging, 2nd edn. Van Nostrand Reinhold, New York, pp 540–563

Masoro EJ (1988) Food restriction in rodents: an evaluation of its role in the study of aging. J Gerontol (Biol Sci) 43:B59–64

Masoro EJ (1989) Nutrition and aging in animal models. In: Munro HN, Danford DE (eds) Nutrition, aging and the elderly. Plenum, New York, pp 25–41

Masoro EJ, Katz MS, McMahan CA (1989) Evidence for glycation hypothesis of aging from the food restricted model. J Gerontol (Biol Sci) 44:B20–22

Masoro EJ, Yu BP, Bertrand HA (1982) Action of food restriction in delaying the aging process. Proc Natl Acad Sci USA 79:4239–4241

McCarter RJ, McGee JR (1989) Transient reduction of metabolic rate by food restriction. Am J Physiol 257:E175–179

McCarter RJ, Masoro EJ, Yu BP (1985) Does food restriction retard aging by reducing the metabolic rate? Am J Physiol 248:E488–492

Pearl R (1928) The rate of living. Knopf, New York

Pongor S, Ulrich PC, Benscath FA, Cerami A (1984) Aging of proteins: isolation and identification of fluorescent chromophore from the reaction of polypeptides with glucose. Proc Natl Acad Sci USA 81:2684–2688

Rubner M (1908) Das Problem der Lebensdauer und seine Beziehungen zu Wachstum und Ernährung. Oldenbourg, Munich

Sabatino F, Masoro E, McMahan CA (1990) Influence of aging and food restriction on plasma corticosterone levels. FASEB J 4:A369

Sacher GA (1977) Life table modification and life prolongation. In: Finch C, Hayflick L (eds) Handbook of the biology of aging. Van Nostrand Reinhold, New York, pp 582–638

Sapolsky RM, Krey LC, McEwen BS (1986) The neuroendocrinology of stress and aging: the glucocorticoid cascade hypothesis. Endoc Rev 7:84–301

Vlassara H, Brownlee M, Cerami A (1986) Nonenzymatic glycosylation: role in pathogenesis of diabetic complications. Clin Chem 32:B37–41

Weindruch R (1985 Aging in rodents fed restricted diets. J Am Geriatr Soc 33:125–132

CHAPTER 12

Aging, Dietary Restriction, and Glucocorticoids: A Critical Review of the Glucocorticoid Hypothesis

R.R. Holson[1], P.H. Duffy[1], S.F. Ali[1], and F.M. Scalzo[1]

Introduction

Philosophers and scientists have long been beguiled by the possibility that, like growth and development, senescence and death are biologically programmed. Since development is under endocrine control, by analogy the control of aging has also been sought in this system. Early research concentrated upon the impact of hypophysectomy on biological measures of aging. While this work has produced intriguing evidence for delay or even reversal of some age-related changes (see Regelson 1983; Cole et al. 1982; and Everitt 1976 for reviews), hypophysectomized subjects seldom outlived intact controls and often died of hormonal abnormalities attendant upon this radical procedure. Similarly, the search for a "death hormone" under pituitary regulation (Denckla 1981) continues to be frustrating. Among the known hormones, at this time the adrenal glucocorticoids (GC) are doubtlessly the best candidate for a role in the endocrine modulation (if not control) of some aspects of aging.

Glucocorticoids and Aging

That GCs can contribute to aging is suggested by evidence of several sorts. Chronic stress, for example, has been associated clinically and experimentally with accelerated aging (Arvay 1976; Paré 1965; Gunderson and Rahe 1974; Selye and Tuchweber 1976), observations which imply but do not require a GC role. Further, hyperadrenocorticism in humans (Cushing's syndrome) is associated with reduced longevity due to arteriosclerosis, osteoporosis, muscle wasting, immunologic dysfunction, and diabetes. These are all common degenerative diseases of the elderly, suggesting that an abnormal elevation of GCs can accelerate

[1]National Center for Toxicological Research, Division of Reproductive and Developmental Toxicology, Jefferson, AR 72079, USA

aging (Findlay 1949; Solez 1952). From a comparative viewpoint, massive adrenal hyperactivity is also implicated in the programmed death of Pacific salmon (Robertson and Wexler 1957; 1960) and of some mice (Diamond 1982).

Wexler and his colleagues have extended their findings of adrenal abnormalities in salmon to rats, in a series of reports spanning several decades. These investigators have shown that repeated breeding in rats produces ". . . hyperlipidemia, hyperglycemia, hypertension, premature arteriosclerosis, and accelerated aging" (Wexler 1964; 1984; Wexler et al. 1964), and that this effect is linked to adrenocortical hyperactivity (Wexler and Kittinger 1965; Iams and Wexler 1977). Subsequent work by this group also demonstrated a "Cushingoid" pathophysiology (inducing thinning of skin, hair loss, diabetes, vascular lesions, and myocardial fibrosis) accompanied by hyperadrenocorticism in a spontaneously obese strain of the SHR rat (Wexler and McMurtry 1983). Adrenalectomy prevented the hyperphagia, obesity, and degenerative disorders seen in these animals (Wexler and McMurtry 1981).

The Glucocorticoid Cascade Hypothesis

The suggestive findings mentioned in the previous section have been largely overshadowed by the work of Landfield, Sapolsky, and their colleagues, work restricted to the relationship between GCs and the degeneration which accompanies aging in the hippocampus. This model is important for several reasons. It focuses upon the probable neural locus of the memory loss which is an invariant accompaniment of human aging, and hence upon a very important aspect of the loss of functional capacity which accompanies age. In adopting the strategy of concentrating narrowly upon a single much-studied area of the brain, this research has been able to make rapid progress. Perhaps most important, these researchers have produced a specific, testable hypothesis, the glucocorticoid cascade hypothesis, linking neural control of adrenal secretion, aging, and cell death in the hippocampus. This paper will concentrate upon this body of work, but the reader is reminded that this topic is a subset of a larger "glucocorticoid hypothesis," linking this hormone to a wide range of the disorders which accompany aging.

The glucocorticoid cascade hypothesis was first clearly enunciated by Landfield et al. (1980), although the term itself is drawn from a recent review by Sapolsky (Sapolsky et al. 1986). As shown in Fig. 12.1, the hypothesis posits a negative feedback cascade between the aging hippocampus and the adrenals. It begins with the observation that the hippocampal neuron is enriched in glucocorticoid receptors, and that these neurons exert an inhibitory influence on the hypothalamic-pituitary-adrenal axis. Over the life span, it is postulated that GC exposure causes cell death in the hippocampus, presumably a receptor-mediated effect. This cell death reduces hippocampal inhibition of adrenal secretion, which further increases hippocampal cell death, resulting in greater GC secretion, and so on, ultimately producing functional memory loss.

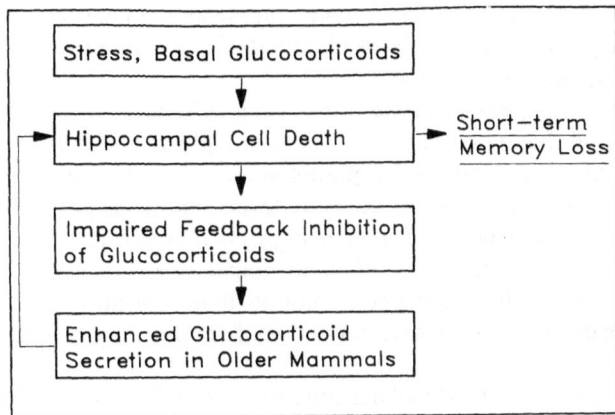

Figure 12.1. Proposed negative feedback cascade in the glucocorticoid cascade hypothesis showing how abnormal regulation of glucocorticoid secretion can produce memory loss

To validate this model, a number of experimentally difficult steps are required. First, it must be demonstrated that over the course of normal aging there is growing neural cell loss in the hippocampus, and that this cell loss occurs at a time and in a quantity sufficient to produce the memory loss of aging. Next (and perhaps most difficult) it must be proven that physiological levels of glucocorticoid secretion do cause the observed incidence of hippocampal cell death, and the mechanism of this hormonal neurotoxicity must be identified. Finally, there is the matter of hippocampal feedback regulation of GC secretion. While this aspect of the hypothesis is not logically required (physiological levels of GC over a lifetime could induce a receptor-mediated excitotoxicity, for example, without postulating any loss of higher control of the adrenals), the strong form of the hypothesis requires that with aging there be a loss of inhibitory control of the adrenals, one directly correlated with the magnitude of hippocampal damage.

Glucocorticoids and Age-Related Degeneration in the Hippocampus

Some progress has been made in each of the areas already mentioned. Starting with the brain, it is now well established that hippocampal damage causes short-term memory loss, that the hippocampus does undergo progressive neuronal cell loss and astrocytic proliferation with age (Landfield et al. 1977), and that the hippocampus has an as yet poorly understood inhibitory influence upon GC secretion (Sapolsky et al. 1983). On the other hand, it is not known to what degree neuronal death in the hippocampus is directly responsible for the memory loss which occurs with aging, it has not been established that GC secretion is predominantly responsible for age-induced hippocampal neurotoxicity, and only a few tantalizing glimpses have been obtained of a possible mechanism for this hypothesized GC neurotoxicity.

Turning first to the question of the link between GC secretion and hippocampal cell death, we find that much remains to be accomplished here. At the very least, it is necessary to demonstrate that a pathologic condition of the hippocampus is correlated over a wide range of GC exposure levels, including physiological levels. Landfield et al. (1978) demonstrated a high correlation at 13 months of age between basal trough plasma GCs and hippocampal astrocytic gliosis in nine male Fischer rats. Of course this report could not attribute causality, so it was unclear whether the obtained GC levels were reflective of or contributory to the hippocampal damage.

In a second paper, Landfield and his colleagues (Landfield et al. 1981) reported that male Fischer rats adrenalectomized at 18 months of age and killed at 27 months showed fewer signs of hippocampal aging in the CA_1 subregion than did same-age controls. However, there was high mortality (42)% in the adrenalectomized group and considerable overlap between hippocampal aging measures in brains of controls and survivors of adrenalectomy. Further, adrenalectomy causes hypersecretion of adrenocorticotropic hormone (ACTH), and inclusion in this experiment of a group of controls treated with an ACTH fragment suggested that part of the protective effect of adrenalectomy might have been ACTH mediated. More important, it has recently been reported that adrenalectomy induced profound cell death in the dentate gyrus of the rat (Sloviter et al. 1989). These findings cast grave doubts on the utility of adrenalectomy for studying GC-induced brain damage.

Attempts have also been made to assess the effect of chronic pharmacological GC exposure on the hippocampus. Aus der Mühlen and Ockenfels (1969) exposed young male guinea pigs to daily pharmacological doses of cortisone for 4 weeks and reported some pyknotic neurons in thalamus, midbrain, and hippocampus, accompanied by neuronal swelling. Similarly, Sapolsky et al. (1985) gave intact 5-month-old male rats daily injections of corticosterone (5 mg daily) for 3 months. This treatment caused a permanent loss of GC receptors in the hippocampus, due both to neuron loss and to receptor downregulation. These changes were seen most prominently in the CA_3 subregion of Ammon's horn. While encouraging, both experiments are subject to the criticism that they used nonphysiological treatments, and hence do not really demonstrate GC toxicity under more typically physiological conditions.

To date this small handful of experiments are the only ones to come to our attention. While encouraging, they clearly need independent replication, especially over a range of chronic physiological stressors, before a prudent observer could accept such data as convincing proof of the glucocorticoid cascade hypothesis.

Although it is still uncertain whether glucocorticoids are directly responsible for age-related hippocampal cell death, there is strong evidence for GC mediation of the effect of neurotoxins on the hippocampus. In a series of studies, Sapolsky and his colleagues (Sapolsky 1985a,b; Sapolsky and Pulsinelli 1985; Sapolsky 1986a,b) have shown that hippocampal damage caused by cerebral ischemia, local hippocampal injection of an excitotoxin (kainic acid) or an antimetabolite (3-acetylpyridine) are all reduced by adrenalectomy and increased by corticosterone injections. This synergistic interaction between GCs and hippocampal neurotoxicity is partly blocked by coadministration of glucose or related brain fuels (Sapolsky 1986b). Again these experiments do not provide the proverbial smoking gun, but they do point toward a possible mechanism of GC-induced hippocampal neurotoxicity, which is the well-known GC inhibitory effect upon cellular glucose uptake (Munck 1971).

Age-Induced Changes in Glucocorticoid Secretion

If the hypothesized GC induction of age-related cell death in the hippocampus remains to be proven, what of age-related changes in GC secretion? If the glucocorticoid cascade hypothesis is correct as stated (Sapolsky et al. 1986), then there should be increases in GC secretion in older animals. Sapolsky (Sapolsky et al. 1983) and Landfield (Landfield et al. 1978) both suggest that this increased GC secretion should manifest at least in part as increased basal (unstimulated) peak and trough GC levels. Additionally, Sapolsky and his collaborators (Sapolsky et al. 1983) have found that stress-induced increases in GC levels may reach identical peaks in young and old rats, but the elderly GC levels then decline to baseline more slowly, resulting in a prolonged elevation of GC stimulation.

Such changes with age should be well documented in the sizeable literature on mammalian pituitary-adrenal function across the life span. Regrettably, however, this literature is extremely, perhaps surprisingly, confused and contradictory. Work in humans has reported no age-related changes in plasma GC levels (Jensen and Blichert-Toft 1971), while rodent research to date has produced very mixed results. A number of investigators report enhanced basal GC levels in older rats, in line with the GC cascade hypothesis (Grad and Khalid 1968; Riegle 1973; Landfield et al. 1978; Sapolsky et al. 1983; Dekosky et al. 1984; Stewart et al. 1988). On the other hand, equally reliable laboratories report no such effect (Hess and Riegle 1970; Tang and Phillips 1978; Sonntag et al. 1987). Since many of the positive reports utilized blood draw techniques which could be stressful (ether + orbital sinus drain, Riegle 1973; tail blood, Sapolsky et al. 1983; N_2 asphyxiation in home cage, DeKosky et al. 1984; tail blood draws every 4 h, Stewart et al. 1988), it may be that elderly rats simply are more reactive to such procedures. In any case, it is certainly not possible to draw any firm conclusions based on this literature.

There may be better reason to suppose that stress-induced GC levels are elevated for a longer time in old rats. In elderly humans and rodents GC plasma clearance rates appear to be slowed (Samuels 1956; Hess and Riegle 1972), a factor which could account for Sapolsky's findings. Moreover, Riegle has reported that elderly rats fail to downregulate responsiveness of the pituitary adrenal axis to chronic ACTH (Hess and Riegle 1972) or restraint stress (Riegle 1973). Thus there may well be more subtle regulatory abnormalities in the aged pituitary-adrenal system.

Studies of ACTH-invoked GC secretion, in vivo and in vitro, certainly support such a possibility. A variety of elderly mammals appear to have an attenuated GC response to ACTH, including humans (Samuels 1957; Dilman 1976), cattle (Riegle and Nellor 1967), goats (Riegle et al. 1968), and rodents (Hess and Riegle 1970). This attenuation of elderly adrenal ACTH sensitivity has also been reported in vitro (Pritchett et al. 1979; Malamed and Carsia 1983; Popplewell et al. 1986) and may extend to an insensitivity to corticotropin release factor (CRF) at the hypothalamic level (Hykla et al. 1984). While such alterations are well documented, this ACTH insensitivity would certainly not suggest a GC hyperactivity in the elderly, contrary to the GC hypothesis.

In summary, at this writing neither leg of the GC cascade hypothesis seems to be firmly grounded in experimental results. The relationship between GCs and hippocampal damage is not established, while the literature on age-induced changes in basal and stimulated GC secretion is in disarray. This is due in large part to the very real experimental difficulties involved in testing this hypothesis. As we have seen, adrenalectomy is strongly contraindicated, while pharmacological manipulation of this system is always open to criticism on the grounds that it introduces nonphysiological influences. What is needed is a less invasive means of chronically altering plasma GC levels, followed by life span histological and functional evaluation of the effects of such intervention on the hippocampus.

The burden of this chapter is to alert the reader to the possibility that chronic caloric restriction may provide one such means. It now appears that this procedure, long known to substantially extend rodent life span, may also chronically elevate plasma GC levels in restricted rats and mice. If so, comparisons between restricted and ad libitum-fed subjects will allow a critical test of the GC cascade hypothesis. The remainder of this paper will be devoted to consideration of this topic.

Dietary Restriction and Glucocorticoid Secretion

Acute starvation is accompanied by increased GC secretion (Chowers et al. 1969; Bouillé and Assenmacher 1970), an effect which is central to the role of this steroid in emergency mobilization and redirection of glucose stores (Munck et al. 1984). Thus there is no doubt that severe dietary restriction (DR) will initially enhance GC secretion. The important questions involve the duration, timing, and size of DR-induced GC secretion in chronic, mild DR.

Answers to this question are complicated by a second phenomenon. GCs in rodents display a well-known circadian rhythmicity, with a pronounced peak at night, during the period of peak activity and food intake. Restriction of food availability to a few hours outside of this natural peak causes a "splitting" of this circadian rhythm, with a primary GC peak now appearing near the time of feeding and a secondary peak during the standard period of night-time activity (Honma et al. 1983; Krieger 1974; Morimoto et al. 1977). Life span DR studies are a great deal of work, and feeding times are typically synchronized with circadian rhythms of human care takers, not rodent subjects. Consequently one must also expect circadian GC rhythms in DR subjects to be initially upset by feeding-driven alterations in GC circadian rhythms. Again the question is how large and enduring such changes are, and what happens if food is made available during the rodent high activity and feeding stage.

Answers to these questions are still actively being sought, but available information indicates that DR-induced alterations in GC levels are predictable from the above considerations, and may be permanent. Thus two recent studies have demonstrated that the "splitting" of the circadian GC rhythm by morning provision of the DR ration persists across the life span (Armario et al. 1987; Stewart et al. 1988). This phenomenon guarantees that sampling of GC levels during the normal morning GC trough in ad libitum-fed rats will reveal a greatly enhanced GC level in DR subjects, even after chronic restriction for many months (Leakey et al., this volume). Conversely, samples drawn in the evening, during the GC peak in ad libitum-fed subjects, suggest that levels are lower in DR than in ad libitum-fed animals. The question is whether total 24-h secretion is enhanced or decreased by DR. Here the published literature is in conflict. Armario et al. (1987) report that total 24-h GC secretion is increased in young rats placed on 65% of ad-libitum food restriction for 34 days. Stewart et al. (1988), on the other hand, report that 24-h GC levels are elevated in 5-month-old rats on 60% DR for 2 months, but that 24-month-old DR rats had lower levels than ad libitum controls, due to a pronounced increase in the GC levels of ad libitum-fed animals. As already noted, these findings are questionable, due to a blood draw technique which involved repeated tail-blood draws every 4 h for 24 h.

A far better technique for resolving this question involves phase shifting the vivarium lighting cycle, so that the standard 10 a.m. feeding of the DR group coincides with the early lights-off phase in the animal room. These experiments have been conducted for DR B6C3F$_1$ mice in this laboratory. Subjects were fed ad libitum or 60% DR from 14 weeks of age. Prior to killing (decapitation within 1 min of removal from the home room) lighting was phase shifted (for a minimum of 3 weeks) so that provision of the daily DR ration occurred just prior to lights-out. At 30 months of age, DR males and females, restricted for 27 months, displayed a sizeable increase in GC levels across the circadian cycle (Fig. 12.2). Younger subjects were sacrificed at this peak GC level at 7 or 15 months of age. Here too a very substantial increase was seen in DR plasma GC levels (Fig. 12.3).

This life span increase in plasma GC levels in phase-shifted DR mice has now also been reported in rats (Masaro, this volume). These findings tempt one to conclude that the predictable acute increase in GC levels triggered by DR persists throughout the circadian cycle and over the life span of DR subjects. However, a cautionary note is in order here. It can be argued that chronic DR rodents spend a lifetime learning just one thing very well. That is that entry of humans into the home room signals feeding time. Consequently repeated experimenter entries into the home vivarium in the course of a timed killing could trigger an anticipatory GC surge in DR but not in ad libitum-fed rodents. When we reanalyzed our 30-month GC data by sacrifice order (comparing differences between the first,

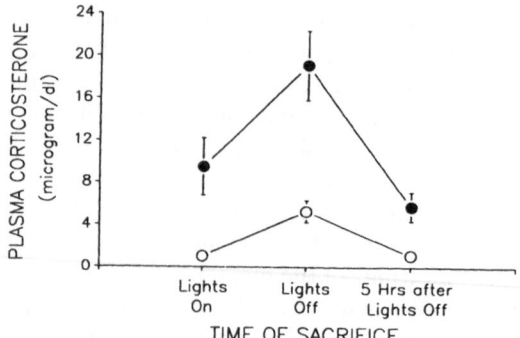

Figure 12.2. Plasma corticosterone levels (mean ± SEM) in restricted (*solid circles*) and ad libitum-fed (*open circles*) B6C3F₁ mice determined across the circadian cycle at 30 months of age (*n* = 8–11 per group)

second, and nth ad libitum and DR pair of mice killed), we found that differences were initially in favor of higher GC levels in the ad libitum member of the pair, but this difference quickly shifted in favor of DR pair members in subsequent pairs (Fig. 12.4).

Thus, while there are compelling theoretical reasons for accepting the finding of a life span GC elevation in DR rodents, such findings should be treated with caution at least until they are replicated in several laboratories using 24-h sampling in isolated DR and ad libitum-fed rodents with indwelling catheters.

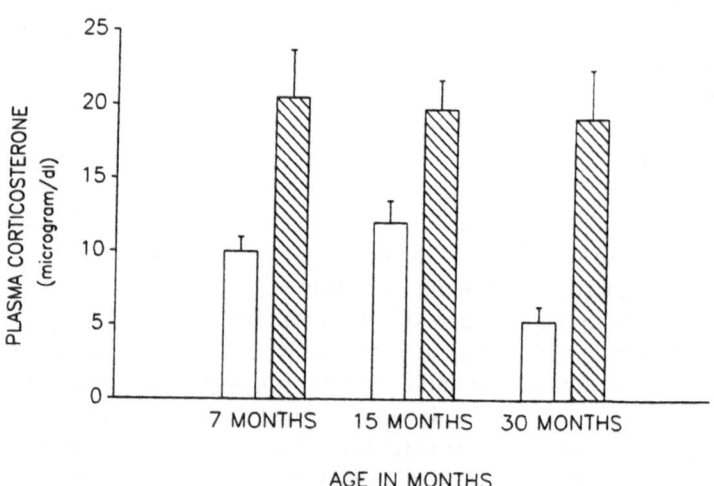

Figure 12.3. Serum corticosterone levels (mean ± SEM) taken at the peak of circadian levels in 7-, 15-, and 30-month restricted (*hatched bars*) and ad libitum-fed (*open bars*) B6C3F₁ mice (*n* = 20–30 per group)

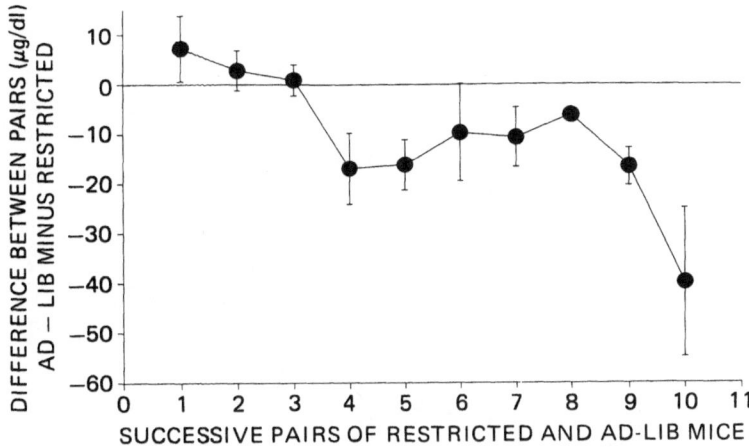

Figure 12.4. Differences in corticosterone levels (mean ± SEM) between pairs of ad libitum-fed and restricted B6C3F₁ mice by order of sacrifice at 30 months

Dietary Restriction and Maze Performance

The ability to learn the location of rewards or escape areas in complex spatial environments declines rapidly with age in rodents. Such behavioral measures are also extremely sensitive to hippocampal damage. Thus if chronic DR restriction does elevate GC levels, and if this elevation results in hippocampal neurotoxicity, DR rodents should be deficient in complex spatial task performance such as the radial eight-arm maze, the Morris water maze or complex T mazes, at least after prolonged DR. To date, no such deficit has been reported, with investigators typically finding no alteration (Beatty et al. 1987) or actual improvements in DR performance on such tasks (Goodrick 1984; Ingram et al. 1987; Idrobo et al. 1987).

A recent experiment in this laboratory produced very similar results. Brown Norway male rats were placed on a lifetime 60% of ad libitum diet at 14 weeks. These DR rats and age-mate ad libitum-fed controls were tested in a complex maze (Fig. 12.5) at 7, 14, or 23 months of age. To control for the substantial behavioral impact of DR and acute hunger, body weights were equated by briefly allowing DR rats to reach 80% of ad libitum body weight, while ad libitum control weights were reduced to the same 80% baseline. All subjects were then tested under 23-h water deprivation, to water reward.

Previous work with this apparatus had revealed that animals with hippocampal damage induced by the neurotoxin trimethyltin (TMT) displayed a characteristic behavioral profile. Such brain-damaged animals eventually made their way to the goal region of the maze (Fig. 13.4), but experienced great difficulty in finding their way back to the start region (Holson et al. 1989, Fig. 12.6). The older (14 and 23 month) brown Norways displayed precisely the same TMT-like deficit relative to younger (7-month-old) Brown Norways (Fig. 12.7). However, at no age did behavior in this complex maze differ as a function of diet. This

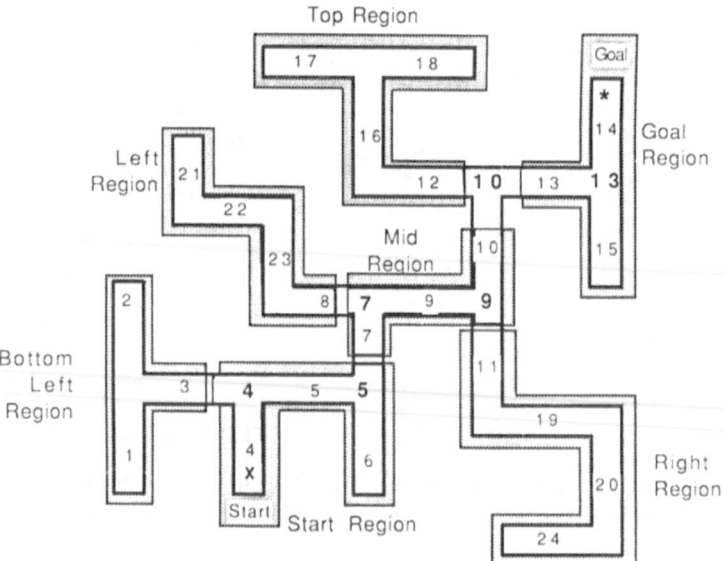

Figure 12.5. Complex maze. The floor plan of the 24-arm complex maze. Each arm is numbered, and major choice points en route to the goal are numbered in bold print. *Shaded lines* indicate the seven maze regions used for purposes of data analysis (adapted from Holson et al. 1989)

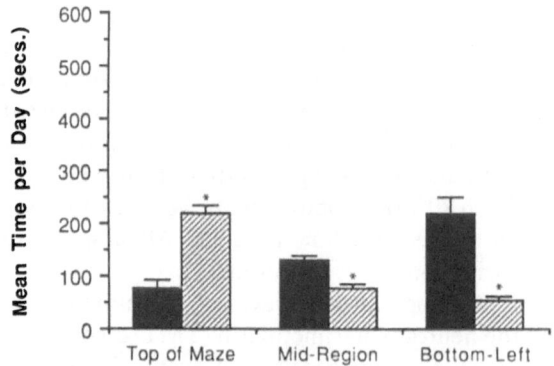

Figure 12.6. Maze performance in 7-month-old Sprague-Dawley rats following a single injection of 7.0 mg/kg trimethyltin (n = 8–10 per group). TMT, *hatched bars*; controls, *black bars*; *, significantly different from controls; $p \leq 0.05$. (Adapted from Holson et al. 1989)

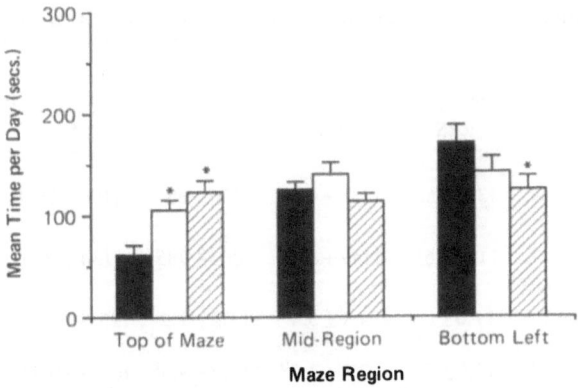

Figure 12.7. Maze performance in 7- (*black bars*), 14- (*open bars*), and 23-month-old (*hatched bars*) ad libitum-fed and diet-restricted Brown Norway rats (data collapsed over diet condition; $n = 10–12$ per group). *, Significantly different from controls; $p \leq 0.05$

experiment suggests that if differences in acute hunger are controlled, chronic caloric restriction neither retards nor accelerates the memory loss seen even in middle-aged rats (14 months) in a complex maze.

Dietary Restriction and Neurochemistry

At the time of killing brains were collected from the same 7-, 18-, and 30-month ad libitum-fed and DR B6C3F$_1$ mice which provided the GC data discussed above (Figs. 12.2, 12.3). Muscarinic binding was measured in hippocampus and caudate, and both D$_1$ and D$_2$ binding was quantified in caudate, with single-point assays in all cases. There was an age-related drop in ^{3}[H]-quinuclidinyl benzylate (QNB) binding in the hippocampus of the 30-month-old mice, relative to both 18- and 14-month-old mice (Table 12.1). Moreover, D$_2$ but not D$_1$ binding sites were reduced in the caudate of 18- and 30-month-old mice, relative to 7-month-old mice (Table 12.1). At no time were differences seen between the brains of ad libitum-fed and DR animals during these measures. Finally, hypothalami from both sets of 30-month-old animals were assayed by high-performance liquid chromatography–electrochemical detection (HPLC-ECD) for content of dopamine, serotonin, and metabolites. Again there were no differences between DR and ad libitum-fed animals (Table 12.2).

These findings do not support earlier reports of reductions in age-related loss of caudate dopamine receptors in DR rats (Levin et al. 1982; Roth et al. 1984), but neither finding suggests an acceleration of brain aging correlates in DR animals.

Table 12.1. Dopaminergic and muscarinic receptor binding in B6C3F$_1$ mice.

	Age					
	7 Months		18 Months		30 Months	
Diet	AL	RES	AL	RES	AL	RES
Hippocampus						
^{3}[H]-QNB[a]	1361±44	1309±53	1333±63	1276±35	1190±26[b,c]	1113±40[b,c]
Caudate						
^{3}[H]-QNB[a]	1136±49	1199±61	1149±85	1067±985	1134±40	1107±68
^{3}[H]-SCH-23390[a]	111.2±3.5	111.0±2.5	104.1±5.4	114.2±7.2	125.8±12.2	133.9±12.5
^{3}[H]-spiro-peridol[a]	189.8±4.2	211.0±7.7	107.3±7.9[b]	110.2±5.4[b]	103.3±9.5[b]	92.3±10.6[b]

AL, ad libitum feeding; RES, 60% calorie restricted diet; QNB, quinuclidinyl benzylate; SCH, Schering.
[a]Mean fmol/mg protein ± SEM.
[b]Significantly different from 7-month group, $p \leq 0.05$.
[c]Significantly different from 18-month group, $p \leq 0.05$.

Conclusion

We began this review by presenting the glucocorticoid hypothesis. In its broadest form, this hypothesis posits that over the course of a lifetime, continual exposure to GCs contributes to the development of a range of the degenerative diseases of aging, from reduced immune competence through cardiovascular and metabolic disorders to memory loss and perhaps to cancer. A narrower, more specific, and, hence, more easily tested variant of this hypothesis, the glucocorticoid cascade hypothesis of Landfield and Sapolsky, has related GC-induced damage to hippocampus to impaired feedback inhibition of GC secretion, in a positive feedback loop which eventually produces substantial hippocampal damage and memory loss in the elderly.

To date, neither form of this hypothesis has been fully tested, and virtually all predictions, drawn from either form of the hypothesis, are under question. Thus

Table 12.2. Dopamine, serotonin and their metabolites in the hypothalamus of 30-month-old male and female B6C3F$_1$ ad libitum-fed and diet-restricted mice (ng/mg tissue ± SEM).

	Ad libitum feeding ($n=16$)	Restricted diet ($n=16$)
Dopamine	0.23 ± 0.01	0.25 ± 0.02
DOPAC	0.11 ± 0.005	0.12 ± 0.007
HVA	0.11 ± 0.004	0.11 ± 0.005
Serotonin	0.77 ± 0.04	0.79 ± 0.04
5-HIAA	0.17 ± 0.02	0.18 ± 0.02

it is unclear whether GC secretion undergoes any substantial alteration with age, and it is certainly not proven that prolonged physiological GC exposure plays any role in the normal deterioration of aging, in hippocampus or anywhere else in the body.

A primary cause of this uncertainty is technical. It is difficult to irrefutably test this hypothesis. Hypophysectomy and adrenalectomy are dubious tools at best, since they trigger a wide range of life-threatening hormonal abnormalities, abnormalities which evidently can, inter alia, cause profound cell loss in the dentate gyrus of the hippocampus. Similarly, pharmacological manipulations of the pituitary adrenal axis are open to many of the same objections—they are nonphysiological interventions, which can tell us little about the normal course of aging.

Clearly what is required is recourse to noninvasive and more "natural" means of elevating or reducing chronic GC secretion within a single experimental species. The thrust of this chapter is that chronic caloric restriction may provide one such treatment. To date, all indications are that GC increases seen classically after acute nutritional deprivation persist around the clock and over the life span in calorically restricted rodents.

This possibility is not a promising one for the glucocorticoid hypothesis. If such a sizeable and persistent GC elevation does indeed occur in the DR rodent, then according to this hypothesis one must expect that damage to the hippocampus and other organ systems would be accelerated. Of course precisely the opposite is the case. Life span DR is the only treatment known to greatly retard the deteriorative diseases of aging; this treatment also does not appear to accelerate the memory loss of aging, and indeed might have the opposite effect. In short, caloric restriction may well totally contradict the glucocorticoid hypothesis, in that elevated glucocorticoids in this condition are associated with a remarkable reduction in the rate of aging. (Parenthetically, glucocorticoids are higher in female than in male rats across the life span, yet female rats on the average outlive males.)

However, it would be altogether premature to sound the death knell for the glucocorticoid hypothesis at this stage. Like the hypothesis itself, the possibility of dietary alterations in GC levels requires much careful research. Just to clearly prove that GC levels are elevated, it will be necessary to carefully evaluate both sexes over the circadian cycle and across the life span. Such evaluations should be conducted on phase-shifted restricted animals, with the daily food ration provided shortly after lights-out. Moreover, the technique of choice for blood draws involves cannulation of rats well adapted to the cannula and housed in sound- and light-proof chambers.

Nor are such experiments adequate. What is required is a demonstration that GC effects on target tissues are accentuated by caloric restriction. Such a demonstration involves evaluation not just of GC secretion, but of the entire cascade of events which ultimately produces a GC effect at the cellular level. Thus it will be necessary to quantify corticosterone-binding globulin levels, GC receptor density and affinity, and postreceptor events in DR animals. Only if such experiments succeed in demonstrating an enhanced glucocorticoid effect on tissues such as the hippocampus in dietary-restricted rodents will this procedure fulfill its promise as a critical test of the glucocorticoid hypothesis.

Acknowledgments. Ms. Patricia Sullivan-Jones is thanked for conducting all the corticosterone assays; Mr. Bobby Gough is likewise thanked for the HPLC-ECD analyses of hypothalamic monoamines. Jay Williams conducted the maze tests, and Ms. Kellye Luckett transformed all of the above into a readable manuscript.

References

Armario A, Montero JL and Jolin T (1987) Chronic food restriction and the circadian rhythms of pituitary-adrenal hormones, growth hormone and thyroid-stimulating hormone. Ann Nutr Metab 31:81–87

Arvay A (1976) Reproduction and aging. In: Everitt AV, Burgess MA (eds) Hypothalamus, pituitary and aging. Thomas, Springfield, pp 363–375

Aus der Mühlen K, Ockenfels H (1969) Morphologische Veränderungen in Diencephalon und Telencephalon nach Störungen des Regelkresises Adenohypophyse-Nebennierenrinde. Z Zellforsch 93:126–141

Beatty WW, Clouse BA, Bierley RA (1987) Effects of long-term restricted feeding on radial maze performance by aged rats. Neurobiol Aging 8:325–327

Bouillé C, Assenmacher I (1970) Effects of starvation on adrenal cortical function in the rabbit. Endocrinology 87:1390–1394

Chowers I, Einat R, Feldman S (1969) Effects of starvation on levels of corticotrophin releasing factor, corticotrophin and plasma corticosterone in rats. Acta Endocrinol 61:687–694

Cole GM, Segall PE, Timiras PS (1982) Hormones during aging. In: Vernadakis A, Timiras PS (eds). Hormones in Development and Aging. Spectrum, New York, pp 477–511

DeKosky ST, Scheff SW, Cotman CW (1984) Elevated corticosterone levels: possible cause of reduced axon sprouting in aged animals. Neuroendocrinology 38:33–38

Denckla WD (1981) Aging, dying and the pituitary. In: Schimke RT (ed) Biological mechanisms of aging. NIH, Baltimore, pp 673–684

Diamond JM (1982) Big bang reproduction and aging in male marsupial mice. Nature 298:115–116

Dilman VM (1976) The hypothalamic control of aging and age-associated pathology. In: Everitt AV, Burgess JA (eds) Hypothalamus, pituitary and aging. Thomas, Springfield, pp 634–667

Everitt AV (1976) Hypophysectomy and aging in the rat. In: Everitt AV, Burgess AV (eds) Hypothalamus, pituitary and aging. Thomas, Springfield, pp 68–75

Findley T (1949) Role of the neurohypothysis in the pathogenesis of hypertension and some allied disorders associated with aging. Am J Med 7:70–84

Goodrick CL (1984) Effects of lifelong restricted feeding on complex maze performance of rats. Age 7:1–2

Grad B, Khalid R (1968) Circulating corticosterone levels of young and old, male and female C57B1/6J mice. J Gerontol 23:522–528

Gunderson EK, Rahe RJ (1974) Life stress and illness. Thomas, Springfield

Hess GD, Riegle GD (1970) Adrenocortical responsiveness to stress and ACTH in aging rats. J Gerontol 25:354–358

Hess GD, Riegle GD (1972) Effects of chronic ACTH stimulation on adrenocortical function in young and aged rats. Am J Physiol 222:1458–1461

Holson RR, Ali SF, Scallet AC, Slikker W Jr, Paule MG (1989) Benzodiazepine like behavioral effects following withdrawal from chronic delta-9-tetrahydrocannabinol administration in rats. Neurotoxicology 10:605–620

Honma K-I, Honma S, Hiroshige T (1983) Critical role of food amount for prefeeding corticosterone peak in rats. Am J Physiol 245:R339–R344

Hykla VW, Sonntag WE, Meites J (1984) Reduced ability of old male rats to release ACTH and corticosterone in response to CRF administration. Proc Soc Exp Biol Med 175:1–4

Iams SG, Wexler BC (1977) Inhibition of spontaneously-developing arteriosclerosis in female breeder rats by adrenalectomy. Atherosclerosis 27:311–323

Idrobo F, Nandy K, Mostofsky DI, Blatt L, Nandy L (1987) Dietary restriction: effects on radial maze learning and lipofuscin pigment deposition in the hippocampus and frontal cortex. Arch Gerontol Geriatr 6:355–362

Ingram DK, Weindruch R, Spangler EL, Freeman JP, Walford RL (1987) Dietary restriction benefits learning and motor performance of aged mice. J Gerontol 42:78–81

Jensen HK, Blichert-Toft M (1971) Serum corticotrophin, plasma cortisol and urinary excretion of 17-ketogenic steroids in the elderly (age group 66–94 years). Acta Endocrinol (Copenh), 66:25–34

Krieger DT (1974) Food and water restriction shifts corticosterone, temperature, activity and brain amines periodicity. Endocrinology 95:1195–1201

Landfield PW, Baskin RK, Pitler TA (1981) Brain aging correlates: retardation by hormonal-pharmacological treatments. Science 214:581–584

Landfield PW, Rose G, Sandles L, Wohlstadter TC, Lynch G (1977) Patterns of astroglial hypertrophy and neuronal degeneration in the hippocampus of aged, memory-deficient rats. J Gerontol 32:3–12

Landfield PW, Sundberg DK, Smith MS, Eldridge JC, Morris M (1980) Mammalian aging: theoretical implications of changes in brain and endocrine systems during mid- and late-life in rats. Peptides 1:185–196

Landfield PW, Waymire JC, Lynch G (1978) Hippocampal aging and adrenocorticoids: quantitative correlations. Science 202:1098–1102

Levin P. Janda JK, Joseph JA, Ingram DK, Roth GS (1982) Dietary restriction retards the age-associated loss of rat striatal dopaminergic receptors. Science 214:561–562

Malamed S, Carsia RV (1983) Aging of the rat adrenocortical cell: response to ACTH and cyclic AMP in vitro. J Gerontol 38:130–136

Morimoto Y, Arisue K, Yamamura Y (1977) Relationship between circadian rhythm of food intake and that of plasma corticosterone and effect of food restriction on circadian adrenocortical rhythm in the rat. Neuroendocrinology 23:212–222

Munck A (1971) Glucocorticoid inhibition of glucose uptake by peripheral tissues: old and new evidence, molecular mechanisms, and physiological significance. Perspect Biol Med 14:265–289

Munck A, Guyre PM, Holbrook NJ (1984) Physiological functions of glucocorticoids in stress and their relation to pharmacological actions. Endocr Rev 5:25–44

Paré WP (1965) The effect of chronic environmental stress on premature aging in the rat. J Gerontol 20:78–84

Popplewell PY, Tsubokawa M, Ramachandran J, Azhar S (1986) Differential effects of aging on adrenocorticotrophin receptors, adenosine 3′,5′-monophosphate response, and corticosterone secretion in adrenocortical cells from Sprague-Dawley rats. Endocrinology 119:2206–2213

Pritchett JF, Sartin JL, Marple DN, Harper WL, Till ML (1979) Interaction of aging with in vitro adrenocortical responsiveness to ACTH and cyclic AMP. Horm Res 10:96–103

Regelson W (1983) The evidence for pituitary and thyroid control of aging: is age reversal a myth or reality? The search for a "death hormone." In: Regelson W, Sinex FM (eds) Intervention in the aging process. Liss, New York, pp 3–52

Riegle GD (1973) Chronic stress effects on adrenocortical responsiveness in young and aged rats. Neuroendocrinology 11:1–10

Riegle GD, Nellor JE (1967) Changes in adrenocortical function during aging in cattle. J Gerontol 22:83–87

Riegle GD, Przekop F, Nellor JE (1968) Changes in adrenocortical responsiveness to ACTH infusion in aging goats. J Gerontol 23:187–190

Robertson OH, Wexler BC (1957) Pituitary degeneration and adrenal tissue hyperplasia in spawning Pacific salmon. Science 125:1295–1296

Robertson OH, Wexler BC (1960) Histological changes in the organs and tissues of migrating and spawning Pacific salmon. Endocrinology 66:222–239

Roth GS, Ingram DK, Joseph JA (1984) Delayed loss of striatal dopamine receptors during aging of dietarily restricted rat. Brain Res 300:27–32

Samuels LT (1956) Effects of aging on steroid metabolism. In: Engle J, Pincus T (eds) Hormones and the aging process. Academic, New York, pp 21–38

Samuels LT (1957) Factors affecting the metabolism and distribution of cortisol as measured by levels of 17-hydroxycorticosteroids in blood. Cancer 10:746–751

Sapolsky RM (1985a) Glucocorticoid toxicity in the hippocampus: temporal aspects of neuronal vulnerability. Brain Res 359:300–305

Sapolsky RM (1985b) A mechanism for glucocorticoid toxicity in the hippocampus: increased neuronal vulnerability to metabolic insults. J Neurosci 5:1128–1232

Sapolsky RM (1986a) Glucocorticoid toxicity in the hippocampus. Neuroendocrin 43:440–444

Sapolsky RM (1986b) Glucocorticoid toxicity in the hippocampus: reversal by supplementation with brain fuels. J Neurosci 6:2240–2244

Sapolsky RM, Pulsinelli WA (1985) Glucocorticoids potentiate ischemic injury to neurons: therapeutic implications. Science 229:1397–1400

Sapolsky RM, Krey LC, McEwen BS (1983) The adrenocortical stress-response in the aged male rat: impairment of recovery from stress. Exp Gerontol 18:55–64

Sapolsky RM, Krey LC, McEwen BS (1985) Prolonged glucocorticoid exposure reduces hippocampal neuron number: implications for aging. J Neurosci 5:1222–1227

Sapolsky RM, Krey LC, McEwen BS (1986) The neuroendocrinology of stress and aging: the glucocorticoid cascade hypothesis. Endocr Rev 7:284–301

Selye H, Tuchweber B (1976) Stress in relation to aging and disease. In: Everitt AV, Burgess JA (eds) Hypothalamus, pituitary and aging. Thomas, Springfield, pp 554–569

Sloviter RS, Valiquette G, Abrams GM, Ronk EC, Sollas AL, Paul LA, Neubort S (1989) Selective loss of hippocampal granule cells in the mature rat brain after adrenalectomy. Science 243:535–538

Solez CA (1952) Aging and adrenal cortical hormones. Geriatrics 7:241–245, 290–294

Sonntag WE, Goliszek AG, Brodish A, Eldridge JC (1987) Diminished diurnal secretion of adrenocorticotropin (ACTH), but not corticosterone, in old male rats: possible relation to increased adrenal sensitivity to ACTH in vivo. Endocrinology 230:2308–2315

Stewart J, Meaney MJ, Aitken D, Jensen L, Kalant N (1988) The effects of acute and life long food restriction on basal and stress-induced serum corticosterone levels in young and aged rats. Endocrinology 123:1934–1941

Tang F, Phillips JC (1978) Some age-related changes in pituitary-adrenal function in the male laboratory rat. J Gerontol 33:377–382

Wexler BC (1964) Spontaneous arteriosclerosis in repeatedly bred male and female rats. J Atheroscl Res 4:57–80

Wexler BC (1984) Hyperlipidemia, hyperglycemia and hypertension in repeatedly bred parents of the obese spontaneously hypertensive rat (obese/SHR) unaccompanied by arteriosclerosis. Atherosclerosis 51:211–222

Wexler BC, Kittinger GW (1965) Adrenocortical function in arteriosclerotic female breeder rats. J Atheroscl Res 5:317–329

Wexler BC, McMurtry JP (1981) Ameliorative effects of adrenalectomy on the hyperphagia, hyperlipidemia, hyperglycemia and hypertension of obese, spontaneously hypertensive rats (obese/SHR). J Exp Pathol 62:146–157

Wexler BC, McMurtry JP (1983) Cushingoid pathophysiology of old, massively obese, spontaneously hypertensive rats (SHR). J Gerontol 38:148–154

Wexler BC, Antony CD, Kittinger GW (1964) Serum lipoprotein and lipid changes in arteriosclerotic breeder rats. J Atheroscl Res 4:131–143

CHAPTER 13

The Effect of Dietary Restriction on the Endocrine Control of Reproduction

B.J. Merry[1] and A.M. Holehan[1]

Introduction

The effect of chronic periods of underfeeding on reproductive performance in rodents has been a subject of contention for a number of years. This situation has arisen partly because of differences in the severity of underfeeding protocols, and secondly because of variability between species in their endocrine response to protracted periods of underfeeding. The early studies reported by McCay and colleagues (McCay et al. 1939), which later formed the basis for a proposed mechanism of action of diet on aging, demonstrated a relationship between the degree of enhanced survival and the length of the prepubertal period. An inverse correlation was observed between the length of retarded growth and the period of life remaining after maturation subsequent to accelerated growth following refeeding. It was demonstrated by Mulinos and Pomerantz (1940) that in adult animals, a hypopituitary condition could be induced by chronic undernutrition, and very severe underfeeding will decrease the secretion of hypothalamic releasing hormones (Campbell et al. 1977). These observations led to the repeated suggestion in the early literature that diet restriction operates through the induction of a "pseudo-hypophysectomized" state, sometimes referred to as dietary hypophysectomy. The extensive studies of Everitt et al. (1980; 1983) in which early surgical hypophysectomy, linked to selected hormone replacement therapy, was shown to mimic many of the effects of chronic underfeeding on pathology and subsequent survival, were interpreted as support for this suggestion. Implicit to this hypothesis is that such animals will remain sexually immature for the greater part of their life span. This prediction was confirmed in A strain mice which were maintained on a calorie-deficient diet for 240 days before returning to ad libitum feeding (Ball et al. 1947). These animals then produced 13 times as many litters after 240 days as their fully fed controls. The consensus from the initial studies on reproduction and chronic dietary restriction was that such feeding regimes induced a state of infertility, but this was reversible on a return to

[1]Institute of Human Ageing, University of Liverpool, P.O. Box 147, Liverpool, L69 3BX, UK

ad libitum feeding. During prolonged periods of underfeeding, reproductive aging appeared to be retarded, refed diet-restricted animals exhibiting greater fecundity than age-matched control animals (Holehan 1984).

Fertility and Reproductive Senescence

Suspended sexual maturation and prolonged periods of infertility are not inevitable consequences of restricted feeding regimes. In a series of studies the authors have shown that a less severe restricted feeding regimen than used in the early studies did not prevent sexual maturation in either male and female rats, but still resulted in a significant increase in longevity (Merry and Holehan 1981; Holehan and Merry 1985a,c). Although puberty was delayed by 63 to 189 days in CFY strain female rats, following sexual maturation more than 70% of underfed animals exhibited normal 5-day oestrous cycles. Chronic underfeeding was observed to abolish the age-related increase in cycle length, a feature of reproductive aging in fully fed rats. Similarly the age-related abnormalities in oestrous cycles which have been well documented with age for fully fed animals, (constant oestrus and persistent di-oestrus), showed a much later onset in underfed animals (Merry and Holehan 1979). The age-related irregularities of the oestrous cycle in fully fed rats reflect changes in ovarian endocrine secretion and in the threshold sensitivity of the hypothalamic-pituitary feedback control (Lu 1983).

In contrast with the rat, even under very mild restricted feeding regimens, mice are unable to sustain normal reproductive physiology. Mice (B6) maintained on 80% normal food intake between 3.5 and 10.5 months were acyclic, being arrested mainly in di-oestrus (Nelson et al. 1985). On return to full feeding at 10.5 months, oestrous cycling resumed at an age when 80% of the fully fed control mice were acyclic. At 12.5 months the follicular reserves of the diet-restricted mice were twice those of the control animals. Restricted feeding will therefore delay reproductive aging in mice as judged by the rate of follicular depletion and oestrous cycling.

While restricted feeding will delay reproductive aging in both rodent species, the retention of normal oestrous cycles and fertility in rats suggests that the endocrine mechanism by which this is achieved may be different. Both species demonstrate cycle lengthening and cessation with age, but the different balance between ovarian and neuroendocrine control of these events within separate species will influence their reaction to underfeeding (Felicio and Nelson 1988).

The progressive neuroendocrine dysfunction of oestrous cycle control can be demonstrated in middle-aged rats (9–11 months) which are still undergoing regular oestrous cycles. The luteinizing hormone (LH) surge is delayed (Cooper et al. 1980; van der Schoot 1976) and its magnitude is significantly decreased (Wise 1982). Serum progesterone is decreased at pro-oestrus in middle-aged rats (Miller and Riegle 1979) and follicle-stimulating hormone (FSH) levels are elevated and sustained on the day of oestrus. The progressive cessation of regular oestrous cycles may be preceded in middle-aged rats exhibiting normal oestrous cycles by enhanced FSH secretion, resulting in premature follicular growth and increased oestrogen production. In a comparison with young animals (4 months old), Lu (1983) reported an early rise in serum oestradiol in relation to the time of ovulation. This observation confirmed the findings of Butcher and Page (1981).

A comparison of the serum profiles for LH, FSH, progesterone, and oestradiol-17β across the 5-day cycle in fully fed and diet-restricted rats 6–7 months of age shows clear differences in the temporal relationship of these hormones. Although the preovulatory peak of LH was reduced in height in diet-restricted rats, it occurred approximately 6 h earlier in the cycle, whereas a delay of 6 h was observed in the rise of oestradiol-17β (Holehan and Merry 1985b). The age-related changes in the endocrine control of oestrous cycles recorded for fully fed animals of this age are not observed in the diet-restricted females.

Experimental Procedures

Follicular Steroidogenesis with Aging

In an attempt to understand the controlling factors which delay reproductive senescence in diet-restricted rats, a series of in vitro studies were undertaken with individual ovarian follicles to determine the effect of age and diet on follicular steroidogenesis. The follicular content and release of oestradiol-17β, progesterone, testosterone, androstenedione and 20α-dihydroprogesterone were measured during unstimulated steroid synthesis and during LH- or testosterone-stimulated synthesis according to the method of Uilenbroek et al. (1981). Nine individual ovarian follicles were removed from control and underfed female rats on each day of the oestrous cycle and were incubated in 0.5 ml of medium 199 for 4 h at 37°C under an atmosphere of 95%O_2/5%CO_2. Four group comparisons were made. Follicular steroidogenesis was compared in 4- and 12-month-old fully fed rats and 12- and 20-month diet-restricted animals.

It was observed that with increasing age, follicles from fully fed rats exhibited an increased release of progesterone in response to 20 μg/ml ovine LH stimulation and enhanced aromatization of available androgen to oestradiol-17β. The overall age change in follicles from fully fed rats was an increase in the activity of the δ⁴ pathway to oestrogen and a concomitant reduced activity of the pathway for progesterone metabolism (Merry and Holehan 1990). In follicles from 4-month-old, ad libitum-fed animals a significant increase in the release of oestradiol-17β was seen only at di-oestrus and pro-oestrus. By 1 year of age in ad libitum-fed rats, significantly increased amounts of oestradiol-17β were released across each stage of the oestrus cycle. The age changes were retarded by underfeeding, and the oestradiol-17β profile across the cycle for 12-month-old diet-restricted animals was very similar to that observed in 4-month-old ad libitum-fed controls. By 20 months follicles from underfed rats did show enhanced oestradiol-17β release but this was restricted to the oestrous and metoestrous phases of the cycle. The inhibition with aging of the progesterone metabolic pathway to 20α-dihydroprogesterone was not as great as observed in follicles from 12-month control rats. Conversely, as the synthesis and release of oestradiol-17β increases with age, so the follicular content and release of testosterone and androstenedione declines.

It is possible therefore to demonstrate retarded aging of the endocrine component of reproduction at the level of the individual follicle. These studies do not provide any indication as to whether the retarded rate of age changes is intrinsic to the follicles and is a direct response of this tissue to underfeeding or whether it is the result of changes induced in the central neuroendocrine control of oestrous cycling.

Heterochronic Orthotopic Ovarian Transplantation

In order to distinguish between these two alternative points of control we have employed the technique of heterochronic, bilateral orthotopic ovarian transplantation, developed for mice by Jones and Krohn (1960). The animal groups employed were similar in age and dietary status to those used in the study of follicular steroidogenesis. Fully fed animals of 4 and 12 months and diet-restricted animals of 12 and 22 months were used to establish the 16 possible transplant combinations in 160 rats. All animals were allowed 50 days to recover from surgery. The functional competence of grafts was determined in all animals by monitoring vaginal cytology over a 30-day period subsequent to the recovery period. In a pilot study, 4-month-old fully fed rats with age-matched ovarian transplants from ad libitum-fed donor animals were mated with proven males. In this study 62% of the females bearing orthotopic transplanted ovaries successfully carried a pregnancy to term and produced live offspring, (Merry and Holehan 1990). Steroidogenesis in individual follicles was studied for each animal bearing bilateral ovarian transplants using the method described above. Because of the 50-day recovery period allowed after surgery and the 30-day period for monitoring vaginal cytology, both host and donor tissue were approximately 2 months older than at transplantation. Follicles were recovered from transplanted ovaries only on the day of oestrus. Follicular oestradiol-17β release over 4 h at oestrus, following orthotopic transplantation between animals of the same age and dietary group, showed no significant differences when compared to oestradiol-17β release recorded in the previous experiment for nontransplanted follicles. These orthotopic ovarian transplants within the same age and dietary group provided the technical control for the experiment. The previously recorded effects of both age and restricted feeding on oestradiol-17β release over 4 h in the in vitro assay were therefore not affected by the trauma of the surgical procedure.

In spite of large interfollicular variance, the transplantation study demonstrated quite clearly that in the rat, the age-associated changes in oestradiol-17β synthesis and release are reversible and dependent upon changes external to the ovary, i.e., neuroendocrine age changes. The reversal of age changes in follicular steroidogenesis was seen in both dietary groups. The release of oestradiol-17β at oestrus for follicles from the ovaries of 12-month-old fully fed and 22-month-old diet-restricted animals transplanted into 4-month-old fully fed or 12-month-old diet-restricted animals was significantly reduced.

When ovaries from 12-month-old fully fed rats were transplanted into 22-month-old diet-restricted rats, follicular oestradiol-17β release was significantly reduced, even though the follicles from 24-month-old diet-restricted and 14-month-old fully fed animals release similar amounts of oestradiol-17β when transplanted within the same age and dietary group. It is unclear why a 24-month-old diet-restricted host cannot maintain oestradiol-17β synthesis in ovarian follicles from a 14-month-old fully fed donor but is able to support follicular steroidogenesis in the ovarian follicles of 24-month-old diet-restricted rats.

Accelerated Aging of Follicular Steroidogenesis

Using this technique it was possible to assess whether, by transplantation into an old host, the release of oestradiol-17β from young follicles could be enhanced to mimic that of old follicles. The only transplantation combination where this was achieved was when the host was a 12-month-old fully fed female. When the ovaries of a 4-month-old fully fed donor were transplanted into a 12-month-old fully fed host, oestradiol-17β release was increased, but the difference was not statistically significant. Conversely, the follicles from the 12-month-old diet-restricted animals showed a significant increase in oestradiol-17β release to levels seen at 24 months of age (Merry and Holehan 1990).

Conclusions

Chronic restricted feeding in the female rat will delay reproductive aging. The age-related changes in steroid hormone secretion can be shown to be retarded at the level of the individual follicle. These age changes in the profile of steroid secretion can be reversed either by transplantation into a chronologically younger host or into the biologically younger diet-restricted animal. Transplantation of young ovaries into control animals showing reproductive aging will show enhanced oestradiol-17β secretion from individual follicles. The delay in reproductive aging in the diet-restricted animal appears, therefore, to be primarily the result of retarded aging in the hypothalamic-pituitary axis and not some intrinsic change induced in the ovarian tissue by the dietary regimen.

References

Ball ZB, Barnes RH, Visscher MB (1947) The effects of dietary caloric restriction on maturity and senescence, with particular reference to fertility and longevity. Am J Physiol 150:511–519

Butcher RL, Page RD (1981) Role of the aging ovary in cessation of reproduction. In: Schwartz NB, Hunzicker-Dunn, M (eds) Dynamics of ovarian function. Raven, New York, pp 253–271

Campbell GA, Kurcz M, Marshall S, Meites J (1977) Effects of starvation in rats on serum levels of follicle stimulating hormone, luteinizing hormone, thyrotropin, growth hormone and prolactin: response to LH releasing hormone and thyrotropin releasing hormone. Endocrinology 100:580–587

Cooper RL, Conn M, Walker RF (1980) Characterization of the LH surge in middle-aged female rats. Biol Reprod 23:611–615

Everitt AV, Seedsman NJ, Jones F (1980) The effects of hypophysectomy and continuous food restriction, begun at ages 70 and 400 days, on collagen aging, proteinuria, incidence of pathology and longevity in the male rat. Mech Ageing Dev 12:161–172

Everitt AV, Wyndham JR, Barnard DL (1983) The anti-aging action of hypophysectomy in hypothalamic obese rats: effects on aging, age-associated proteinuria development and renal histopathology. Mech Ageing Dev 22:233–251

Felicio LS, Nelson JF (1988) Regulation of age-related changes in hypothalamo-pituitary-ovarian function in the mouse. In: Everitt AV, Walton JR (eds) Regulation of neuroendocrine aging. Karger, Basel pp 90–97 (Interdisciplinary topics in gerontology, vol 24)

Holehan AM (1984) The effect of ageing and dietary restriction on reproduction in the female CFY Sprague-Dawley rat. Thesis, University of Hull, Hull

Holehan AM, Merry BJ (1985a) The control of puberty in the dietary restricted female rat. Mech Ageing Dev 32:179–191

Holehan AM, Merry BJ (1985b) Modification of the oestrous cycle hormonal profile by dietary restriction. Mech Ageing Dev 32:63–76

Holehan AM, Merry BJ (1985c) Lifetime breeding studies in fully fed and dietary restricted female CFY Sprague-Dawley rats. I. Effect of age, housing conditions and diet on fecundity. Mech Ageing Dev 33:19–28

Jones EC, Krohn PL (1960) Orthotopic ovarian transplantation in mice. J Endocrinol 20:135–146

Lu JKH (1983) Changes in ovarian function and gonadotropin and prolactin secretion in aging female rats. In: Meites J (ed) Neuroendocrinology of aging, Plenum, New York, pp 103–122

McCay CM, Maynard LA, Sperling G, Barnes LL (1939) Retarded growth, lifespan, ultimate body size and age changes in the albino rat after feeding diets restricted in calories. J Nutr 18:1–13

Merry BJ, Holehan AM (1979) Onset of puberty and duration of fertility in rats fed a restricted diet. J Reprod Fertil 57:253–259

Merry BJ, Holehan AM (1981) Serum profiles of LH, FSH, testosterone and 5α-DHT from 21 to 1000 days of age in ad-libitum-fed and dietary-restricted rats. Exp Gerontol 16:431–444

Merry BJ, Holehan AM (1990) Effects of caloric restriction on endocrine function during aging in rats. In: Ingram D (ed) The potential for nutritional modulation of age processes. Food and Nutrition, Westport (in press)

Miller AE, Riegle GD (1979) Endocrine factors associated with the initiation of constant estrous in aging female rats. Fed Proc 38:1248

Mulinos MG, Pomerantz L (1940) Pseudo-hypophysectomy, a condition resembling hypophysectomy produced by malnutrition. J Nutr 19:493–504

Nelson JF, Gosden RF, Felicio L (1985) Effect of dietary restriction on estrous cyclicity and follicular reserves in aging C57BL/6J mice. Biol Reprod 32:515–522

Uilenbroek JTJ, Van Der Schoot P, Woutersen PJA (1981) Changes in steroidogenic activity of preovulatory rat follicles after blockage of ovulation with Nembutal. In: Schwartz NB, Hunzicker-Dunn M (eds) Dynamics of ovarian function. Raven, New York, pp 41–46

Van Der Schoot P (1976) Changing pro-oestrous surges of luteinizing hormone in ageing 5-day cyclic rats. J Endocrinol 69:287–288

Wise PM (1982) Alterations in pro-oestrous LH, FSH and prolactin surges in middle-aged rats. Proc Soc Exp Biol Med 169:348–354

Chronic Energy Intake Restriction: Influence on Longevity, Autoimmunity, Immunodeficiency, and Cancer in Autoimmune–Prone Mice

R.A. Good[1], E. Lorenz[1], R.W. Engelman[1], and N.K. Day[1]

Introduction

The role of diet in the prevention of diseases of aging and the maintenance of good health has been the subject of great interest and much investigation in recent years. Over the past two decades we have investigated this complex interrelationship, beginning with observations of the immunological damage caused by protein-calorie malnutrition accompanied by deficiencies of zinc and essential nutrients and moving on to investigations of the effects of dietary manipulations such as calorie restriction on immune function, autoimmune disease, and spontaneously occurring cancers in inbred short-lived mouse strains prone to develop age-associated autoimmune disorders and spontaneously occurring cancers.

Chronic Energy Intake Restriction: Effects on Immune Function

Our earliest research on the effects of total food restriction on immune function (Jose and Good 1971; 1973a,b) showed that protein or protein-calorie restriction in mice and other animals regularly produced deficient antibody responses and decreased lymphocyte counts but resulted in enhanced cell-mediated immune functions, including allograft rejection, resistance to numerous viruses, and delayed hypersensitivity responses. However, these experimental findings stood in stark contrast to our field observations (Good and Fernandes 1979) and those of other investigators (Aref et al. 1970; Abbassy et al. 1974; Jose et al. 1970) of the ravaging results of protein-calorie malnutrition on immune function, particularly cellular immunity to infections, in protein-calorie malnourished children and adults. These paradoxical findings could be explained in part by discoveries we and others made about the role in immune response of a most important trace

[1]Department of Pediatrics, All Children's Hospital, University of South Florida, St. Petersburg, FL 33701, USA

element, namely, zinc. Dietary zinc deficiency can profoundly affect numerous immune functions. These include involution of thymus and reduced primary and secondary antibody responses (Prasad 1988; Fraker et al. 1977; Lueke et al. 1978), depressed T-cell counts, depressed T-cell proliferation in response to stimulation with phytomitogens, depressed natural killer cell function, and depressed T-suppressor function and decreased T4:T8 ratios (Fernandes et al. 1979; Schloen et al. 1979; Prasad et al. 1988), and decreased serum thymulin levels and decreased IL-2 activity (Prasad et al. 1987, 1988). In young mice, many of these functions cannot be restored to normal values even after adequate dietary zinc levels are established (Schloen et al. 1979). Recently evidence has been presented that the effect of zinc on thymulin activity may be the mechanism by which this essential element influences T-cell function (Prasad et al. 1987, 1988; Licastro 1986).

Effects on Longevity, Autoimmunity, and Immunodeficiency

Since our early experimental findings demonstrated that dietary restriction wherein all essential nutrients and micronutrients, including zinc, were provided at adequate levels could in fact improve numerous aspects of immune response, we have focused our analyses on the effects of dietary restriction on immune function, autoimmune disease, and cancer in experimental systems. We have studied the influence of energy (calorie) restriction in a number of genetically short-lived strains of mice and have established that restriction of energy intake is the most important nutritional variable in the extension of life span and the prevention of diseases of aging. We have shown that in mice of each of the autoimmune-prone strains studied, chronic energy intake restriction (CEIR) imposed at the time of weaning by means of diet that is reduced by 40% in calories but still provides all essential nutrients greatly prolongs life span and impressively forestalls or inhibits the development of diseases associated with aging (Fernandes and Good 1984; Fernandes et al. 1976a, 1978a,b, 1983). CEIR regularly doubles and triples and can even quadruple life span in these auto-immune-prone strains, including the B/W, NZB, MRL/1pr 1pr, kdkd and BXSB strains. CEIR also dramatically delayed or prevented the development of progressive immunologically based hyalinizing renal disease, vascular lesions, lymphoproliferative disease and even certain malignancies to which mice of these inbred strains are most susceptible (Fernandes and Good 1984; Fernandes et al. 1976a, 1978a,b, 1983).

To ascertain which is the more crucial variable—calorie source or total calorie intake—in this salubrious influence of calorie restriction, we tested diets varying in composition from very high fat/no-carbohydrate to very low fat/high-carbohydrate and moderate-fat/moderate-carbohydrate diets and found no significant difference in the influence of the different diets on life span (Kubo et al. 1977; Gajjar et al. 1987). However, when each experimental diet was fed at a restricted calorie level, the median and maximal longevity of mice of each strain studied

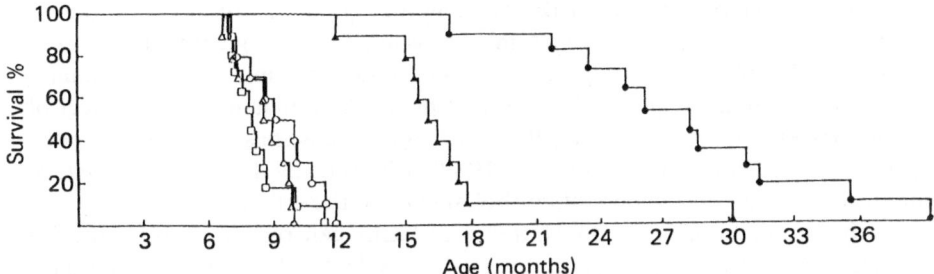

Figure 14.1. Influence of CEIR and dietary composition on life span and autoimmune disease in B/W mice. Survival curves are shown for mice fed from time of weaning a high-calorie/high-carbohydrate (○), high calorie/high-fat (△), a low-calorie/high carbohydrate (●), a low-calorie/high carbohydrate (○) or low-calorie/high-fat diet (▲). Control animals (□) consumed ad libitum a purified diet consisting 73% dextrin plus glycerol, 20% protein, and 4.5% fat. Mice fed the higher calorie intakes died of autoimmune disease and accompanying progressive renal disease before 1 year of age. Life span was greatly extended by the imposition of the low-calorie diet, although a low-calorie/high-fat diet did not offer as great a degree of protection from autoimmune and renal disease as did the low-calorie/high-carbohydrate diet ($p < 0.01$).

was dramatically increased (Fig. 14.1). The most significant shift in longevity curves occurred with a diet relatively low in fat and relatively high in carbohydrate fed at a restricted (40%) calorie level (Kubo et al. 1977; Gajjar et al. 1987).

Cellular and Molecular Bases for Effects

The cellular and molecular bases for this beneficial influence of calorie restriction are only now beginning to be understood. We have shown that CEIR impedes the immunological involution that occurs with aging (Kubo et al. 1984), reduces the formation of circulating immune complexes and deposition of gp70-anti-gp70 immune complexes in a capillary distribution within the glomeruli (Izui et al. 1981), inhibits the age-associated decline of interleukin-2 production and responsiveness (Jung et al. 1982), enhances the activity of the radical scavenging enzymes superoxide dismutase and catalase (Kubo et al. 1987), and augments the levels and activity of intestinal alkaline phosphatase as well as gene expression for this enzyme (Dao et al. 1990). In studies with long-lived animals, various reports have shown that calorie restriction increases intestinal absorption of vitamin A (Hollander et al. 1986), increases catalase activity (Koizumi et al. 1987), enhances NK cell function (Weindruch et al. 1985), retards the age-related decline in DNA repair by lymphoid cells (Licastro and Walford 1985), and augments synthesis of $\alpha_{2\mu}$-globulin as well as mRNA expression and transcription of a $\alpha_{2\mu}$-microglobulin by hepatocytes (Richardson et al. 1987).

We have also reported that CEIR can slow the rate of cell turnover of each of the rapidly replicating cellular systems we have studied, e.g., thymus, spleen, mesenteric lymph nodes, and gastrointestinal tissues in numerous autoimmune-prone mouse strains (Ogura et al. 1987, 1989a). A similar inhibitory effect of CEIR on the proliferative rate of cells of the colon mucosa has been demonstrated by Albanes and Winick (Albanes et al. 1989), who showed that CEIR can reduce the ultimate mass or numbers of epithelial cells in the colon.

Furthermore, we have demonstrated that CEIR can reduce the growth of a special subset of B lymphocytes: the Ly-1+ or CD5+ B lymphocytes that have been closely linked to autoimmune disease and autoantibody production and which are thought to be at increased risk for malignant transformation (Ogura et al. 1989b, 1990; Wang et al. 1980; Ledbetter et al. 1981; Herzenberg et al. 1986; Raveche et al. 1988; Stall et al. 1988). Given that we have also shown that CEIR can slow the rate of cellular proliferation, our finding that CEIR can limit the growth of a cell population so closely linked to hyperplastic and neoplastic clonal expansion points to the need for further studies of dietary prevention of cancer from this perspective.

Effect of CEIR on Cancer Development

In our analyses of the relationship of energy intake and diet on cancer we have focused one group of investigations over the last 15 years on the occurrence of mammary adenocarcinoma in inbred strains of C3H mice which develop and die of mammary malignancy as a result of latent retroviral infection with the murine mammary tumor virus (MMTV). Beginning with initial observations that CEIR inhibited the development of mammary adenocarcinoma in C3H/Umc mice (Fernandes et al. 1976b), we sought to elucidate the mechanisms underlying this prevention of disease. We found that calorie restriction delayed the development of mammary tumors, reduced the levels of circulating prolactin, and delayed the development of precancerous minimal alveolar lesions of the breast (or so-called hyperplastic precancerous nodules of the breast) (Sarkar et al. 1982). In addition, CEIR decreased the expression of both type A and B MMTV particles.

To determine whether development of cancer was influenced more by total caloric intake rather than dietary composition, as has proved the case for development of the renal disease and lymphoproliferative disease which characterized the autoimmune disease of the genetically short-lived mice, we imposed low-calorie/low-fat, low-calorie/high-fat, high-calorie/low-fat, and high-calorie/high-fat feeding regimens in C3H/OuJ mice (Engelman et al. 1990). Here, too, we demonstrated that calorie intake rather than calorie source was the most critical variable in the prevention of mammary adenocarcinoma as a consequence of retroviral infection (Fig. 14.2) (Engelman et al. 1990). This seems to us an especially important observation, since fat rather than total calorie intake is the dietary factor that has been most frequently cited as contributing to breast tumorigenesis (Doll and Peto 1981). However, the close association of fat intake with total

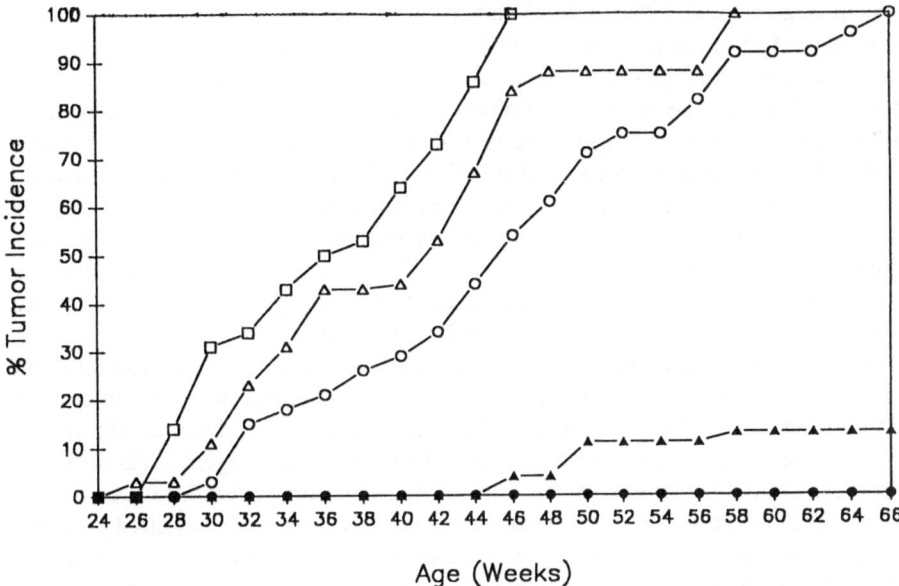

Figure 14.2. Influence of CEIR and dietary composition on mammary adenocarcinoma in C3H mice. Incidence of mammary adenocarcinoma among C3H mice fed from time of weaning a high-calorie/high carbohydrate (○) diet, high-calorie/high fat (△), low-calorie/high-carbohydrate (●), or low-calorie/high-fat diet (▲). Control animals (□) were fed a nonpurified commercial diet ad libitum. Although the mice fed the CEIR diet in which calories were derived principally from fat developed mammary adenocarcinoma more rapidly than did mice fed a CEIR diet in which calories were derived principally from carbohydrate, the overall tendency to develop mammary cancer did not differ between the full-fed isocaloric diets when survival analysis was calculated using Weibull distribution (Chen et al. (1990).

calorie intake in self-selected human diets must be considered in this equation. We have also found that mammary cancer can be delayed by calorie restriction even when such dietary manipulation is not imposed until maturity, i.e., after one successful breeding and associated lactation (Shao et al. 1990).

We have also demonstrated that, in C3H mice, MMTV proviral messages in both mammary gland and liver were suppressed and that expression of the cellular proto-oncogenes int-1, int-2, and v-ha-ras was decreased as a function of CEIR (Chen et al. 1989). Moreover, CEIR apparently decreased the frequency of reintegration of MMTV DNA adjacent to the int-1 gene and appears to thereby inhibit mammary tumorigenesis by inhibiting expression of int-1 (Chen et al. 1990).

The relation of these influences of diet on development of breast cancer and on immune function requires that the role of prolactin receive more than cursory attention. This is the case because calorie restriction exercises such an impressive influence on proliferation of cells of the intestinal epithelium, thymus, and spleen (Ogura et al. 1987, 1989a) and, by inference, even on precursor cells of mammary tumors (Sarkar et al. 1982) and the fact that much recent literature links prolactin to a possible essential role in cell proliferation. For example, prolactin has been shown to be comitogenic with phytomitogens (Buckley et al. 1988), to exert a crucial influence in the regenerative proliferation of hepatic cells after partial hepatectomy (Russell and Snyder, 1968), and to influence protein kinase C translocation in the preproliferative state for epithelial cells (Buckley et al. 1987). Evidence also suggests that a prolactinlike molecule immunoreactive with prolactin antiserum, is produced by lymphoid cells as a consequence of transcription from activated genes in lymphoid cells which are impressively homologous to prolactin (Russell 1988).

Recently, with Hamada et al. we carried out experiments which linked CEIR and prolactin more intimately to the control of breast cancer development. In both ad libitum-fed and CEIR mice of the C3H/Ou strain, we manipulated circulating prolactin levels by employing either the dopaminomimetic agent CV205–502 to suppress circulating prolactin levels in full-fed mice or grafted isologous adeno-hypophyses to elevate serum prolactin levels in CEIR mice. We measured prolactin using an enzyme immunosorbent assay developed in our laboratory by Tomita et al. 1990. In mammary tissue samples taken from the full-fed mice treated with CV205–502 or in the untreated CEIR mice, MMTV messenger RNA could not be detected by northern blot analysis at 17, 21, and 25 weeks of age. By contrast, in the untreated full-fed mice and in the CEIR mice in which serum prolactin was consistently elevated as a consequence of adenohypophysectomy, MMTV mRNA was detected in more than 80% of mammary tissue samples taken from 17-, 21-, and 25-week-old mice. Taken along with our previous findings that CEIR that prevented mammary tumorigenesis consistently lowered circulating prolactin levels, these new reports further support a regulatory role for prolactin in the control of breast cancer by dietary restriction. Future investigations must seek to clarify how CEIR acts to suppress prolactin levels, which seem so important in mediating the nutritional prevention of cancer.

Conclusions

Experimental models provide useful systems for analyses of the respective roles of dietary components in the pathogenesis of autoimmune diseases and cancers. Studies in genetically short-lived mice revealed that calorie intake, rather than calorie source, is the most critical variable in the dietary control of development of autoimmune diseases, including autoimmune-based renal disease, vascular and cardiovascular disease, and lymphoproliferative diseases. Calorie restriction also delays or prevents development of cancers to which these inbred mice are most susceptible. A diet in which calories are restricted by 40% and in which carbohydrate intake is relatively low and fat intake relatively high was most effective in preventing disease and extending longevity. Studies of C3H mice prone to develop mammary adenocarcinoma due to latent retroviral infection showed that such chronic energy intake restriction dramatically inhibited development of mammary tumors, in association with reduction in hyperplastic alveolar nodule formation and suppression of MMTV proviral messages in mammary gland and other organs. Recent evidence suggests that the beneficial effects of CEIR may be mediated by a possibly crucial influence of prolactin on cellular proliferation.

References

Abbassy AS, Badr El-din MK, Hassan A, Aref GH, Hammad SA, El-Araby II, Badr El-Din AA (1974) Studies of cell-mediated immunity and allergy in protein-energy malnutrition. J Trop Med Hyg 77:13–21

Albanes D, Salbe AD, Winick M, Levander OA, Taylor PR (1989) Early caloric restriction and cellular growth of colonic mucosa in rats. FASEB J:3047

Aref GH, Badr El-din MK, Hassan A, El-Araby I (1970) Immunoglobulins in kwashior-kor. J Trop Med Hyg 73:186–191

Buckley AR, Putnam CW, Evans R, Laird HE II, Shah GN, Montgomery DW, Russell DH (1987) Hepatic protein kinase C: translocation stimulated by prolactin and partial hepatectomy. Life Sci 41:2827–2834

Buckley AR, Putnam CW, Russell DH (1988) Prolactin as a mammalian mitogen and tumor promoter. Adv Enzyme Regul 27:371–391

Chen RF, Good RA, Hellerman G, Engelman RW, Tanaka A, Nonoyama M, Day NK (1989) Gene expression of mammary tumor virus proviral DNA and proto-oncongene Int-1 during nutritional regulation of mammary tumor development. FASEB J 3:A1051

Chen RF, Good RA, Engelman RW, Hamada N, Tanaka A, Nonoyama M, Day NK (1990) Suppression of mouse mammary tumor proviral DNA and proto-oncogene expression: association with nutritional regulation of mammary tumor development. Proc Natl Acad Sci USA 87:2385–2389

Dao ML, Shao R, Risley J, Good RA (1989) Influence of chronic energy intake restriction on intestinal alkaline phosphatase. J Nutr 119:2017–2022

Doll R, Peto R (1981) The causes of cancer. Oxford University Press, London

Engelman RW, Day NK, Chen RJ, Tomita Y, Bauer-Sardiña I, Dao ML, Good RA (1990) Calorie consumption level influences development of C3H/Ou breast adenocarcinoma with indifference to calorie source. Proc Soc Exp Biol Med 193:23–30

Fernandes G, Alonso DR, Tanaka T, Thaler HT, Yunis EJ, Good RA (1983) Influence of diet on vascular lesions in autoimmune-prone B/W mice. Proc Natl Acad Sci USA 80:874–877

Fernandes G, Friend P, Yunis EJ, Good RA (1978a) Influence of dietary restriction on immunologic function and renal disease in (NZB×NZW)F1 mice. Proc Natl Acad Sci USA 75:1500–1504

Fernandes G, Good RA (1984) Inhibition by diet of lymphoproliferative disease and renal damage in MRL/1p mice. Proc Natl Acad Sci USA 81:6144

Fernandes G, Nair M, Onoe K, Tanaka T, Floyd R, Good RA (1979) Impairment of cell-mediated immunity functions by dietary zinc deficiency in mice. Proc Natl Acad Sci USA 76:457–461

Fernandes G, Yunis EJ, Good RA (1976a) Influence of diet on survival of mice. Proc Natl Acad Sci 73:1279–1283

Fernandes G, Yunis EJ, Good RA (1976b) Suppression of adenocarcinoma by the immunological consequences of calorie restriction. Nature 263:504–506

Fernandes G, Yunis EJ, Miranda M, Smith J, Good RA (1978b) Influence of dietary restriction on immunologic function and renal disease in (NZB×NZW)F1 mice. Proc Natl Acad Sci USA 75:2888–2892

Fraker PJ, Haas S, Leuke RW (1977) Effect of zinc deficiency on the immune response of young adult A/Jax mouse. J Nutr 107:1889–1895

Gajjar A, Kubo C, Johnson BC, Good RA (1987) Influence of extremes of protein and energy intake on survival of B/W mice. J Nutr 117:1136–1140

Good RA, Fernandes G (1979) Nutrition, immunity, and cancer — a review. I. Influence of protein or protein-calorie malnutrition and zinc deficiency on immunity. Clin Bull MSKCC 9:3–12

Hamada N, Engelman RW, Tomita Y, Chen RF, Iwai H, Bauer-Sardiña, Good RA, Day NK (1990) Effects of controlled prolactin on mouse mammary tumor proviral DNA expression. Proc Natl Acad Sci USA :6733–6737

Herzenberg LA, Stall AM, Lalor PA, Sidman C, Moore WA, Parks DR, Herzenberg LA (1986) The Ly-1 B cell lineage. Immunol Rev 93:81–102

Hollander D, Dadufalza V, Weindruch R, Walford RL (1986) Influence of life-prolonging dietary restriction on intestinal vitamin A absorption in mice. Exp Gerontol 9:57–60

Izui S, Fernandes G, Hara I, McConahey PJ, Jensen FC, Dixon JF, Good RA (1981) Low-calorie diet selectively reduces expression of retroviral envelope glycoprotein gp70 in sera of (NZB×NZW)F1 hybrid mice. J Exp Med 154:1116–1124

Jose DG, Good RA (1971) Absence of enhancing antibody in cell-mediated immunity to tumour heterografts in protein deficient rats. Nature 321:323–325

Jose DG, Good RA (1973a) Quantitative effects of nutritional essential amino acid deficiency upon immune responses to tumors in mice. J Exp Med 137:1–9

Jose DG, Good RA (1973b) Quantitative effects of nutritional protein and calorie deficiency upon immune responses to tumors in mice. Cancer Res 33:807–812

Jose DG, Welch JS, Doherty RL (1970) Humoral and cellular immune responses to streptococci, influenza, and other antigens in Australian aboriginal school children. Aust Paediatr J 6:191–202

Jung LKL, Palladino MA, Calvano S, Mark DA, Good RA, Fernandes G (1982) Effect of calorie restriction on the production and responsiveness to interleukin 2 in (NZB×NZW)F1 mice. Clin Immunol Immunopathol 25:295–301

Kaplan J, Hess JW, Prasad AS (1988) Impaired interleukin-2 production in the elderly: association with mild zinc deficiency. J Trace Elements Exp Med 1:3–8

Koizumi A, Weindruch R, Walford RL (1987) Influences of dietary restriction and age on liver enzyme activities and lipid peroxidation in mice. J Nutr 117:361–367

Kubo C, Day NK, Good RA (1984) Influence of early or late dietary restriction on life span and immunological parameters in MRL/Mp-1pr/1pr mice. Proc Natl Acad Sci USA 81:5831–5835

Kubo C, Johnson BC, Gajjar A, Good RA (1977) Crucial dietary factors in maximizing life span and longevity in autoimmune-prone mice. J Nutr 117:1129–1135

Kubo C, Johnson BC, Misra HP, Dao ML, Good RA (1987) Nutrition, longevity and hepatic enzyme activities in mice. Nutr Rep Int 35:1184–1194

Ledbetter JA, Evans RL, Lipinski M, Cunningham-Rundles C, Good RA, Herzenberg LA (1981) Evolutionary conservation of surface molecules that distinguish T lymphocyte subpopulations in mouse and man. J Exp Med 153:310–323

Licastro F (1986) Down's Syndrome as a model of accelerated aging. Immunopathol Immunother Lett 2(1):6–7

Licastro F, Walford Rl (1985) Aging, proliferative potential and DNA repair capacity in short-lived and long-lived mice. Mech Ageing Dev 31:171

Lueke RW, Simonel CE, Fraker PJ (1978) The effect of reactivated dietary intake on the antibody mediated response of the zinc deficient A/Jax mouse. J Nutr 108: 881–887

Ogura M, Ogura H, Pahwa R, Johnson BC, Good RA (1987) Decrease by chronic energy intake restriction (CEIR) of proliferation of epithelial cells of the gut in (NZB×NZW)F1 mice. Fed Proc 46:151

Ogura M, Ogura H, Ikehara S, Dao ML (1989a) Decrease by chronic energy intake restriction of cellular proliferation in the intestinal epithelium and lymphoid organs in autoimmunity-prone mice. Proc Natl Acad Sci USA 86:5918–5922

Ogura M, Ogura H, Ikehara S, Good RA (1989b) Influence of dietary energy restriction on the numbers and proportions of Ly-1[+] B lymphocytes in autoimmunity-prone mice. Proc Natl Acad Sci USA 86:4225–4229

Ogura M, Ogura H, Lorenz E, Ikehara S, Good RA (1990) Undernutrition without malnutrition restricts the numbers and proportions of Ly-1 B lymphocytes in autoimmune (MRL/1 and BXSB) mice. Proc Soc Exp Biol Med 193:6–12

Prasad AS (1988) Zinc in growth and development and spectrum of human zinc deficiency. J Am Coll Nutr 7:377–384

Prasad AS, Dardenne M, Abdallah J, Meftah S, Brewer GJ, Bach JF (1987) Serum thymulin and zinc deficiency in humans: serum thymulin and zinc deficiency in humans. Trans Assoc Am Physicians 1987:222–231

Prasad AS, Meftah S, Abdallah J, Kaplan J, Brewer GJ, Bach JF, Dardenne M (1988) Serum thymulin in human zinc deficiency. J Clin Invest 82:1202–1210

Raveche ES, Lalor P, Stall A, Conroy J (1988) In vivo effects of hyperdiploid LY-1[+] cells of NZB origin. J Immunol 141:4133–4139

Richardson A, Butler JA, Rutherford MS, Semsei I, Gu MZ, Feranades G, Chiang WH (1987) Effect of age and dietary restriction on the expression of $\alpha_{2\mu}$-globulin. J Biol Chem 262:1282–1285

Russell DH: New aspects of prolactin and immunity: a lymphocyte-derived prolactin-like product and nuclear protein kinase C activation (1988) Trends in Pharmaceutical Sciences 10:44

Russell D, Snyder SH (1968) Amine synthesis in rapidly growing tissues ornithine decarboxylase activity in regenerating rat liver, chick embryo and various tumors. Proc Natl Acad Sci USA 60:1420–1427

Sarkar NH, Fernandes G, Telang NT, Kourides IA, Good RA (1982) Low-calorie diet prevents the development of mammary tumors in C3H mice and reduces circulating prolactin level, murine mammary tumor virus expression, and proliferation of mammary alveolar cells. Proc Natl Acad Sci USA 79:7758–7762

Schloen LH, Fernandes G, Garofalo JA, Good RA (1979) Nutrition, immunity and cancer —a review. II. Zinc, immune function and cancer. Clin Bull MSKCC 9: 63–75

Shao R, Dao ML, Day NK, Good RA (1990) Dietary manipulation of mammary tumor development in adult C3H/Bi mice. Proc Soc Exp Biol Med 193:313–318

Stall AM, Farinas MC, Tarlinton DM, Lalor PA, Herzenberg LA, Strober S, Herzenberg LA (1988) Ly-1 B cell clones similar to human chronic lymphocytic leukemias routinely develop in older normal mice and young autoimmune (New Zealand Black-related) animals. Proc Natl Acad Sci USA 85:7312–7316

Tomita Y, Engelman RW, Iwai H, Hamada N, Bauer-Sardiña, Day NK and Good RA (1990) Enzyme immunosorbent assay for circulating mouse prolactin application to study of nutritional influences on murine mammary tumor development. (Abstr.) FASEB J 4(4):A1170

Wang CY, Good RA, Ammirati P, Dymbort G, Evans RL (1980) Identification of a p69, 71 complex expressed on human T cell sharing determinants with B-type chronic lymphatic leukemic cells. J Exp Med 151:1539–1544

Weindruch R, Devens BH, Raff HV, and Walford RL (1985) Influence of dietary restriction and aging on natural killer cell activity in mice. J Immunol 130:993–996

CHAPTER 15

Behavioral and Neurobiological Effects of Dietary Restriction in Aged Rodents

D.K. Ingram[1]

Introduction

The detrimental effects on brain and behavior resulting from protein-calorie deprivation have long been of central interest in nutritional research particularly involving early development (Morgane et al. 1978). Therefore, it is surprising that relatively few gerontological studies of dietary restriction (DR) have addressed such questions (Forbes 1980) even though concerns have been expressed about possible detrimental effects of this nutritional treatment (Moment 1978). With respect to brain and behavioral function at an advanced age, does DR in rodents produce detrimental effects or beneficial effects as it does for numerous other physiological and biochemical parameters, or no effects at all?

A behavioral and neurobiological analysis of DR has important implications from several perspectives. First, from the perspective of quality-of-life issues, it would be mandatory to determine whether the effects were detrimental or beneficial. If the effects were detrimental, this evidence would argue against the efficacy of DR as a prolongevity treatment. If age-related declines in brain and behavioral function are indeed retarded by DR, then additional evidence would be provided that aging rate can be altered by such treatments. Furthermore, these investigations would provide clues to possible neural mechanisms involved in the prolongevity effects observed in various DR regimens.

Thus, the objective of this chapter is to review studies concerning the effects of DR on brain and behavioral function. Previous reviews (Ingram 1990, 1991) can be consulted for additional detail. The rationale and objectives for a behavioral analysis of aging and putative interventions have also been reviewed (Ingram 1983, 1984; Ingram and Reynolds 1986).

[1]Molecular Physiology and Genetics Section, Laboratory of Cellular and Molecular Biology, Nathan W. Shock Laboratories, Gerontology Research Center,* National Institute on Aging, Francis Scott Key Medical Center, Baltimore, MD 21224, USA
*The Gerontology Research Center is fully accredited by the American Association for the Accreditation of Laboratory Animal Care.

Table 15.1. Effect of intermittent feeding on life span in male rodents.

Strain	Starting age (months)	Increase in mean life span over ad libitum fed controls (%)
Wistar rat	1.5	82*
	10.5	34*
	18.0	11*
C57BL/6J mice	1.5	50*
	6.0	11*
	10.0	0
A/J mice	1.5	12*
	6.0	2
	10.0	−14*
B6AF$_1$ mice	1.5	20*
	6.0	19*
	10.0	−1
C57BL/6J ob/ob mice	1.5	26*
C57BL/6J A^y/a mice	1.5	79*

*$p < 0.05$, according to Lee-Desu survival analysis

Survival Effects of Dietary Restriction

Numerous dietary regimens have been used to implement DR that have increased survival of laboratory rodents and other species (Weindruch and Walford 1988). In our laboratory a regimen of intermittent feeding using a conventional diet (NIH-07; 4.2 kcal/g) has produced robust effects on life span in various rodent strains when implemented at several ages (Table 15.1; Ingram and Reynolds 1987). Compared to ad libitum (AL) intake, this regimen of every other day (EOD) feeding reduces total food consumption 15–30% depending on age and markedly improves survival most notably when initiated near weaning. The EOD regimen has also been used in other laboratories to assess effects of DR on brain and behavioral function (e.g., Beatty et al. 1987; Campbell and Gaddy 1987).

Behavioral Effects of Dietary Restriction

Psychomotor Performance

As observed in Table 15.2, several studies have investigated the effects of DR on performance in psychomotor tests including sensorimotor abilities, strength, locomotor activity, coordination, and balance. In general, the results clearly indicate that DR is not detrimental to psychomotor performance but rather appears beneficial in several tasks. However, the interpretation of performance differences between control and DR mice is complicated by the consideration of other factors, such as differences in body weight, motivation, or pathology.

Table 15.2. Effects of dietary restriction on indices of psychomotor aging in rodents.

Task	Species	Strain	Sex	Diet effect	Reference
Grip strength	Mice	C57BL/6J	M	0	Ingram (1984)
	Mice	B6AF$_1$	M	0	Ingram (1990)
Tightrope	Mice	C57BL/6J	M	+	Ingram (1984)
				+	Harrison and Archer (1987)
	Mice	B6AF$_1$	M	+	Ingram (1990)
	Mice	B6CBAF$_1$	M	+	Harrison and Archer (1987)
	Rats	F344	M	+	Campbell and Gaddy (1987)
Rotorod	Mice	C57BL/6J	M	+	Ingram (1984)
	Mice	B6AF$_1$	M	+	Ingram (1990)
	Mice	C3B10RF$_1$	F	+	Ingram et al. (1987)
Pole descension	Rats	F344	M	+	Campbell and Gaddy (1987)
Elevated bridge	Rats	F344	M	+	Campbell and Gaddy (1987)
Acoustic startle response	Rats	F344	M	0	Campbell and Gaddy (1987)
Stereotyped turning and other behaviors	Rats	Wistar	M	+	Joseph et al. (1983)
Exploratory activity	Mice	C57BL/6J	M	+	Ingram (1984)
				+	Harrison and Archer (1987)
	Mice	B6AF$_1$	M	0	Ingram (1990)
	Mice	C3B10RF$_1$	F	0	Ingram et al. (1987)
	Mice	B6CBAF$_1$	M	+	Harrison and Archer (1987)
Runwheel activity	Mice	C56BL/6J	M	+	Ingram (1984)
	Mice	B6AF$_1$	M	+	Ingram (1990)
	Mice	C3B10RF$_1$	F	+	Ingram et al. (1987)

+, improved performance; 0, no effect

Campbell and Gaddy (1987) examined the effects of an EOD regimen begun in male Fischer 344 (F-344) rats at 12 months of age. When followed longitudinally beginning at 15 months of age, the performance of restricted rats was superior to that of AL controls in several tasks involving strength, balance, and coordination. The wire suspension, or tightrope, task measures the ability to remain suspended by the forepaws from a taut wire or string over several trials. The pole descension task measures the ability to locomote down a wire mesh cylinder. The elevated bridge measures the ability to locomote along a narrow pathway and avoid falling. Although EOD-fed rats demonstrated superior performance in these tasks, the rate of age-related loss in these abilities was similar to that of controls. In addition, in measures of sensorimotor reaction to visual and auditory stimuli, no difference in the acoustic startle response was noted between EOD- and AL-fed groups.

The Joseph et al. (1983) study concerned the behavioral responses of aged male Wistar rats maintained on an EOD regimen since weaning. The relevance of the analysis of stereotyped behaviors will become evident in later discussion of neurobiological studies.

Examining the effects of the EOD regimen begun at 6 mo of age, Ingram (1984, 1990, 1991) investigated the performance of male C57BL/6J and B6AF$_1$/J mice in a battery of simple psychomotor tests. In tests of locomotor activity, aged (26–28 months) C57BL/6J mice fed EOD exhibited greater exploratory activity in an open-field and higher levels of wheel-running activity compared to AL-fed controls. This observation was repeated in aged (26–28 months) B6AF$_1$/J mice for wheel-running activity but not for exploratory activity. Aged mice from both genotypes also demonstrated superior performance compared to AL controls in two tests of strength and coordination. The tightrope test was similar to that described above for rats. Rotorod performance reflects the ability to remain balanced on a rotating cylinder. However, in a direct measure of forepaw grip strength, no evidence of a diet effect was obtained in either genotype.

Harrison and Archer (1987) tested adult (9–13 months) and aged (25–27 months) C57BL/6J and B6CBAF$_1$/J mice begun on DR (about 33% of control level) at weaning. Compared to AL-fed controls, restricted mice showed higher levels of open-field activity and better tightrope performance. These observations were made despite the fact that the DR regimen used in this study actually reduced mean but not maximum lifespan in C57BL/6J mice.

The interpretation of all these findings must be considered in light of intervening variables. When comparing groups of DR and AL control animals, differences in body weight, motivational levels, and pathology can possibly intervene to alter the interpretation that the differences in behavioral performance are due to altered aging rate (Ingram 1991). Lighter animals might perform better in tightrope tests or balance better on rotorods or elevated bridges. DR animals might be more motivated to explore open fields or run in activity wheels. Pathology unique to control animals might hinder their performance in certain tasks. How can the investigator provide an experimental protocol to reduce some of these possibilities?

To control for motivational differences, animals can be tested only after a feeding period. To control for body weight differences, perhaps weights could be added to DR animals. To control for pathology, only animals with no obvious physical impairment could be used, and then a postmortem necropsy might be used to eliminate animals with obvious pathology after the fact of behavioral analysis. All these approaches might still have limitations that would not eliminate intervening variation completely.

In a study using female C3B10RF$_1$ mice, Ingram et al. (1987) used a protocol that attempted to reduce the influence of body weight and motivational factors. Several weeks prior to the behavioral analysis, the experimental group of mice restricted from weaning (about 42% relative to control intake) were placed on the control diet, which was slightly less than an ad libitum level. Through this measure, the body weights of experimental and control groups differed by less than 5%, and the animals were tested while they were on the identical diet. The comparison of performance indicated that middle-aged (11–15 months) and aged (31–35 months) DR mice had higher levels of runwheel activity but did not differ in exploratory activity in an open field. The rotorod performance of aged DR mice was also superior to that of controls.

Table 15.3. Effects of dietary restriction on indices of cognitive aging in rodents.

Task	Species	Strain	Sex	Diet effect	Reference
Stone maze	Mice	C3B10RF$_1$	F	+	Ingram et al. (1987)
	Rats	Wistar	M	+	Goodrick (1984)
Radial maze	Mice	C57BL/6J	F	+	Idrobo et al. (1987)
	Rats	Sprague-Dawley	M	0	Beatty et al. (1987)
	Rats	Wistar	M	0	Bond et al. (1989)
	Rats	F344	M	0	Stewart et al. (1989)
Swim maze	Rats	F344	M	+	Stewart et al. (1989)

+, improved performance; 0, no effect

Further investigation is needed that attempts to control for possible intervening variation in an efficient manner. In addition, age-related decline in psychomotor performance should be followed in a longitudinal fashion, such as attempted by Campbell and Gaddy (1987), to determine if the rate of aging is altered by DR.

Learning and Memory Performance

Several studies have assessed the effects of DR on performance in conventional tests of learning and memory used with laboratory rodents (Table 15.3). The findings here are more mixed than those observed in studies of psychomotor performance. Although it is clear that DR had no detrimental effects on learning and memory performance of aged rodents, beneficial effects were reported in only certain paradigms. Again the issue of intervening variation should not be ignored in interpreting these results.

In the radial maze, the task is to retrieve food rewards located at the distal ends of maze arms radiating out from a center platform. The most efficient performance during a session is to select different arms and thus not to revisit an arm from which a food reward has been obtained. By utilizing visuospatial cues available to them in the room housing the maze, young rats are very efficient at remembering which arms have been visited, and thereby rarely repeat visits. Age-related declines have been reported in the ability of rats to learn this task (e.g., Ingram et al. 1981). In studies of three different rat strains shown in Table 15.3, no effect of DR on the performance of aged (21–30 months) males was noted (Bond et al. 1989; Beatty et al. 1987; Stewart et al. 1989). Thus, it would appear that DR does not alter the age-related decline in the type of short-term, working memory required for efficient performance in this task.

However, one positive result in radial maze performance has been observed. Idrobo et al. (1987) reported that 15-month-old female C57BL/6J mice maintained on DR since 3 months of age performed more efficiently in a radial arm maze than AL-fed controls. Whether this finding might be evidence of a species difference in response to DR cannot be determined. No age comparisons were made in the Idrobo et al. (1987) study. A major procedural difference was also evident between the rat DR studies and the Idrobo et al. (1987) study. The latter used water as the motivating reinforcement compared to food in the rat studies. The use of food-motivated tasks would appear to be problematic in DR studies, even though the diet of control rats is reduced substantially prior to the learning experience. If anything, however, one would expect that DR animals might be more highly motivated than controls. Another problem concerns the use of tasks, like the radial maze, that heavily tax visual abilities. Aged animals, particularly from albino strains, are more likely to have retinal degeneration and lens opacities. The use of pigmented strains, like the C57BL/6J strain, may offset this problem to some degree. Interestingly, Bernstein et al. (1985) found no age effect on performance of male C57BL/6J mice in this task.

Goodrick (1984) used food motivation in his study of performance in a 14-unit T-maze. In this maze, the rat must learn 14 successive positions (left-right) discriminations to locomote from a start area to a goal box where it obtains a food reward. This paradigm has produced robust evidence of age-related performance declines in a variety of rodent species with various motivational manipulations, and performance likely does not depend on the use of visual cues for accurate responding (Ingram 1988). An EOD feeding regimen since weaning was associated with enhanced learning ability (fewer errors) compared to AL-fed controls that had been food-deprived for several weeks prior to testing. Was this difference due to an altered aging rate or to a difference in motivation for food? Ingram et al. (1987) attempted to overcome this potential motivational confound by using a shock-motivated paradigm and DR mice that had been placed on the control diet several weeks prior to maze testing. Aged (33 months) female C3B10RF$_1$ mice on DR (about 40%) since weaning demonstrated superior learning compared to controls. The performance requirement was to avoid the onset of foot shock by moving deliberately and continually through the maze.

A different type of paradigm using negative reinforcement also produced evidence of beneficial effects of DR on learning performance in aged rodents. Stewart et al. (1989) assessed the performance of male F344 rats in a swim maze task in which the animal must use visuospatial cues to locate a submerged platform to escape from swimming. An age-related decline in learning ability (swim distance to find platform) was observed. Superior performance was noted in 8-, 16-, and 24-month-old DR animals in comparison to AL-fed controls. This task also uses visual cues, but the intervening role of this factor would be questioned somewhat because of the lack of DR effects observed in the radial maze task by the same investigators.

The memory processing requirements of the swim maze and 14-unit T maze differ operationally from that required in the radial maze. In the former paradigms the task is to remember responses across sessions, and the task demands do not change across sessions; thus, interaction with more long-term *reference* memory stores is being demanded. In the radial maze, however, the task is to store responses in a short-term *working* memory within a session which should be reset at the beginning of any new session. Thus, the differential results with respect to these reference memory and working memory tasks might be due to

Table 15.4. Effects of dietary restriction on neurochemical indices of aging in rodents.

Parameter	Species	Strain	Sex	Diet effect	Reference
Striatal DA receptor	Rats	Wistar	M	+	Levin et al. (1981)
concentration	Rats	Wistar	M	+	Roth et al. (1984)
Striatal ACh receptor	Rats	Wistar	M	+	London et al. (1985)
concentration					
Striatal, cerebellar, and	Rats	Wistar	M	+	London et al. (1985)
hippocampal ChAT					
activity					
Striatal, cerebellar, and	Rats	Wistar	M	0	London et al. (1985)
hippocampal TH and GAD					
activity					
Striatal, hypothalamic, and	Rats	F344	M+F	–	Kolta et al. (1989)
olfactory bulb NE content					
Striatal and hypothalamic	Rats	F344	M+F	–	Kolta et al. (1989)
DA and 5-HT content					

DA, dopamine; ACh, acetylcholine; ChAT, choline acetyltransferase; TH, tyrosine hydroxylase; GAD, glutamic acid decarboxylase; NE, norepinephrine; DA, dopamine; 5-HT, 5-hydroxytryptamine; +, retarded age effect; –, reversed age effect; 0, no effect

differential effects of DR on different memory processes. This suggestion, however, would run counter to demonstrations that aging affects both types of memory processes in rats (Lowy et al. 1985).

Bond et al. (1989) also tested rats in a different type of memory task, referred to as a habituation paradigm. Compared to young controls, aged rats required more trials to reach an asymptotic level of consumption of a novel saccharin solution; however, no significant difference in habituation was observed between aged DR and control rats.

Neurobiological Effects of Dietary Restriction

Few investigations have attempted to make direct links between the behavioral effects of DR and specific neurobiological mechanisms. However, several studies have produced indirect evidence to indicate that DR can retard age-related decline in brain function that has indirect bearing on the behavioral findings.

Neurochemical Parameters

The age-related declines in levels and receptors of neurotransmitters and in their synthetic enzymes have provided an area for examining the effects of DR on neurochemical parameters (Table 15.4). Again no detrimental effects have been observed, and several studies have produced evidence that DR can retard specific age-related losses.

The concentration of dopamine (DA) receptors in the striatum has been the focus of intense research because of the robust nature of the age-related decline in this neurochemical parameter and its relationship to motor dysfunction (Joseph and Roth 1988; Morgan and Finch 1988). Levin et al. (1981) observed that aged (24 months) male Wistar rats on an EOD regimen since weaning had a higher concentration of striatal DA receptors than AL-fed controls. In a study comparing this parameter in several age groups of Wistar rats, Roth et al. (1984) confirmed and extended this observation. The age-related loss of striatal DA receptors was retarded by DR. This effect was due to the chronic nature of the treatment since there was no DR effect in 3-month-old rats or in 24-month-old rats that had been placed on the EOD diet for 2 weeks prior to killing.

Joseph et al. (1983) produced behavioral findings that directly linked the effects of DR to DA receptors. Unilateral injections of the DA toxin, 6-hydroxydopamine, were made in the substantia nigra (the origin of striatal DA fibers) given to young (6–12 months) and aged (24 months) Wistar rats. This treatment normally results in an upregulation of DA receptors in the intact striatum. During the test session a week later, amphetamine, a DA agonist, is injected peripherally, and the rats show a stereotyped turning response that is correlated with DA receptor concentration in the intact striatum (Joseph et al. 1981). An age-related decrease in this locomotor response was observed in AL-fed rats. However, the responses of aged rats fed EOD since weaning was similar to that of young AL-fed rats. In another experiment the cholinergic antagonist atropine was given to stimulate DA release. With this treatment young AL-fed rats again displayed increased stereotyped behavior compared to aged AL-fed rats. The response of aged EOD-fed rats was greater than that of aged AL rats. However, when comparing the responses of aged DR rats and young AL rats to this treatment, aged DR rats exhibited greater responding at low doses and reduced responding at higher doses of atropine, which paralleled the responses of their aged AL counterparts. These observations of neurotransmitter receptor cross talk indicated that the cholinergic system may exhibit different responses to DR compared to the dopaminergic system.

London et al. (1985) provided additional evidence of differential effects of DR on neurotransmitter systems. Neurotransmitter synthetic enzyme activities and cholinergic receptor concentrations were examined in different brain regions of young (6 months) and aged (24 months) male Wistar rats fed AL or EOD since weaning. The regional activities of choline acetyl-transferase (ChAT), glutamic acid decarboxylase (GAD), and tyrosine hydroxylase (TH) were used as cellular markers for the integrity of the cholinergic, gamma-aminobutyric acid-ergic (GABA), and adrenergic systems, respectively. Regarding TH activity, no DR effect was observed except that cortical TH activity was lower among DR rats compared to controls at 6 months of age. No significant DR effect was observed in GAD activities, except that cortical GAD activity was lower among DR rats compared to controls at 6 months of age. DR effects were most apparent for cholinergic markers. ChAT activity was notably higher in striatum, cerebellum, and hippocampus of 24-month-old DR rats than in AL controls. At 6 months of age, hippocampal ChAT activity was also higher among EOD animals compared to controls. Regarding cholinergic receptors, diet effects were observed only in the striatum, where muscarinic receptor concentrations were markedly higher (>40%) among EOD groups compared to AL controls.

The increased sensitivity to low doses of atropine observed in the Joseph et al. (1983) study may have been due to the higher muscarinic receptor concentration as reported in the London et al. (1985) study. Effects of DR on hippocampal cholinergic receptor function should also be considered with respect to the age-related impairment observed in learning the 14-unit T maze. Cholinergic antagonists and lesions to hippocampal cholinergic systems have been used to disrupt learning in this maze (Ingram 1988). However, in the London et al. study (1985), no DR effect was observed with respect to the age-related decline in hippocampal muscarinic receptor concentration observed, although hippocampal ChAT activity was higher among DR rats at 6 and 24 months of age.

Kolta et al. (1989) examined several monoamine parameters in selected brain regions of aged (22 months) male and female F344 rats fed AL or on DR (40% reduction) since the age of 3 months. Relative to control values, decreased concentrations of norepinephrine were observed in the striatum, hypothalamus, and olfactory bulbs of DR rats. A similar reduction was observed for serotonin in striatum and hypothalamus. No other DA metabolite was affected by DR, except 5-hydroxyindoleacetic acid, which was reduced in nucleus accumbens and hypothalamus of female DR rats, but increased in olfactory bulb of male DR rats. No age comparisons were made in this study. However, rather than viewing the data with respect to alterations of age-related change in neurochemical parameters, these investigators viewed the lower levels of norepinephrine and serotonin among aged DR rats as evidence that catecholamines had been reduced chronically by this treatment and that this effect might have a wide range of implications for brain aging. In drawing this conclusion, they further recognized the need for more complete analysis including the effects of acute DR treatments.

Neuromorphological Parameters

Applying various types of histological analysis, several laboratories have begun to analyze the effects of DR on selected neuromorphological parameters (Table 15.5). Similar to the neurochemical findings, an impression thus far emerging from this limited effort is that the effects of DR may be selective to specific systems.

Moroi-Fetters et al. (1989) analyzed Golgi-stained neurons to quantify the morphology of dendritic spines from the parietal cortex of young (6 months) and aged (24 months) male Wistar rats. An age-related decline in spine density was observed in AL-fed groups while little or no decline was indicated in 24- and 30-month-old rats fed EOD since weaning or in 24-month-old rats on EOD for only 5 months prior to killing. Additional morphological characterization of L-type (lollipop) and N-type (nubbin) spines also indicated DR effects. Among AL rats, the proportion of L-type spines declined with age, whereas all EOD rats showed a retention of L-type spines. Thus, these data suggested that specific aspects of neuronal aging could be retarded by DR.

Table 15.5. Effects of dietary restriction on neurohistological indices of aging in rodents.

Parameter	Species	Strain	Sex	Diet effect	Reference
Hippocampal and frontal cortical lipofuscin	Mice	C57BL/6J	F	+	Nandy (1985)
Whole-brain lipofuscin	Mice	Swiss albino	F	+	Chipalkatti et al. (1983)
Cerebellar Purkinje cell lipofuscin	Rats	Sprague-Dawley	M	0	Forbes (1980)
Brain-reactive antibodies	Mice	C57BL/6J	F	+	Nandy (1982)
Parietal cortical dendritic spine density	Rats	Wistar	M	+	Moroi-Fetters et al. (1989)
Striatal medium spiny neuronal dendritic spine parameters	Mice	C57BL/6J	M	0	McNeill et al. (1991)
Auditory cortical morphological parameters	Rats	Sprague-Dawley	M	0	West et al. (1984)

+, retarded age effect; 0, no effect

McNeill et al. (1991) examined other neuromorphological features of the dendritic tree in aged (26 months) male C57BL/6J mice fed AL or intermittently (3/7 days per week) since weaning. Focusing on the medium spiny neuron of the striatum, these investigators found no DR effects related to the number of dendritic arbors, mean total dendritic length per cell, mean total number of segments, or mean segment length. However, they found no evidence of a significant age effect either; therefore, the selection of these neurons might not have been as sensitive as that of parietal cortical neurons observed by Moroi-Fetters et al. (1989). These striatal spiny neurons are hypothesized to contain either GABA, substance P, or enkephalin as their neurotransmitters. If examined, striatal cholinergic interneurons might have been more affected by aging and thus more responsive to DR.

West et al. (1984) found no evidence of a significant DR effect on neuromorphological parameters in 24-month-old male Sprague-Dawley rats either. However, their analysis was not quantitative and involved small numbers of rats. Age-related morphological alterations in the auditory cortex, including abnormal myelinated axons, intradendritic membranous bodies, and increased lipofuscin, did not differ between AL controls and rats restricted since weaning. Examining the brains of 48-month-old DR rats, these investigators reported several phenomena rarely or never observed in younger controls, including thickening of basal lamina of blood vessels, abnormalities in neuroglia, and evidence of neuritic plaques appearing to have an amyloid core. These observations, particularly the report of neuritic plaques, suggest that certain neuropathology, not usually seen in AL rats with shorter life spans, was permitted to emerge in very long-lived DR rats.

Lipid Peroxidation Products

What mechanisms might be involved in producing selected retardation of age-associated alterations in brain aging through DR? A major hypothesis of DR action has focused on the retardation of damage associated with normal by-products of oxidative metabolism (Weindruch and Weindruch 1988). DR might retard lipid peroxidation, which is one potential destructive by-product of oxidative metabolism. Several neurohistological techniques can be used to assess this potential directly and indirectly.

Nandy (1985) examined the production of lipofuscin in the brains of female C57BL/6J mice as one indirect measure of lipid peroxidation. Based on its autofluorescent quality, lipofuscin deposition can be measured in neurons and represents an ubiquitous marker of cytological aging (Nandy 1985). Analyzing lipofuscin autofluorescence in hippocampus and frontal neocortex, he observed that mice fed about 50% of AL control levels since the age of 3 months showed much less evidence of lipofuscin accumulation than control brains. When DR was initiated at 24 months of age, no differences between DR mice and control levels of lipofuscin in these brain regions were observed 6 months later. Chipalkatti et al. (1983) confirmed the DR effects on brain lipofuscin when initiated early in life in whole-brain preparations extracted from female Swiss albino mice at 4.5 and 12 months of age. Lipofuscin accumulation was greater among AL-fed controls than brains of mice on DR (about 50% of control intake) since weaning. This finding was not repeated for lipofuscin accumulation in cerebellar Purkinje cells of 7-month-old male Sprague-Dawley rats fed either AL or restricted (2-7.5 g per day) since weaning. Forbes (1980) noted a trend towards reduced lipofuscin in the DR rats; however, possibly in this study the diet treatment had not been implemented for a sufficient period of time to observe effects on this parameter.

Chipalkatti et al. (1983) also analyzed the activities of lysosomal enzymes (acid phosphatase and cathepsin) as measures of lipid peroxide-induced damage to lysosomes. Compared to control brains, these indices were reduced in brains of DR mice at 3, 6, and 9 months of age. In addition, these investigators measured the activity of superoxide dismutase (SOD). Levels of this endogenous antioxidant can index protection against lipid peroxidation. At 6 months of age, brain SOD was higher among DR mice, whereas at 12 months of age, this measure was lower in DR mice.

Immunological Parameters

Immunological mechanisms may also be involved in neuronal aging, and these may be sensitive to manipulation by DR (Weindruch and Walford 1988). The formation of brain-reactive antibodies (BRA) is suggested to provide an index of autoimmune susceptibility. Nandy (1982) examined the incidence of BRA in the frontal cortex and hippocampus of female C57BL/6J mice. A high frequency of BRA was observed among AL-fed controls at 12 months of age, whereas there was little evidence of BRA among mice on DR (50% of control level since weaning) at this age. No difference in BRA between control and DR groups was observed in 27-month-old mice on the DR regimen for only 3 months (Nandy 1981). Therefore, DR appears to retard the development of autoimmunity in brain when the treatment is initiated early but not when introduced late in life.

Summary and Conclusions

A clear conclusion derived from studies focusing on the effects of DR on behavior of aged rodents is that such nutritional treatments are not detrimental to performance. Indeed, beneficial effects were observed in several psychomotor and learning tasks in both rats and mice. On the other hand, some types of behavioral performance did not appear significantly affected by DR. Whether DR has selective effects on behavioral performance or whether methodological factors might account for the negative results will require further investigation. Few studies have addressed the issue of adequate control over intervening variables (e.g., body weight, motivation, sensory abilities, pathology). This concern must be expressed in future analysis of behavioral performance to design appropriate control groups and to utilize meaningful tasks.

Analysis of neurochemical effects of DR has also produced some inconsistencies, but similar to the behavioral results, little evidence of detrimental effects has emerged. DR appears to retard the age-related declines in striatal DA and cholinergic receptor concentrations. Other markers of cholinergic function also appear sensitive to manipulation by DR. In addition, reduction in monoamine levels in the hypothalamus and striatum also occur with chronic DR. Whether there is differential sensitivity of certain neurochemical parameters to DR also remains to be determined.

Regarding neuromorphological analysis of DR effects, again, mixed results appear. Aspects of dendritic morphology in the parietal cortex appear to be affected in aged rats on DR but not affected in a particular striatal neuron of mice. Neuromorphological evaluation of neurons in the auditory cortex of aged rats also did not reveal a DR effect, although rare pathology was observed at advanced ages among DR rats. When the accumulations of lipofuscin and BRA were evaluated in aged mice, there was evidence that DR could retard such parameters if initiated shortly after weaning but not when started at a late age. Further investigation of these parameters will be important as they possibly relate to mechanisms of brain aging pertaining to neurochemical and neuromorphological measures.

In conclusion, investigation of the effects of DR on brain and behavioral function has only recently begun to be conducted systematically. This research will continue because of its importance in assessing the impact of DR on function and because neurobiological mechanisms may underlie the phenomenon. Future research should focus on a specific system that can be assessed at behavioral, physiological, cellular, and molecular levels. Little attention has been given to hypothalamic systems. These should be of particular interest because of their involvement in regulation of peripheral neuroendocrine events. Finally, additional emphasis should be given to early effects of DR on brain and behavior. In this fashion, rather than assessing effects late in life, findings emerging from these investigations of early life might provide clues to mechanisms, such as reduced monoamine production (Kolta et al. 1989).

Acknowledgments. This work was made possible by the valuable technical assistance of John Freeman, Edward Spangler, Albert Ellis, Kathleen Schrieber, Nancy Cider, John Whitaker, and Brain Sievers, and the contributions of valued collaborators, including Elaine Bresnahan, Barbara Davis, James Joseph, Edythe London, Thomas McNeill, Ronald Mervis, George Roth, Mark Reynolds, Roy Walford, Steven Waller, and Richard Weindruch, and the continued inspiration of Charles Goodrick. I also wish to thank the International Life Sciences Institute for sponsoring and helping to organize the conference for which this paper was prepared.

References

Beatty WW, Clouse BA, Bierley RA (1987) Effects of long-term restricted feeding on radial maze performance by aged rats. Neurobiol Aging 8:325–327

Bernstein D, Olton DS, Ingram DK, Waller SB, Reynolds MA, London ED (1985) Radial maze performance in young and aged mice: neurochemical correlates. *Pharmacol Biochem Behav* 5:301–307

Bond NW, Everitt AV, Walton J (1989) Effects of dietary restriction on radial-arm maze performance and flavor memory in aged rats. *Neurobiol Aging* 10:27–30

Campbell BA, Gaddy JR (1987) Rate of aging and dietary restriction: sensory and motor function in the Fischer 344 rat. *J Gerontol* 42:154–159

Chipalkatti S, De AK, Aiyar AS (1983) Effect of diet restriction on some biochemical parameters related to aging in mice. *J Nutr* 113:944–950

Forbes WB (1980) Dietary restriction, longevity, and CNS aging. In: Stein D (ed) *The psychobiology of aging: problems and perspectives.* Elsevier, Amsterdam, pp 145–160

Goodrick CL (1984) Effects of lifelong restricted feeding on complex maze performance in rats. *Age* 7:1–2

Harrison DE, Archer JR (1987) Genetic effects on responses to food restriction in aging mice. *J Nutr* 117:376–382

Idrobo F, Nandy K, Mostofsky DI, Blatt L, Nandy L (1987) Dietary restriction: effects on radial maze learning and lipofuscin pigment deposition in the hippocampus and frontal cortex. *Arch Gerontol Geriatr* 6:355–362

Ingram DK (1983) Toward the behavioral assessment of biological aging in the laboratory mouse: concepts, terminology, and objectives. *Exp Aging Res* 9:225–238

Ingram DK (1984) Biological age: a strategy for assessment. In: Cherbotarev F, Tokar AV, Voitenko VP (eds) Gerontology and geriatrics yearbook (USSR). Institute for Gerontology, Kiev, pp 30–38

Ingram DK (1988) Complex maze learning in rodents as a model of age-related memory impairment. *Neurobiol Aging* 9:475–485

Ingram DK (1990) Perspectives on genetic variability in behavioral aging of mice. In: Harrison DE (ed) *Genetic effects on aging*. Telford, Caldwell, NJ, pp 205–231

Ingram DK (1991) Effects of dietary restriction on brain and behavioral function in aging rodents. In: Ingram DK, Baker GT III, Shock NW (eds) *The potential for nutritional modulation of aging processes*. Food and Nutrition Press, Westport, CN, pp 289–310

Ingram DK, Reynolds MA (1986) Assessing the predictive validity of psychomotor tests as measures of biological age in mice. *Exp Aging Res* 12:155–162

Ingram DK, Reynolds MA (1987) The relationship of body weight to longevity within laboratory rodent species. In: Woodhead AD, Thompson KH (eds) Evolution of longevity in animals. Plenum, NY, pp 247–282

Ingram DK, London ED, Goodrick CL (1981) Age and neurochemical correlates of radial maze performance in rats. *Neurobiol Aging* 2:41–47

Ingram DK, Weindruch R, Spangler EL, Freeman JR, Walford RL (1987) Dietary restriction benefits learning and motor performance of aged mice. *J Gerontol* 42:78–81

Joseph JA, Roth GS (1988) Upregulation of striatal dopamine receptors and improvement of motor performance in senescence. In: Joseph J (ed) *Central determinants of age-related declines in motor function*. *Ann NY Acad Sci* 515:355–362

Joseph JA, Filburn CR, Roth GS (1981) Development of dopamine receptor denervation supersensitivity in the neostriatum of the senescent rat. *Life Sci* 29:575–584

Joseph JA, Whitaker J, Roth GS, Ingram DK (1983) Life-long dietary restriction affects striatally mediated behavioral responses in aged rats. *Neurobiol Aging* 4:191–196

Kolta MG, Holson R, Duffy P, Hart RW (1989) Effect of long-term caloric restriction on brain monoamines in aging male and female Fischer 344 rats. *Mech Aging Dev* 48:191–198

Levin P, Janda JK, Joseph JA, Ingram DK, Roth GS (1981) Dietary restriction retards the age-associated loss of rat striatal dopaminergic receptors. *Science* 214:561–562

London ED, Waller SE, Ellis AT, Ingram DK (1985) Effects of intermittent feeding on neurochemical markers in aging rat brain. *Neurobiol Aging* 6:199–204

Lowy AM, Ingram DK, Olton DS, Waller SB, Reynolds MA, London ED (1985) Discrimination learning requiring different memory components in rats: age and neurochemical comparisons. *Behav Neurosci* 99:638–651

McNeill TH, Davis BJ, Hamill RW, Bresnahan EL, Ingram DK (1991) Assessment of lifelong dietary restriction on neostriatal morphology and peripheral neuroendocrine hormones in aged C57BL/6J mice. In: Ingram DK, Baker GT III, Shock NW (eds) *The potential for nutritional modulation of aging processes*. Food and Nutrition, Westport, CN, pp 311–326

Moment G (1978) The Ponce de Leon trail today. In: Behnke J, Finch C, Moment G (eds) *The biology of aging*. Plenum, New York, pp 1–18

Morgan DG, Finch CE (1988) Dopaminergic changes in the basal ganglia: a generalized phenomenon of aging in mammals. In: Joseph J (ed) *Central determinants of age-related declines in motor function*. *Ann NY Acad Sci* 515:145–160

Morgane PJ, Miller M, Kemper T, Stern W, Forbes W, Hall R, Bronzino J, Kissane J, Hawrylewicz E, Resnick O (1978) The effects of protein malnutrition on the developing central nervous system in the rat. *Neurosci Biobehav Rev* 2:137–230

Moroi-Fetters SE, Mervis RF, London ED, Ingram DK (1989) Dietary restriction suppresses age-related changes in dendritic spines. *Neurobiol Aging* 10:317–322

Nandy K (1981) Effects of caloric restriction on brain-reactive antibodies in sera of old mice. *Age* 4:117–121

Nandy K (1982) Effects of controlled dietary restriction on brain-reactive antibodies in sera of aging mice. *Mech Ageing Dev* 18:97–102

Nandy K (1985) Lipofuscin as a marker of impaired homeostasis in aging organisms. In: Davis BB, Wood WG (eds) *Homeostatic function and aging*. Raven, New York, pp 139–148

Roth GS, Ingram DK, Joseph JA (1984) Delayed loss of striatal dopamine receptors during aging of dietarily restricted rats. *Brain Res* 300:27–32

Stewart J, Mitchell J, Kalant N (1989) The effects of life-long food restriction on spatial memory in young and aged Fischer 344 rats measured in the eight-arm radial and the Morris water mazes. *Neurobiol Aging* 10:669–675

Weindruch R, Walford RL (1988) *The retardation of aging and disease by dietary restriction*. Thomas, Springfield, IL

West CD, Volicer L, Vaughan DW (1984) Aging in the food-restricted rat: body temperature, receptor function, and morphologic changes in the brain. In: Armbrecht HJ, Pendergast JM, Coe RM (eds) *Nutritional intervention in the aging process*. Springer-Verlag, Berlin, pp 78–92

CHAPTER 16

Prevention in the Decline of Membrane-Associated Functions in Immune Cells During Aging by Food Restriction[*]

G. Fernandes[1], J.T. Venkatraman[1], E. Flescher[1], S. Laganiere[1], H. Iwai[1], and P. Gray[1]

Introduction

In the past two decades, there has been considerable interest in the role of food restriction (FR) in delaying the progression of immune deficiency and in preventing rise in chronic diseases with aging (Walford et al. 1973; Fernandes et al. 1976, 1990b; Yu et al. 1985; Good and Gazzar 1986; Weindruch and Walford 1988; Masoro 1988). Though FR has been well accepted as an approach for prolonging life span and delaying the aging process in rodents, the precise mechanisms through which FR delays aging and, particularly, the loss of immune function, are still not clearly understood.

Aging is generally associated with a decline in immune functions, particularly the T cell-mediated functions (Walford 1969; Good et al. 1980; Fernandes 1984; Hausman and Weksler 1985; Miller and Harrison 1985; Miller et al. 1989). We and others have proposed that FR acts effectively on the immune system and on T cell functions, in particular, which generally decline earlier with aging than B-cell functions. The original immunological theory of aging proposed that aging is due to decline in immune response capacity which was associated closely with failure in homeostatic mechanisms (Walford 1969; Walford et al. 1987). Now, changes in membrane fatty acid composition of spleen cells (Fernandes et al. 1987, 1990a), alterations in calcium distribution, transport, and binding (Miller 1987), decrease in interleukin-2 (IL-2) production and IL-2 receptor expression (Weindruch and Walford 1988; Iwai and Fernandes 1989), and poor proliferative response to mitogens (Fernandes et al. 1978) are some of the age-associated changes observed in lymphoid cells which may be closely linked to the failure of the immune system with age.

The present study focuses on the effect of FR (40%) on changes in the immune cell functions of Fischer 344 male rats during aging (3-24 months), especially

*This work was supported by N.I.H. Grants AG01188 and AG03417
[1]Department of Medicine, The University of Texas Health Science Center at San Antonio, 7703 Floyd Curl Drive, San Antonio, TX 78284-7874, USA

with regard to T cell-mediated functions. We have begun to investigate the impact of FR on the fatty acid composition of immune cells and their response to mitogens, membrane fluidity, intracellular Ca^{2+} levels (in response to concanavalin A, ConA), IL-2 production and IL-2 receptor expression during aging in Fischer 344 rats. The results gathered from our ongoing studies suggest that FR has a dramatic effect on the above functions and also strongly indicates that these may be some of the mechanisms through which FR may prevent the loss of immune functions and possibly prevent changes in gene expression in T-cell and B-cell repertoire during aging.

Materials and Methods

Animals and Diets. Weanling male Fischer 344 rats obtained from Charles River Breeding Laboratories were maintained in a barrier facility from 4 weeks of age. The FR group was fed 60% of the amount of semipurified diet consumed by the ad libitum (AL) group from 6 weeks of age as described previously (Yu et al. 1985).

Cell Culture. Spleens were aseptically collected in RPMI containing 5% fetal bovine serum (FBS), gently homogenized and washed, and red cells were lysed with tris-buffered ammonium chloride. T cells and B cells were separated by passing spleen cells through nylon wool column to obtain enriched T cells as described previously (Iwai and Fernandes 1989).

Measurement of IL-2 Receptor. Purified T cells were cultured in 15% FBS containing medium (2.5 µg/ml ConA). After culturing cells for specified days, 5×10^6 cells were incubated with serial dilutions of ^{125}I-labeled IL-2 in a total volume of 200 µl for 20 min at 37°C (Robb et al. 1981). Bound IL-2 was separated from free, and the cell pellets containing the bound IL-2 were counted in a gamma counter.

Measurement of Intracellular Ca^{2+}. Spleen mononuclear cells were prepared by Histopaque density gradient centrifugation of nonadherent cells. Cells were resuspended in RPMI-1640 at 5×10^6/ml and loaded with Fluo-3, and $[CA]_i$ was measured by the use of fluorescence-activated flow cytometry (Rabinovitch et al. 1986). Analyses were performed using a FACS IV system. Resting levels of Ca^{2+} and Ca^{2+} concentration after adding ConA were calculated.

Fluorescence Polarization. Spleen cells (25×10^6) in 10% FBS containing RPMI were loaded with 10 µM 6-diphenylhexatriene (DPH) probe by incubating for 2-3 h at 37°C in a CO_2 incubator. The cells were sedimented and washed to remove excess label. Fluorescence of the DPH-loaded cells was measured at 25°C in a Farrand MK-2 spectrofluorometer with excitation and emission polarizers at 365 (excitation) and 440 (emission) wavelengths (Schachter and Shinitzky 1977).

Fatty Acid Analysis. Lipids were extracted from adherent and/or nonadherent spleen cells with chloroform-methanol (2:1), and fatty acid esters were deriva-

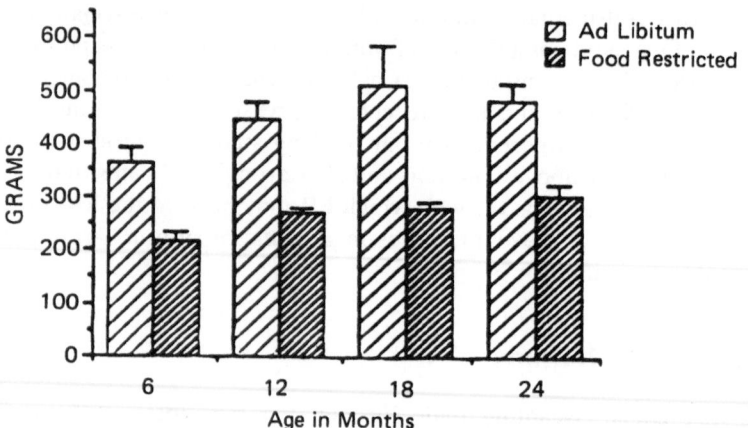

Figure 16.1. Influence of AL feeding and FR on the body weights of Fischer 344 rats. AL-fed animals continued to gain weight until 18 months of age, after which a slight reduction in body weight occurred. FR animals weighed significantly less than the AL-fed animals and consistently maintained their body weights.

tized and analyzed by gas chromatography using a fused silica capillary column (Laganiere and Yu 1987).

Mitogen Response. Spleen cells (5×10^5 cells/100 µl) without removing red cells were cultured in microtiter plates in 1% FBS-RPMI in the presence of ConA (2.5 µg/ml), phytohemagglutinin (PHA; 25 µg/ml), or lipopolysaccharide (LPS; 40 µg/ml). After 48 h incubation, 0.5 µCi of [³H]-methyl thymidine was added and incubation continued for 16 h. The cells were harvested and [³H]thymidine incorporation was counted in a liquid scintillation counter.

Results

Body Weights and Survival. Figure 16.1 indicates that the body weights of Fischer 344 rats increased with age up to 18 months in AL-fed groups. The FR animals weighed significantly less and were able to maintain their low body weights throughout their life span. FR dramatically prolongs the life span of Fischer 344 rats and survival remained similar to that previously reported (Masoro et al. 1982; Yu et al. 1985).

Proliferative Response of Spleen Cells to Mitogens. Proliferative response of spleen cells to optimum concentrations of all three mitogens decreased dramatically in AL-fed animals at 24 months of age (Fig. 16.2). In general, the spleen cells were much more responsive to ConA than to PHA or B cell mitogen (BCM). In contrast, FR maintained or enhanced proliferative response of splenocytes to mitogens for all age groups, although a significant decline was also noted in 24-month-old rats (Fig. 16.2).

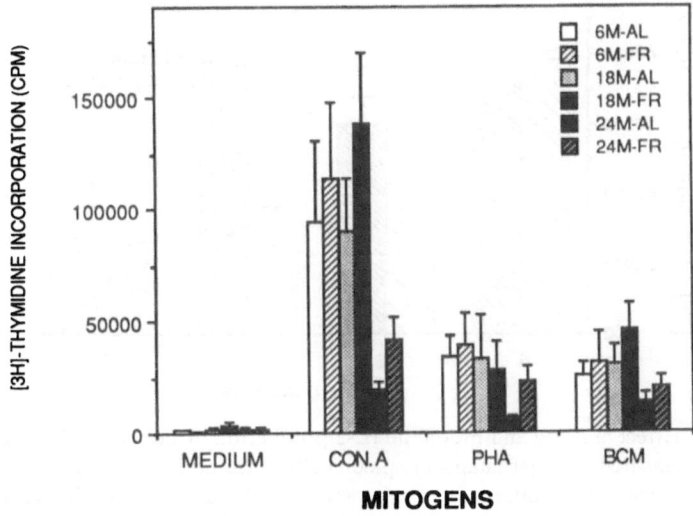

Figure 16.2. Effect of aging and FR on the proliferative response of spleen cells to mitogens in 6-, 18-, and 24-month old animals. Values are mean±SE (n=6-8) of cpm [³H]-methyl thymidine incorporated into 5x10⁵ cells in 24-h culture. *PHA*, phytohemagglutinin; *ConA*, concanavalin A; *BCM*, B cell mitogen

Production of IL-2 by Spleen Cells. Supernatants collected from spleen cells cultured in the presence of ConA for 48 h revealed higher IL-2 production in FR rats at 12, 18, and 24 months (Fig. 16.3).

Ca²⁺ Influx in the Splenic Lymphocytes. The resting Ca²⁺ level was similar in splenocytes of 24-month-old AL and FR rats (Fig. 16.4). The addition of ConA, however, markedly increased the intracellular Ca²⁺ concentration in the spleen cells of FR animals ($\sim$ 240 nM) compared to the AL group ($\sim$ 156 nM).

Fluorescence Polarization. FR results in more fluid spleen cell membranes (decreased anisotropy, r) than in the rats fed AL when anisotropy values were determined with DPH probe (Fig. 16.5). Lipid order with DPH appears to reach deep within the hydrophobic core of the membrane where the most probable compositional determinants of fluidity values are cholesterol content and/or unsaturation states or chain lengths of the fatty acyl groups of phospholipids.

Fatty Acid Composition of Splenic Nonadherent Cells. Although none of the saturated fatty acyls were modulated either by age or by FR, proportions of 18:2ω6, 20:4ω6, and 22:4ω6 were markedly altered in 24-month-old AL-fed rat lymphocytes. The 18:2ω6 levels decreased in the lymphocytes of 24-month-old AL-fed animals while 22:4ω6 content increased in AL-fed animals (Table 16.1). FR significantly prevented these changes. More specifically, 18:2ω6 level was significantly higher at 24 months of age (p<0.01) and 20:4ω6 and 22:4ω6 were significantly lower in lymphocytes of 24-month-old FR rats.

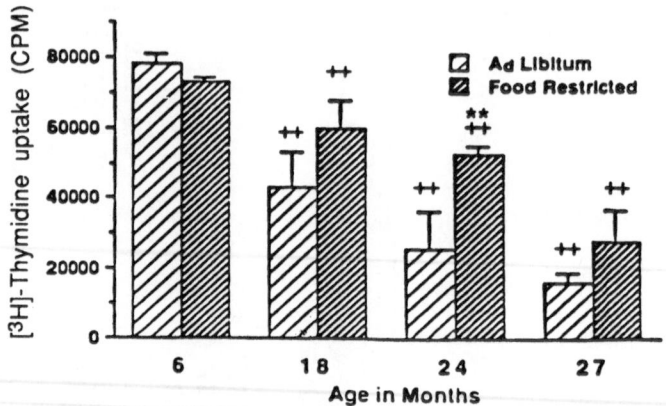

Figure 16.3. Effect of aging and FR on the IL-2 production by cultured spleen cells. IL-2 levels were measured in supernatants of spleen cells cultured for 48 h in the presence of ConA by measuring the proliferative response of CTLL cells. AL feeding resulted in a rapid reduction in the production of IL-2 during aging. ++, Significantly different from 6-month-old AL-fed control rats at $p < 0.05$ level; **, significantly different from age-matched AL-fed animals at $p < 0.05$ level

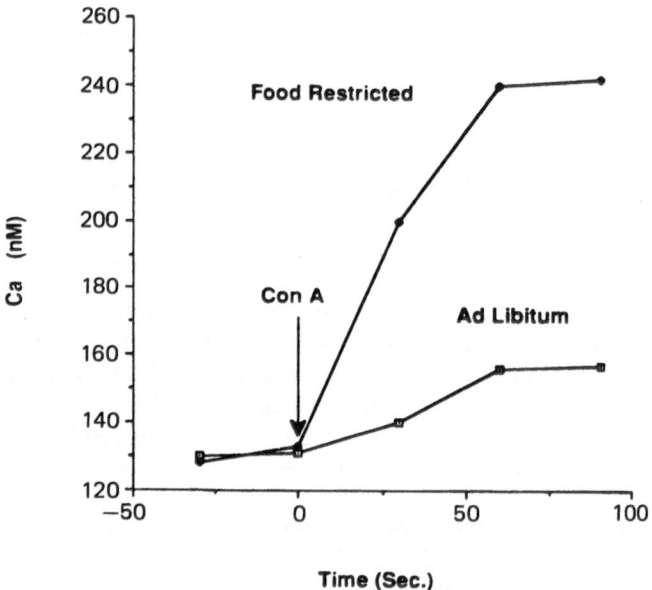

Figure 16.4. Effect of FR on calcium influx in the splenic lymphocytes of 24-month-old rats. Lymphocytes from FR and AL-fed rats were stimulated with ConA. FR(*circles*) markedly enhanced the rise in intracellular Ca^{2+} of 24-month-old rat spleen lymphocytes in response to ConA compared to the AL group (*squares*)

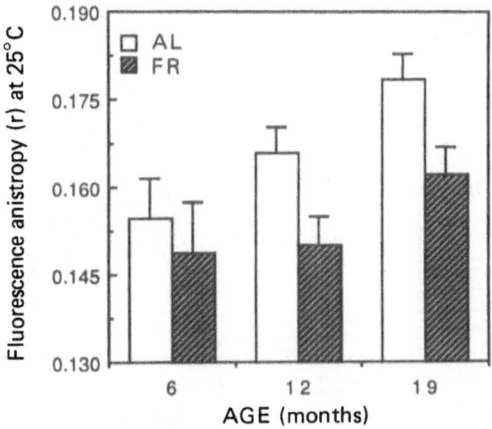

Figure 16.5. Fluorescence polarization measurements in the spleen cells of AL (*open bars*) and FR (*hatched bars*) rats. Membrane fluidity was measured by steady-state fluorescence polarization using 1,6-diphenylhexatriene (DPH). Anisotropy (r) values were higher for the spleen cells from AL-fed than FR animals, suggesting that the fluidity was higher in the splenocytes of FR animals

High-Affinity IL-2 Receptors on Splenic T Cells. The expression of HA receptors on splenic T cells of 10-month-old AL and FR rats was measured. The expression of IL-2 receptors (molecules per cell) reached a peak at 72 h in both the groups of rats; however, the number of IL-2 receptor molecules/cell was greater in FR rats than in the AL group (Fig. 16.6).

Table 16.1. Effect of FR and aging on the fatty acid composition of splenic lymphocytes in Fischer 344 rats.

Fatty acid	6 Months of age		24 Months of age	
	AL	FR (%w/w)	AL	FR (%w/w)
16:0	19.12 ± 1.28	20.63 ± 0.98	18.75 ± 1.12	20.71 ± 1.31
16:1	1.42 ± 0.24	1.26 ± 0.20	0.67 ± 0.040	0.71 ± 0.07
18:0	13.16 ± 0.58	16.15 ± 1.50	15.29 ± 0.75	15.72 ± 1.46
18:1	10.03 ± 1.04	10.05 ± 0.50	9.83 ± 0.51	11.00 ± 0.29
18:2ω6	9.54 ± 1.48	9.94 ± 0.63	8.07 ± 0.16	9.22 ± 0.36
18:3ω6	0.16 ± 0.04	0.00	0.82 ± 0.02	0.00 ± 0.03
20:2ω6	1.37 ± 0.53	2.39 ± 0.19	1.45 ± 0.20	0.97 ± 0.92
20:4ω6	19.36 ± 0.67	20.26 ± 1.18	21.58 ± 1.05	16.75 ± 1.63
22:4ω6	3.59 ± 0.18	3.75 ± 0.50	6.39 ± 1.11	4.24 ± 1.13
22:5ω6	0.97 ± 0.06	0.78 ± 0.08	2.40 ± 0.47	0.82 ± 0.20
22:6ω3	0.57 ± 0.07	0.56 ± 0.15	0.71 ± 0.08	0.55 ± 0.12

AL, ad libitum feeding; FR, food restriction

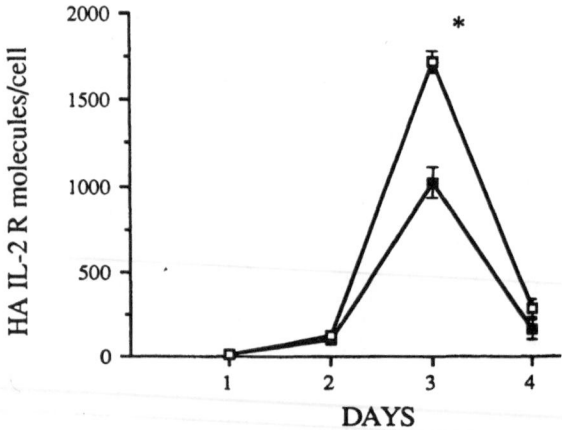

Figure 16.6. High-affinity (HA) IL-2 receptors on splenic T cells of AL and FR rats. HA IL-2 receptors in purified T cells from AL (■) and FR (□) rats were measured using recombinant IL-2 (rIL-2) and [^{125}I]-IL-2. The expression of IL-2 receptor (R) molecules reached a peak at 72 h in AL and FR rats and the number of HA IL-2 receptor molecules per cell was higher in FR rats. (* Significantly higher from AL group at $p < 0.001$.) The response of spleen cells to mitogens decreased with age. FR improved the response of splenocytes to mitogens.

Discussion

FR is known to prolong life span and delay the aging process in rodents. Aging is generally associated with decline in immune function, particularly the T cell-mediated immune functions. The present study indicates that FR markedly alters the fatty acid composition of splenic nonadherent cells and increases membrane fluidity, intracellular Ca^{2+} levels (in response to ConA), IL-2 receptor numbers on T cells, and proliferative response of spleen cells to mitogens in aging Fischer 344 rats. The above observed alterations induced by lifelong FR may be partly responsible for delaying the aging process of the immune system in Fischer 344 male rats.

Several studies in the past have shown that FR strongly influences the immune system, but the exact mechanism is still not known. However, the aging process of the immune cells may depend on extracellular homeostatic and integrative mechanisms provided by neuroendocrine, immune, and intracellular communicating systems. Indeed, the T-cell cellular defects have been associated with the loss of immune function in a large proportion of T cells during aging but not in all of the T cells. Overall, many aging T cells or T-cell subsets do not respond well to mitogens or antibodies when compared to cells obtained from young animals. These changes are currently linked to the failure of T cells from old mice to generate adequate Ca^{2+} signals in response to ConA (Miller et al. 1987, 1989). Our present results clearly indicate that FR maintains higher Ca^{2+} response to

ConA in enriched splenocytes stimulation, which appears to aid in the better preservation of immune functions during aging. However, our new findings also suggest that the increased Ca^{2+} influx may be due to fewer changes occurring both in fatty acid composition and membrane fluidity of lymphocytes with age by FR.

We have recently observed higher IL-2 production and increase in the number of HA IL-2 receptors in the T cells of FR animals (Iwai and Fernandes 1989). The binding of IL-2 to HA IL-2 receptor is thought to provide a proliferation-promoting signal, and the internalization of IL-2 appears to be mediated through HA IL-2 receptor (Robb et al. 1981; Weissman et al. 1986) while the biological functions of low-affinity IL-2 receptors are still not well understood. However, our results indicate that fewer changes with age in membrane fatty acid composition in FR animals could maintain a well-functioning IL-2R for a longer period of time.

Alternatively, faulty metabolism of calcium has been implicated by Miller (1987) in decline in proliferative response during aging, and FR seems to enhance Ca^{2+} levels as well as response of splenocytes of mitogens. FR also retards age-related changes in T-cell subsets in long- and short-lived strains of mice (Fernandes 1987; Weindruch et al. 1982). FR may maintain higher precursor T cells and raise the proportion of $CD8^+$ and $CD4^+$ cells, which secrete and utilize growth factors (e.g., IL-2 by T-helper cells, which are activated by ConA), thereby maintaining improved proliferative responses with age (Iwai and Fernandes 1989; Weindruch and Walford 1988; Miller and Harrison 1985).

Our investigations on the influence of FR on immunological functions suggest both enhanced T-cell functions and B-cell functions, such as production of antibodies to thymus-dependent antigens and response to alloantigens, most likely by elevating IL-2 production and response via enhanced expression of HA IL-2 receptors on T cells (Iwai and Fernandes 1989). Hence, prevention of early loss of T-cell function may be a principal mechanism, whereby FR may delay aging and maintain a prolonged disease-free life span.

Finally, it is quite apparent that AL feeding induces marked alterations in the fatty acid content of all lymphoid cells during aging. However, how FR could induce alterations in the biochemical and structural changes of lymphoid cells in the immune system is not yet elucidated. The fatty acid composition of thymocytes and splenic adherent and nonadherent cells is also found to be strikingly modulated by FR in our earlier pilot studies (Fernandes et al. 1987). For many years, it has been well known that the reactivity of the immune system can be modulated by diet- or lipid-induced changes in the fatty acid composition of membranes for effector lymphoid cells (Meade and Mertin 1978; Trotter et al. 1982). However, our new findings suggest that the decline in immune response during aging may be closely associated with age-related alterations in lymphoid cell fatty acids, and FR appears to reverse some of these alterations. Bilayer properties like fluidity, mobility of cell-surface receptors, and transduction of information through the bilayer may be perturbed by AL food intake which appears to be the most critical event occurring during aging. Subcellular membrane fatty acid changes and membrane rigidity may result in impairment of immune regulation along with increasing free radical production and/or other immunosuppressive products such as prostaglandins (PGs) or other cytokines. Indeed, FR is already known to inhibit free radical formation by modulating fatty acid composition of mitochondrial and microsomal membranes obtained from liver tissues of animals of various ages (Laganiere and Yu 1989a,b), and a similar mechanism may be present in the spleen cells, which needs to be confirmed in spleen cells of aging rats. We also would like to point out, though, that the source of lipids used in the diet may have a profound effect not only on the fatty acid composition of membranes but also on lymphoid cells for their immunological functions during the aging process. The diet we used in the present study contained 10% corn oil, which elevates PG production in the AL-fed groups compared to FR animals because of elevated arachidonic acid levels in the membranes (unpublished data). It appears that although FR animals consume the same amount of food per gram of body weight, some yet unknown mechanisms appear to dramatically alter the membrane fatty acid composition of spleen cells, which lowers PGE_2 levels in FR animals.

In conclusion, the present study suggests that during aging, FR prevented loss of T-cell-mediated immune functions through alterations in fatty acid composition of lymphoid cells of the immune system, increased membrane fluidity and intracellular Ca^{2+} levels increased IL-2 production, IL-2 receptor numbers, and proliferative response of spleen cells to mitogens. However, further investigations are urgently required to elucidate and to unravel some of the more precise molecular mechanisms through which FR may delay the aging process and retard the development of age-related diseases. Indeed, some progress is already underway to compare changes occurring at the level of gene expression by FR. For instance, both α_{2u}-globulin and senescence marker protein-2 expression, which are regulated by hormones, were found to be modulated by FR during aging in Fischer 344 male rats (Richardson et al. 1987; Chatterjee et al. 1989). In addition, cytochrome P-450 IIB-1 and IIB-2 expression, which is known to decline with age in AL-fed rats, was also prevented by FR, particularly the loss of drug-inducible cytochrome P-450 b/e mRNAs in the livers of 20-month-old Fischer 344 male rats (Horbach et al. 1990). Thus FR not only modulates the membrane-associated cellular functions but also appears to show changes in several genes expressing in different tissues with age in rats as well as in short-lived autoimmune disease-prone mice (Fernandes 1990; Fernandes et al. 1990b).

Acknowledgments. The authors wish to thank Dr. E. Masoro and Dr. B. P. Yu for their valuable collaboration, Vikram Tomar for his expert assistance, and Kay Meerscheidt for her help in typing this manuscript.

References

Chatterjee B, Fernandes G, Yu BP, Song C, Kim JM, Demyan W, Roy AK (1989) Calorie restriction delays age-dependent loss in androgen responsiveness of the rat liver. FASEB J 3:169–173

Fernandes G (1984) Nutritional factors: modulating effects on immune function and aging. Pharmacol Rev 36:123S–129S

Fernandes G (1987) Influence of nutrition on autoimmune disease. In: Goidl EA (ed) Aging and the immune response. Dekker, New York, pp 225–242

Fernandes G (1990) Modulation of immune functions and aging by Omega-3 fatty acids and/or calorie restriction. In: Ingram DK (ed) Nutritional modulation of aging process. Food and Nutrition, Bridgeport (in press)

Fernandes G, Yunis EJ, Good RA (1976) Influence of diet on survival of mice. Proc Natl Acad Sci USA 73:1279–1283

Fernandes G, Friend P, Yunis EJ, Good RA (1978) Influence of dietary restriction on immunological function and renal disease in (NZBxNZW)F$_1$ mice. Proc Natl Acad Sci USA 75:1500–1504

Fernandes G, Khare A, Laganiere S, Yu B, Sandberg L, Friedrichs WE (1987) Effect of food restriction and aging on immune cell fatty acids, functions and oncogene expression in SPF Fischer-344 rats. Fed Proc 46:567A

Fernandes G, Flescher E, Laganiere S, Gray P, Venkatraman JT (1990a) Effect of food restriction on cellular immunity, membrane fluidity and Ca^{2+} influx in aging Fischer-344 rats. FASEB J 4:A1040

Fernandes G, Venkatraman J, Khare A, Horbach GJMJ, Friedrichs W (1990b) Modulation of gene expression in autoimmune disease and aging by food restriction and dietary lipids. Proc Soc Exp Biol Med 193:16–22

Good RA, West A, Fernandes G (1980) Nutritional modulation of immune responses. Fed Proc 39:3098–3104

Good RA, Gazzar AJ (1986) Diet, immunity and longevity. In: Hutchison ML, Munro HN (eds) Nutrition and aging, Academic, New York, pp 235–249

Hausman PB, Weksler ME (1985) Changes in the immune response with age. In: Finch CE, Schneider EL (eds) Handbook of the biology of aging, 2nd edn. Van Nostrand Reinhold, New York, pp 414–432

Horbach GJMJ, Venkatraman JT, Fernandes G (1990) Food restriction prevents the loss of isosafrole inducible cytochrome P-450 mRNA and enzyme levels in aging rats. Biochem Internatl 20:725–730

Iwai, H, Fernandes G. (1989) Immunological functions in food-restricted rats: enhanced expression of high-affinity interleukin-2 receptors on splenic T cells. Immunol Lett 23:125–132

Laganiere S, Yu BP (1987) Anti-lipoperoxidation action of food restriction. Biochem Biophys Res Commun 145:1185–1191

Laganiere S, Yu BP (1989a) Effect of chronic food restriction in aging rats. I. Liver subcellular membranes. Mech Aging Dev 48:207–219

Laganiere S, Yu BP (1989b) Effect of chronic food restriction in aging rats. II. Liver cytosolic antioxidants and related enzymes. Mech Aging Dev 48:221–230

Masoro EJ (1988) Food restriction in rodents: an evaluation of its role in the study of aging. J Gerontol 43:B59–B64

Masoro EJ, Yu BP, Bertrand HA (1982) Action of food restriction in delaying the aging process. Proc Natl Acad Sci USA 79:4239–4241

Meade CJ, Mertin J (1978) Fatty acids and immunity. In: Paoletti R, Kritchevsky D (eds) Advances in lipid research. Academic, London, pp 127–165

Miller RA (1987) Calcium and the aging immune system. Neurobiol Aging 8:368–370

Miller RA, Harrison DE (1985) Delayed reduction in T cell precursor frequencies accompanies diet-induced life span extension. J Immunol 134:1426–1429

Miller RA, Jacobson B, Weil G, Simons ER (1987) Diminished calcium influx in lectin-stimulated T cells from old mice. J Cell Physiol 132:337–342

Miller RA, Philosophe B, Tinis I, Weil D, Jacobson B (1989) Defective control of cytoplasmic calcium concentration in T-lymphocytes from old mice. J Cell Physiol 188:175–182

Rabinovitch PS, June CH, Grossmann A, Ledbetter JA (1986) Heterogeneity among T-cells in intracellular free calcium responses after mitogen stimulation with PHA or anti-CD3. Simultaneous use of indo-1 and immunofluorescence with flow cytometry. J Immunol 137:952–961

Richardson A, Butler JA, Rutherford MS, Semsei I, Gu MZ, Fernandes G, Chiang WH (1987) Effect of age and dietary restriction on the expression of α_{2u}-globulin. J Biol Chem 262:12821–12825

Robb RJ, Munck A, Smith KA (1981) T-cell growth factor receptors. Quantitation, specificity and biological relevance. J Exp Med 154:1455–1474

Schachter D, Shinitzky M (1977) Fluorescence polarization studies of rat intestinal microvillus membranes. J Clin Invest 59:536–548

Trotter J, Flesch I, Schmidt B, Ferber E (1982) Acyl transferase-catalyzed cleavage of arachidonic acid from phospholipids and transfer to lysophosphatides in lymphocytes and macrophages. J Biol Chem 257:1816–1823

Walford R (1969) The immunologic theory of aging. Munksgaard, Copenhagen

Walford RL, Liu RK, Gerbase-Delima M, Mathies M, Smith GS (1974) Long-term dietary restriction and immune function in mice: response to sheep red blood cells and mitogenic agents. Mech Aging Dev 2:447–452

Walford RL, Harris S, Weindruch R (1987) Dietary restriction and aging: historical phases, mechanisms, current directions. J Nutr 117:1650–1654

Weindruch R, Kristie JA, Naeim F, Mullen B, Walford RL (1982) Influence of weaning-initiated dietary restriction on response to T-cell mitogens and on splenic T-cell levels in a long-lived mouse hybrid. Exp Gerontol 17:49–64

Weindruch R, Walford RL (eds) (1988) The retardation of aging and disease by dietary restriction. Thomas, Springfield

Weissman AM, Harford JB, Svetlik PB, Leonard WL, Depper JM, Waldmann TA, Greene WC, Klausner RD (1986) Only HA receptors for interleukin-2 mediated internalization of ligand. Proc Natl Acad Sci USA 83:1463–1466

Yu BP, Masoro J, McMahan CA (1985) Nutritional influences on aging of Fischer-344 rats: 1. physical, metabolic and longevity characteristics. J Gerontol 40:657–670

Part IV
Effects and Mechanisms
of Dietary Restriction:
Biochemical Consequences

CHAPTER 17

Caloric Restriction and Its Effects on Molecular Parameters, Especially DNA Repair

A. Turturro[1] and R.W. Hart[2]

Introduction

Caloric restriction is a paradigm which inhibits degenerative diseases, chronic induced toxicity, and aging in every species tested (Allaben et al., Chapter 4). An obvious candidate for a mechanism which can be effective in producing such changes in so many diverse biological systems is one that affects a basic biological process involved in disease and toxicity. DNA damage has been postulated to be a common cause or effector for aging, cancer, and a number of degenerative diseases (Turturro et al. 1984). The major modulator of DNA damage is DNA repair, a series of processes which eliminates various kinds of DNA damage, or minimizes the damage to the basic genetic templates (Sancar et al. 1988). Stimulation of DNA repair by caloric restriction, either directly or indirectly, could be responsible for many of the effects seen in caloric restriction.

DNA Repair

There are two major types of excision repair, base excision repair and nucleotide excision repair (Sancar et al. 1988). These two types are usually measured by using different agents predominantly repaired by one set of enzymatic reaction. Base excision repair is often measured using methylmethanesulfonate (MMS). Nucleotide excision repair is often estimated from repair associated with damage from UV irradiation. In addition, there is an O^6-methylguanine repair system important to the removal of the mutagenic adduct O^6-methylguanine often occurring after alkylation damage. This process occurs through the O^6-methylguanine acceptor protein (MGAP), whose activity is fairly linearly associated with capacity to repair (Lindahl et al. 1988). By evaluating these different processes a more complete picture can be given of the repair status of a cell than with any one system.

[1]Office of the Director and [2]Director, National Center for Toxicological Research, Jefferson, AR 72079, USA

Table 17.1. Age and UV-induced repair in cells isolated from male F-344 Rats.[a]

Age (Months)	Hepatocytes[b]		Kidney cells[c]	
	Ad libitum	Restricted	Ad libitum	Restricted
5	3252/563[4]	–	–	–
13	2419/494[4]	2924/511[4]	–	–
22	1397/458[5]	2009/499[5]	1466/731[4]	1707/722[4]
28	960/362[3]	1343/405[5]	563/421[3]	1035/537[6]
34	–	1264/418[3]	–	766/574[4]

[a]Restriction is 60% of ad libitum diet. All values are dpm per µg DNA for irradiated/unirradiated cells after 1 h. Numerical superscripts are number of rats evaluated. Ratios of restricted and ad libitum diets are significantly different ($p < 0.01$) at all ages and in both type of cells. (Adapted from Weraarchakull et al. 1989).
[b]Irradiated with 877 J/m².
[c]Irradiated with 100 J/m².

There have been four major studies of DNA repair with caloric restriction. The first used livers, spleens, brains, and kidneys from Sprague-Dawley rats derived from a colony of animals in which body size, not food intake, was restricted, as is more common. Woodhead et al. (1985) found neither an effect with age nor caloric restriction on MGAP activity. This was followed, in 1986, by Licastro et al. (1988), using B10C3F mice, who showed a significant inverse linear correlation in lymphocyte UV DNA repair with age. The decline was less steep in calorically restricted animals, leading to higher levels in restricted animals when same-aged mice were compared.

Two more recent studies will be considered in more detail. Weraarchakul et al. noted (1989), in cells isolated from kidney and liver of Fischer 344 rats, that there was a decline in both control and UV-stimulated DNA repair with age (Table 17.1). The extent of repair was higher in hepatocytes than nephrocytes, and the time pattern of decline occurred differently in the different tissues. However, for both tissues, consistent with Licastro et al. (1988), caloric restriction resulted in less of a decline with age.

Lipman et al. (1989) used both mice and rats from the NCTR/NIA specific-pathogen-free colonies in order to utilize the differences in age-specific pathologies in the two species to screen out possibilities of interaction of pathology, age, and caloric restriction. Skin fibroblasts were derived from two rat genotypes, Brown Norway (BN) and Brown Norway × Fischer 344 F1 hybrid BNF1 and the B6C3F$_1$ mouse.

It was found, after an 18-hour repair period, that there was an increase in both rat genotypes in skin cells of both types of excision repair (Table 17.2) (Lipman et al. 1989). Also increased were the levels of MGAP activity in BN rats.

In order to find some basis for this increase and extend the paradigm to another species, skin fibroblasts from the mice were evaluated at different times of day, and their response was correlated with circadian changes seen in body temperature. One hour after lights on (an inactive phase for rodents) there were no sig-

Table 17.2. DNA repair and caloric restriction in rat skin cells.[a]

Inducer of DNA damage	Brown-Norway		BN × F-344	
	Ad libitum	Restricted	Ad libitum	Restricted
MMS (0.5 mM)[b]	2583/2234[8]	3547/3007[8]	6603/4704[7]	7625/4763[11]
UV (20 J/m²)[b]	3072/2234[8]	4247/3007[8]	9745/4704[7]	13044/4763[11]
Spontaneous[c] (MGAP levels)	0.37[8]	0.64[8]	–	–

[a]Restriction is 60% of ad libitum diet. All rats are 18 months of age. Numerical superscripts are number of rats evaluated. Ad libitum vs. restricted repair values are significantly different ($p < 0.01$) in all cases. (Adapted from Lipman et al. 1989).
[b]All values are cpm per 10^6 cells for exposed or irradiated/control cells after 18 h.
[c]Values are fmol/µg DNA of O^6-methylguanine-acceptor protein activity.

nificant differences in MGAP activity. At lights off (waking period) MGAP levels increased 79% and 147% in ad libitum (AL) and dietary restriction (DR) mice with levels of MGAP being significantly greater in DR mice (38%; $p < 0.01$). At 3 hours after feeding, MGAP activity in restricted mice was still significantly higher than the AL controls (19%; $p < 0.01$) (Table 17.3). In both rats and mice, therefore, caloric restriction results in higher levels of DNA repair in a number of systems when compared to age-matched controls. Recent data from Busbee et al. (personal communication) are consistent with the findings of both an age-associated decline and the elevation of DNA repair in restricted animals when compared to ad libitum animals.

Mechanism

Although a number of mechanisms may be important in accounting for the observed elevation of DNA repair in older animals, two of the simplest mechanisms to consider are temperature and hormonal effects.

Table 17.3. MGAP activity.

Time	Diet	Body temperature (°C)	MGAP (fmol/µg DNA)
4:00 a.m.	AL	38.0 ± 0.01	0.32 ± 0.02
	CR	36.6 ± 0.15*	0.38 ± 0.02*
7:00 a.m.	AL	36.5 ± 0.04	0.19 ± 0.02
	CR	35.7 ± 0.15*	0.19 ± 0.04
6:00 p.m.	AL	37.8 ± 0.05	0.34 ± 0.02
	CR	36.6 ± 0.15*	0.48 ± 0.02*

MGAP, O^6-methylguanine acceptor protein; AL, ad libitum feeding; CR, caloric restriction.
*Significantly different, $p < 0.01$.
[a]In B6C3F$_1$ mice (adapted from Lipman et al. 1989).

Table 17.4. DNA repair and temperature in human fibroblasts.[a]

Temperature (°C)	Radiosensitivity[b]	Gamma-irradiation (% DNA SSB rejoined[c])		UV-irradiation[d]
		15 min	60 min	
34	0.236 ± 0.03[11]	86.4 ± 7.2	100.7 ± 9.1	5.21 ± 0.9[7]
37	0.290 ± 0.04[10]	81.3 ± 12.3	92.7 ± 10.1	4.98 ± 0.7[6]
39	0.234 ± 0.02[10]	91.8 ± 11.8	104.0 ± 17.3	6.98 ± 0.6[7]

[a]IMR-90 human fibroblasts. Numerical superscripts are number of samples evaluated (adapted from Mayer et al. 1987). [b]Radiosensitivity (single-strand breaks, SSB) is given in elution rates at time = 0 after 6 Gy of gamma irradiation.
[c]Percentage of SSB rejoined at 15 and 60 min after 6 Gy of gamma-irradiation.
[d]Grains in nucleus after UV irradiation of 22.5 J/m^2 and incubation for 3 h with radioactive thymidine.

Temperature Effects

One possibility is that lowering temperature may directly increase DNA repair. As an enzymatic process, it is not unexpected that DNA repair could be affected by temperature. It has been known for some time that extensive hyperthermia can inhibit DNA repair in a number of systems. It appears that this is mediated by affecting the DNA polymerase alpha (Kampinga and Konings 1987), an enzyme important to both DNA repair and DNA replication. In vitro, the effects of temperature were evaluated in fibroblasts grown under mildly hypothermic or very mild hyperthermic conditions (Mayer et al. 1987). When damage is induced by gamma irradiation or UV light, minor effects are seen. The most significant are listed in Table 17.4. Cells are less radiosensitive (i.e., develop less damage) for DNA single-strand breaks (SSB) and appear to repair these breaks faster in hypothermic than normothermic cells. Consistent, but certainly not striking, results are also seen for UV-induced repair, at a dose approximately equal to that used by Lipman et al. (1989).

This in vitro experiment certainly does not give definitive information on the DNA repair of cells maintained almost lifelong at a particular body temperature in vivo. However, the little information that exists, such as this study, suggests that there is presently little evidence for significant positive effects of DNA repair induced by temperature alone.

Hormone Effects

One of the most consistent effects of caloric restriction is the decrease in the levels of a number of circulating hormones, especially the gonadotropin (Merry and Holehan, Chapter 13). Available information suggests that hormones can significantly affect DNA repair. An example of this is the affect of the estrous cycle on DNA repair in mammary gland and uterine DNA repair after insult by N-methylnitrosourea (MNU). Since NMU is not metabolically activated, hormonal

Table 17.5. Estrous stage and DNA repair in Sprague-Dawley rats.

Tissue	Time (h)	Diestrous	Proestrous	Estrous
Mammary epithelium	1	0.150*	0.150	0.160
	8	0.123	0.150	0.162
Uterus	1	0.122*	0.072	0.073
	8	0.079	0.086	0.062
Liver	1	0.111	0.077	0.088
	8	0.122	0.097	0.102

Virgin female Sprague-Dawley rats, 50–53 days old, were given a 50 mg/kg body weight dose of N-methylnitrosourea. Values are ratios of O^6-methyl-guanine to N-methylguanine (from Braun et al. 1987 and Ratko et al. 1988).
*Repair differences between stages are significant at $p < 0.05$.

effects on metabolism are not responsible for any effects. It can be seen that the time in the cycle when gonadotropin effect is lowest (diestrus) is the time when DNA repair is highest (Table 17.5) (Braun et al. 1987; Ratko et al. 1988). How this phenomenon decreases DNA repair is unknown. However, since circulating hormones are significantly decreased in a number of instances in caloric restriction, this factor may account for the increase in DNA repair found when the animals are compared to ad libitum animals. Interestingly, this area remains fairly unexplored.

Conclusion

DNA repair appears to be preserved by caloric restriction as animals age, in a number of different tissues. The mechanism for this is unknown, although present evidence suggests some hormonal mechanism. The ultimate effect of DNA repair is not in isolation, but a complex mixture of DNA damage, DNA repair, and either proliferation, allowing DNA damage to express itself in proliferative cells as a mutation, or transcription of a damaged template, expressed in postreplicative cells as a loss in function. That the result is complex is indicated by some work from Randerath et al. (to be published), which indicates that I compounds, i.e., compounds which increase with age in rats, are increased with caloric restriction. By consideration of these molecular changes in the appropriate context, we can increase our understanding of their role in caloric restriction.

References

Braun RJ, Ratko T, Pezzuto J, Beattie C (1987) Estrous cycle modification of rat uterine DNA alkylation by N-methylnitrosourea. Cancer Lett 37:345–352

Kampinga H, Konings W (1987) Inhibition of repair of X-ray-induced DNA damage by heat: the role of hyperthermic inhibition of DNA polymerase activity. Radiat Res 112:86–98

Licastro F, Weindruch R, Davis LJ, Walford RL (1988) Effect of dietary restriction upon the age-associated decline of lymphocyte DNA-repair activity in mice. Age 11:48–52

Lindahl T, Sedgwick B, Sekiguchi M, Nakabeppu Y (1988) Regulation and expression of the adaptive response to alkylating agents. Ann Rev Biochem 57:133–157

Lipman J, Turturro A, Hart RW (1989) The influence of dietary restriction on DNA repair in rodents: a preliminary study. Mech Ageing 135–143

Mayer PJ, Bradley MO, Nichols WW (1987) The effect of mild hypothermia (34°C) and mild hyperthermia (34°C) on DNA damage, repair and aging of human diploid fibroblasts. Mech Ageing 39:203–222

Randerath E, Hart RW, Turturro A, Reddy R, Danna TF, Randerath K (to be published) Effects of caloric restriction on I-compounds (putative indigenous DNA modifications) in rat liver and kidney. Proc Am Assoc Cancer Res

Ratko T, Braun RJ, Pezzuto J, Beattie C (1988) Estrous cycle modification or rat mammary gland DNA alkylation by N-methyl-N-nitrosourea. Cancer Res 48:3090–3093

Sancar A, Sancar G (1988) DNA repair enzymes. Ann Rev Biochem 57:29–67

Turturro A, Hart RW (1984) DNA repair and aging. In: Sciapelli DG, Migake G (eds) Comparative biology of major age-related diseases. pp 19–45, Alan Liss, N.Y.

Weraarchakull N, Strong R, Wood WG, Richardson A (1989) The effect of aging and dietary restriction on DNA repair. Exp Cell Res 181:197–204

Woodhead AD, Merry BJ, Cao EH, Holehan AM, Grist E, Carlson C (1985) Levels of O^6 methylguanine-acceptor protein in tissues of rats and their relationship to carcinogenicity and aging. J Natl Cancer Inst 75:1141–1145

CHAPTER 18

Prevention of Free Radical Damage by Food Restriction

B.P. Yu[1], D.-W. Lee[1], and J.-H. Choi[1]

Introduction

It is a popular notion that oxidative damage of cellular constituents by free radicals is the underlying cause of many chronic diseases and physiological dysfunctions (Oberley and Oberley 1986; Harman 1986). This premise was formalized by D. Harman as the free radical theory of aging (Harman 1956). For the last three decades, many attempts to establish the quantitative relationship between free radical activity and age-related dysfunction have produced only marginal supporting data. The difficulty came from two major shortcomings: (1) the lack of techniques for in vivo measurement of free radicals and tissue damage, and (2) the unavailability of suitable experimental paradigms in which free radical activity can readily be modified. Although no technical breakthrough has been made, dietary restriction does provide a paradigm in which free radical metabolism can be effectively modulated. Recent data from several laboratories (Koizumi et al. 1987; Semsei et al. 1989) point to the fact that dietary restriction is probably the best free radical modulator that we know of at present (Yu 1990).

In the present study, we report the quantification of modulation of free radicals by food restriction by assessing age-related free radical damage on membrane structure.

Materials and Methods

Animals and Diet. Male, specific-pathogen-free Fischer 344 rats from Charles River Lab were used. Upon arrival, rats (28±2 days old) were housed singly in a barrier facility. They were fed ad libitum until 6 weeks of age, at which time they were divided into the ad libitum (AL) fed control group and 40% food-restricted (FR) group. The detailed information on diet composition and

[1]The University of Texas Health Science Center at San Antonio, Department of Physiology, 7703 Floyd Curl Drive, San Antonio, TX 78284, USA

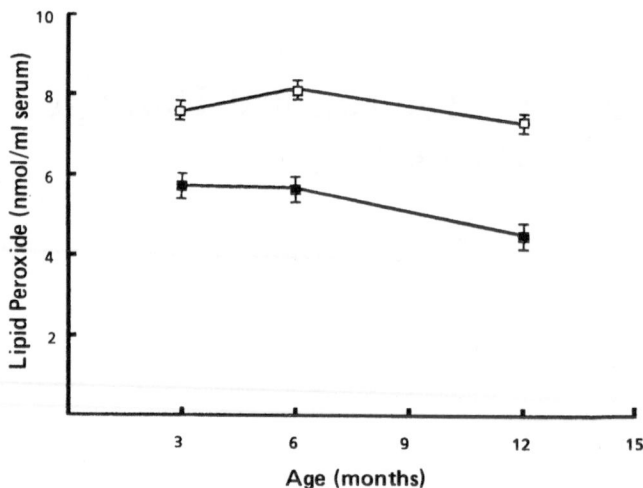

Figure 18.1. Determination of lipid peroxide in serum. Data represent the mean ± SEM; *n* = 9 for each dietary and age group. *Open squares*, ad libitum feeding; *solid squares*, food restriction.

the procedures for the operation of the barrier were the same as reported previously (Yu et al. 1985).

Preparation of Membranes and Cytosols. The preparation of mitochondrial and microsomal membranes was carried out for liver (Laganiere and Yu 1989a) and kidney (Choi and Yu, 1989), respectively. Cytosols from homogenates were prepared by the method of Laganiere and Yu (1989b).

Induction of In Vitro Lipid Peroxidation. In vitro lipid peroxidation was carried out enzymatically in the presence of NADPH/ADP-Fe according to the method of Laganiere and Yu (1987).

Determination of Serum Lipid Peroxide. The serum lipid peroxide was analyzed by the method of Yagi (1987).

Determination of Cytochrome P-450 Destruction. Microsomal cytochrome P-450 destruction during lipid peroxidation was measured by the method of Schacter et al. (1972). Enzymatic lipid peroxidation was induced in microsomes by addition of ADP-Fe (5 mM/0.2 mM) and 1 mM NADPH. Cytochrome P-450 destruction was determined by measuring the remaining P-450 content after a 10-min incubation at 25°C. Cytochrome P-450 destruction by cumene hydroperoxide was performed with 0.5 mM cumene hydroperoxide in place of ADP-Fe and NADPH. All other conditions were the same as those of enzymatic lipid peroxidation. The content of cytochrome P-450 was measured from the spectral difference in absorbance at 450 nm and 490 nm.

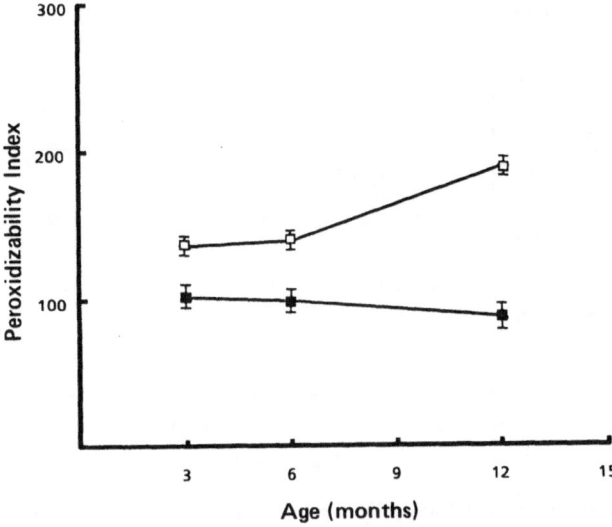

Figure 18.2. Changes in peroxidizability of kidney mitochondria. Data represent the mean ± SEM; $n = 9$ for each dietary age group. The peroxidizability index was calculated by the method of Witting and Horwitt (1964). *Open squares*, ad libitum feeding; *solid squares*, food restriction.

Determination of Protective Factors Against Lipid Peroxidation and Cytochrome P-450 Destruction. Microsomes (1 mg/ml) were preincubated with appropriate protectors to be tested in 0.1 M potassium phosphate buffer (pH 7.6) at 25°C according to the modified method of Iba and Mannering (1987). The lipid peroxidation was induced by addition of ADP-Fe (5 mM/0.2 mM) or cumene hydroperoxide (0.5 mM). The reaction mixture was incubated for 10 min and the inhibitory effect of protective factors was estimated by MDA formation and P-450 destruction.

Results

A widespread phenomenon of lipid peroxidation of rat tissues was shown by data on serum and kidney mitochondria (Figs. 18.1, 18.2) . The serum lipid peroxide level was consistently elevated throughout the age periods studied, while a reduced level of lipid peroxide in serum of restricted rats was remarkably well sustained. It should be noted that as early as 3 months of age (i.e., only 1½ months of restriction), the lipid peroxide value of FR rats already showed dietary effect of reduction at approximately 25% below that of the control.

The structural fragility of microsomal membranes due to oxidative stress with age was tested by monitoring the cytochrome P-450 degradation. As shown in

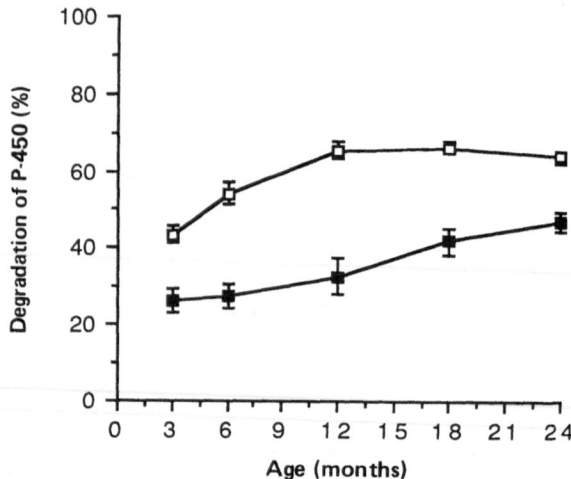

Figure 18.3. Destruction of P-450 by lipid peroxidation; $n = 6$ for each dietary and age group. Experiments were carried out as described in "Materials and Methods." *Open squares*, ad libitum feeding; *solid squares*, food restriction.

Fig. 18.3, the induction of in vitro lipid peroxidation in the presence of NADPH/ADP-Fe led to the disturbance of membrane structure, causing a substantial degradation in AL rats compared to FR rats (Fig. 18.3). The second test on the membrane instability was carried out by exogenous cumene hydroperoxide, a known potent oxidizer. Figure 18.4 shows that the addition of 0.5 mM cumene hydroperoxide caused more than 75% destruction of cytochrome P-450 of the control rats. Much less destruction of P-450 was sustained by microsomal membranes of restricted rats, indicating the strong membrane intactness and stability which, interestingly, appears to increase progressively with age (Fig. 18.4).

The antioxidant protective activity in cytosols was assessed by measuring the ability to suppress the in vitro lipid peroxidation. The suppressive action of cytosols from FR rats consistently showed a higher potency compared to that of the control (Fig. 18.5).

To define possible components responsible for such potent inhibitory actions, several well known antioxidant free radical scavenging reagents and enzymes were tested (Table 18.1). Addition of 2 mM reduced glutathione into the incubation medium inhibited lipid peroxidation and cytochrome P-450 destruction by 95% and 60%, respectively, making glutathione the most potent inhibitor tested. Interestingly, superoxide dismutase and catalase provided little protection against lipid peroxidation and P-450 destruction. Equally interesting were the absence of protective actions of other well-known free radical scavengers, such as mannitol, benzoate, and thiourea (Table 18.1).

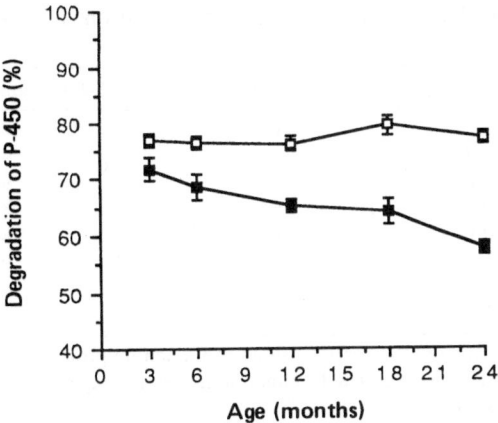

Figure 18.4. Destruction of P-450 by cumene hydroperoxide. The concentration of cumene hydroperoxide was 0.5 mM; $n = 6$ for each dietary and age group. Experiments were carried out as described in "Materials and Methods." *Open squares*, ad libitum feeding; *solid squares*, food restriction.

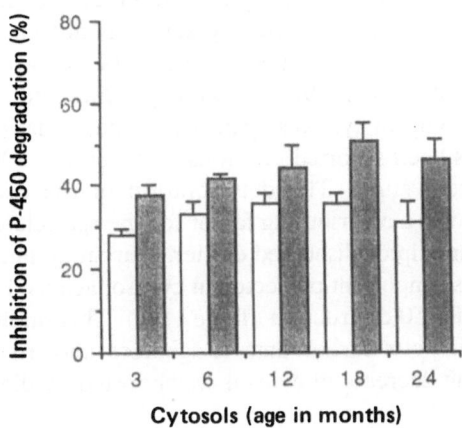

Figure 18.5. Protection of P-450 by cytosols against lipid peroxidation. Aliquots of cytosols prepared from rats of different groups at various ages were added into lipid peroxidation incubation as described in "Materials and Methods." *Open bars*, ad libitum feeding; *shaded bars*, food restriction.

Table 18.1. Factors affecting lipid peroxidation and cytochrome P-450 degradation.

Scavengers	MDA formation (%)	P-450 degradation (%)
None	100	100
Mannitol (5 mM)	109	98
Benzoate (5 mM)	95	98
Thiourea (2 mM)	95	82
Catalase (400 U)	98	84
SOD (200 U)	93	91
GSH (2 mM)	8	42
DTT (2 mM)	4	36

Values are relative to the control, which contained no protector.
MDA, malondialdehyde; GSH, glutathione; SOD, superoxide dismutase; DTT, dithiothreitol

Discussion

In the present report, data reinforcing the antiradical action of dietary restriction are documented. The findings on serum lipid peroxide and kidney mitochondrial lipid strongly indicate that the increased oxidative activity is a general phenomenon occurring in a variety of aging tissues or organs. The clues on the biochemical basis for the increased peroxidative tone came from the use of the peroxidizability index (Fig. 18.2). The increased peroxidizability of lipids strongly suggests that the membrane lipids become more oxidized with age, leading to an increased tissue susceptibility to oxidative stress. The biochemical process by which this age effect is counteracted by dietary restriction is the reduction of the amount of polyunsaturated fatty acids in the membrane. Although the precise regulatory mechanism for this dietary action is not known at this time, it should be noted that a similar mechanism seems to be operational in membranes from liver (Laganiere and Yu 1989a) and kidney (Choi and Yu 1989).

The antioxidant activity of cytosols against lipid peroxidation and cytochrome P-450 destruction has been reported previously (Yu et al. 1990) and was further explored in the present study. The deterioration of the cytosolic antioxidant activity with age could be a contributing factor for the age-related increased peroxidizability of membrane lipids discussed earlier. Our study indicates that reduced glutathione is the most important protector in cytosol against both lipid peroxidation and cytochrome P-450 destruction (Table 18.1). This finding and its interpretation are in line with our earlier data (Laganiere and Yu 1989b) where the age-related glutathione decrease in cytosol was boosted by dietary restriction.

Conclusions

In summary, our study on the antioxidant action of dietary restriction produced strong data supporting that age-related oxidative damages of the subcellular membranes and the deterioration of cytosolic protective components are prevented by dietary restriction. It strengthens the notion that at present dietary restriction is the only antiradical interventive means with such diversity and efficacy (Yu 1990).

References

Choi JH, Yu BP (1989) The effect of food restriction on kidney membrane structures of aging rats. Age 12:133–136

Harman D (1956) A theory based on free radical and radiation chemistry. J Gerontol 11:298–300

Harman D (1986) Free radical theory of aging: role of free radical in the origination and evolution of life, aging, and disease processes. In: Johnson JE Jr, Walford R, Harman D, Miquil J (eds) Free radicals, aging, and degenerative diseases. Liss, New York, pp. 3–49

Iba MM, Mannering GJ (1987) NADPH-and linoleic acid hydroperoxide-induced lipid peroxidation and destruction of cytochrome P-450 in hepatic microsomes. Biochem Pharmacol 36:1447–1455

Koizumi A, Weindruch R, Walford RL (1987) Influence on dietary restriction on liver enzyme activities and lipid peroxidation in mice. J Nutr 117:361–367

Laganiere S, Yu BP (1987) Anti-lipoperoxidation action of food restriction. Biochem Biophys Res Commun 145:1185–1191

Laganiere S, Yu BP (1989a) Effect of chronic food restriction in aging rats. I. Liver subcellular membranes. Mech Ageing Dev 48:207–219

Laganiere S, Yu BP (1989b) Effect of chronic food restriction in aging rats. II. Liver cytosolic antioxidants and related enzymes. Mech Ageing Dev 48:221–230

Oberley LW, Oberley TD (1986) Free radical, cancer, and aging. In: Johnson JE Jr, Walford R, Harman D, Miquil J (eds) Free radicals, aging, and degenerative diseases. Liss, New York, pp 325–371

Schacter BA, Marver HS, Meyer UA (1972) Hemoprotein catabolism during stimulation of microsomal lipid peroxidation. Biochem Biophys Acta 279:221–227

Semsei I, Rao G, Richardson A (1989) Changes in the expression of superoxide dismutase and catalase as a function of age and dietary restriction. Biochem Biophys Res Commun 164:620–625

Witting LA, Horwitt MK (1964) Effect of degree of fatty acid unsaturation in tocopherol deficiency-induced creatinuria. J Nutr 82:19–33

Yagi K (1987) Lipid peroxides and human diseases. In: Paltauf F, Schmid HHD (eds), Chemistry and physics of lipids, vol. 45. Elsevier, Shannon, pp 337–351

Yu BP (1990) Food restriction research: past and present status. In: Rothstein M (ed), Review of biological research in aging, vol. 4. Liss, New York, pp 349–371

Yu BP, Lee DW, Marler CG, Choi JH (1990) Mechanism of food restriction: protection of cellular homeostasis. Proc Soc Exp Biol Med 193:13–15

Yu BP, Masoro EJ, McMahan CA (1985) Nutritional influences on aging of Fischer rats. I. Physical, metabolic, and longevity characteristics. J Gerontol 40:657–670

CHAPTER 19

Modifications in Regulation of Intermediary Metabolism by Caloric Restriction in Rodents

R.J. Feuers[1], D.A. Casciano[1], J.G. Shaddock[1], J.E.A. Leakey[1], P.H. Duffy[1], R.W. Hart,[1] J.D. Hunter,[2] and L.E. Scheving[3]

Introduction

It is well established that caloric restriction extends life span (McCay et al. 1935; Sacher 1977; Masoro 1985; Holehan and Merry 1986) and significantly retards the rate of occurrence of most age-associated degenerative disease (Walford et al. 1974; Harrison and Archer 1983; Kritchevsky et al. 1984; Loyd 1984; Pollard et al. 1984; Pollard and Luckert 1985; Johnson et al. 1986; Ruggeri et al. 1987). The search for a mechanism(s) through which these significant benefits occur has begun, but remains elusive. However, some progress has been made and it appears that numerous processes may be involved (Walford 1983; Weindruch and Walford 1988). Changes in endocrine function (Frolkis et al. 1970; Andres and Tobin 1977; Finch 1977; Everitt 1980; Masoro 1988) and resultant regulation of certain metabolic processes have been implicated (Heuson and Legros 1972; Kmiec and Mysliwski 1983; Feuers et al. 1989).

There has been considerable speculation concerning the effect of caloric restriction on overall metabolic rates, and it now appears that numerous important changes occur. Caloric restriction has been shown to rapidly change circadian rhythms of body temperature, oxygen consumption, carbon dioxide production, and respiratory quotient relative to feeding, whereas these changes are less significant when animals are fed ad libitum (AL; Duffy et al. 1989). Reduction in circulating insulin has been implicated in the mediation of certain caloric restriction effects (Heuson and Legros 1972; Kmiec and Mysliwski 1983), and such hormonal responses may be important in the coordination of metabolic rates. It has also been found that enzymes associated with glycolysis and lipid metabolism are reduced, while those associated with gluconeogenesis and amino acid metabolism are increased in response to caloric restriction (Ruggeri et al. 1987; Feuers et al. 1989).

[1]Food and Drug Administration, Division of Genetic Toxicology, National Center for Toxicological Research, Jefferson, AR 72079, USA
[2]University of Texas at El Paso, El Paso, TX 79968, USA
[3]University of Arkansas for Medical Sciences, Little Rock, AR 72205, USA

A large body of evidence, which has accumulated over many years, demonstrates that insulin stimulates glycolysis and lipid metabolism while rates of gluconeogenesis and glucagon act in opposition. Thus, it appears that caloric restriction results in a reduced role for insulin and an increased role for glucagon in the regulation of intermediary metabolism. This conclusion has been supported by findings concerning the regulation of relative activity of pyruvate kinase (a key regulated enzyme in glycolysis). We have found in preliminary experiments that upon aging under conditions of AL feeding, this enzyme is maintained at high concentrations in a predominately phosphorylated (inactive) form relative to calorically restricted animals (Feuers et al. 1989). This finding further suggests that caloric restriction produces significant changes in the ratio of insulin to glucagon, in a direction which modifies metabolic pathways to increase efficiency.

A more detailed understanding of the effects of caloric restriction on the regulation of intermediary metabolism should provide additional insight into the mechanisms through which caloric restriction elicits its life span-increasing effects. With detailed understanding of how caloric restriction alters regulation of specific key metabolic reactions, it may be possible to back-extrapolate to the effects of caloric restriction at higher centers. We have undertaken a systematic investigation of the effects of caloric restriction on specific hormonal events and their impact on regulation of enzymatic activities and levels of substrates of intermediary metabolism in rodents.

Materials and Methods

Animal Husbandry

A detailed description of husbandry procedures and environmental conditions has been reported elsewhere (Duffy et al. 1989). Briefly, male Fischer 344 rats or male B6C3F$_1$ mice were raised at the National Center for Toxicological Research (NCTR) in a specific-pathogen-free (SPF) environment at 23°C and were maintained on a light-dark cycle so that the lights were on from 0600 to 1800 hours. The animals were housed singly in plastic cages with metal tops and hardwood chip bedding. The animals were divided into a control group that was fed AL and a caloric restricted (CR) group that received 60% of the AL ration starting at 14 weeks of age and continuing for the remainder of their lives. When the test animals were transferred from the SPF facility to the animal testing area, they were allowed to acclimate for at least 3 weeks to specific photoperiods and feeding schedules (see specific methods for individual experiments). The rats used for this in vivo study and all mice were fed just prior to the onset of the dark span. Dark onset feeding was utilized to cause restricted animals to eat at times similar to those at which AL animals normally eat (Duffy et al. 1989).

Insulin Binding and Blood Glucose

Three dozen male $B6C3F_1$ mice of three different age groups ($Y=4$ months, $M=12$ months, and $O=25$ months) on each diet were used. Six different circadian time intervals were established for analysis of binding of 0.4 µCi/0.15 µl to which 6 U of regular, unmodified beef insulin was added to moderate and stabilize the uptake and binding kinetics. Just prior to insulin injection, a drop of tail blood was obtained for basal glucose determinations using the One Touch glucose analyzer (by Lifescan, Johnson and Johnson). ^{125}I-insulin was injected subcutaneously into groups of 6 mice at 4-h intervals at 0230, 0630, 1030, 1430, 1830, and 2230 hours. After exactly 20 min, another blood sample was taken for postinsulin glucose analysis, and mice were killed. The left lobe of the liver was removed, weighed, and cpm/gm measured in a Packard Auto-Scintillator Spectrograph standardized for [^{125}I] assays (Hunter et al. 1990).

In Vivo Pyruvate Kinase Regulation

Male Fischer 344 rats were maintained as outlined above. Three rats from each age group (young, adult, and old) were killed at either 16 h after lights on (HALO) or 4 HALO. Thus, 16 HALO represents 4 h postfeeding and 4 h into the dark span. Livers were quickly excised, prepared, and analyzed for kinetic parameters of pyruvate kinase by our standard procedures (Feuers et al. 1980; Feuers and Casciano 1985).

In Vitro Pyruvate Kinase Regulation

Hepatocyte primary cultures were established from the livers of adult male Fischer 344 rats via the in situ collagenase perfusion technique of Oldham et al. (1979). The viability of the cells, as determined by trypan blue exclusion, was greater than 88%. Monolayer cultures were established by plating 1.25×10^6 viable hepatocytes in 2 ml Williams' Medium E (WE) containing 17% fetal bovine serum (FBS) and supplemented with 10 µm dexamethasone, 2 mU/ml insulin, 10 mM HEPES, 2 mM L-glutamine, and 0.02 mg/ml gentamycin per well in Costar 6-well tissue culture dishes. Following a 1- to 2-h attachment period, cells were incubated for 24 h at 37°C, 95% air/5% CO_2 humidified atmosphere in WE containing 10% FBS.

After hepatocytes had been allowed to attach and had been maintained in culture for 24 h they were exposed to serum-free medium which contained either 1×10^{-10} M insulin or 3×10^{-10} M glucagon for 20 min (Feuers and Casciano 1985). After this incubation period, reactions were stopped by applying 1 ml of our standard homogenization buffer to which 0.5% Triton-X 100 had been added. The lysed cells were carefully removed from the culture dishes and sonicated to assure complete disruption. Pyruvate kinase (PK) kinetic studies were done according to our standard procedures (Feuers and Casciano 1985).

Table 19.1. Altered regulation of hepatic pyruvate kinase under conditions of caloric restriction during the mid-light and mid-dark in young, middle-aged, and old rats.

Age/diet	V_{max}[a]	In vivo		In vitro	
		$K_{.5}$		$K_{.5}$	
		Mid-dark	Mid-light	Insulin	Glucagon
Young CR	6200	.25	.64	.14	.82
Young AL	4050	.39	.45	.35	.70
Adult CR	6902	.24	.98	.14	.83
Adult AL	7215	.48	.39	.34	.71
Old CR	5822	.31	.56	.21	.75
Old AL	6432	.64	.41	–	–

CR, calorically restricted (60% of ad libitum feeding); AL, ad libitum feeding
[a]Hill equation: $v = V_{max}[S]^n/K' + [S]^n$ and $K_{.5}$ = square root of K'.

Results and Discussion

Pyruvate kinase catalyzes a key regulated and essentially irreversible step of glycolysis. Pancreatic beta cells respond to increasing blood glucose levels by secreting insulin, which then binds to its receptor in various tissues. A primary liver response is the uptake of glucose. As blood glucose decreases, the insulin receptor is downregulated. Among a cascade of events which result from insulin binding in the liver is stimulation of rates of glycolysis. One mechanism through which glycolysis is stimulated involves the insulin-mediated synthesis and phosphatase-catalyzed dephosphorylation of PK, which increases its activity. This increase in activity is the result of increased enzyme concentration and, more importantly, an increase in affinity for its substrate (lower K_m), phosphoenolpyruvate (PEP). When glucose levels become low, glucagon is secreted from alpha cells of the pancreas. At the liver, glucagon binds to its receptor, stimulating adenylate cyclase activity, which results in increased cAMP levels. Cyclic-AMP stimulates a protein kinase-dependent phosphorylation of PK, which raises its K_m, thereby inactivating the enzyme. The net result is an inhibition of glycolysis with a concurrent stimulation of gluconeogenesis.

When rats were maintained on a reversed light-dark schedule, fed just prior to the onset of the dark span, and tissues taken during the mid-dark, K_m and PK values were lower than when tissues were taken during the mid-light in all cases except adult and old AL rats (Table 19.1). Because the enzyme is a tetramer with two binding sites, the Hill equation with $n=2$ was employed to obtain more accurate estimates of binding affinity ($K_{.5}$) for this more complex system. Mid-dark $K_{.5}$ values were similar for young, adult, and old CR rats, while the values for all ages of AL fed rats were higher (Table 19.1). The change in $K_{.5}$ values from mid-dark to mid-light was greater for CR rats regardless of age, and $K_{.5}$ values were also higher for CR rats. Mid-dark V_{max} values, which are a measure of enzyme concentration, were lower in young AL rats when compared to young CR rats. However, as animals aged, it appears that enzyme concentration becomes

Table 19.2. ^{125}I-insulin binding in young, middle-aged, and old calorically restricted or ad libitum fed B6C3F$_1$ male mice at different times of day expressed in cpm per gram of tissue.

Diet	Age	0200 hours	0600 hours	1000 hours	1400 hours	1800 hours	2200 hours
AL	Y	–	9200	–	9900	–	5225
	M	6885	6622	6518	7100	8312	6720
	O	3522	2101	2620	2680	2610	4808
CR	Y	5600	9406	11680	12222	10800	7420
	M	9185	9240	11200	10580	15800	8850
	O	6680	4200	4250	5180	5320	4185

AL, ad libitum fed; CR, calorically restricted (60% of ad libitum fed); Y, young (4 months); M, middle-aged (12 months); O, old (>24 months)

higher in AL rats. This suggests the enzyme concentration increases to compensate for the higher K_m (lower relative activity) seen in aging AL rats when circadian glucose levels are highest.

This interpretation was investigated in detail using hepatocytes from AL or CR rats of various ages. When hepatocytes are exposed to insulin for extended times, the reactions leading to dephosphorylation can be driven essentially to completion. Alternatively, glucagon saturation leads to complete phosphorylation of the enzyme (Feuers and Casciano 1985). These two treatments simulate the conditions encountered during mid-dark (high insulin) and mid-light (high glucagon). The $K_{.5}$ obtained from hepatocytes of all aged CR rats was below 0.21 mM PEP when treated with insulin. When AL rats were evaluated, the $K_{.5}$ values were greater than 0.34 mM PEP for young and adult rats. Hepatocytes from old AL rats were not successfully cultured. When hepatocytes from CR animals of each age group were exposed to glucagon, $K_{.5}$ values rose to greater than 0.75 mM PEP. These data demonstrate that a broad range exists for regulation of PK activity in CR rats that is not compromised by aging. AL mice did not demonstrate this range. When hepatocytes of young and adult rats were given glucagon, the $K_{.5}$ increased to only 0.70 and 0.71, respectively.

The role of insulin and its receptor in the compromised regulation of PK due to aging was also evaluated in AL and CR rats. ^{125}I-insulin binding at the liver was higher among 4-, 12-, or over-24-month-old CR mice (Table 19.2), indicating either binding affinity or receptor concentration is increased by CR. Although binding levels were higher in young CR mice, maximum binding occurred in young AL or CR mice just prior to dark onset. As animals eat, glucose levels increase and insulin secretion is stimulated. High insulin stimulates glucose uptake and receptor downregulation. Since the receptor upregulates during the light span as glucose levels diminish, maximum insulin response would be expected when receptor concentration or binding affinity is maximum. Regardless of diet, the time of maximum binding was found to occur at later times

Table 19.3. Effect of exogenous insulin on blood glucose of young, middle-aged, and old B6C3F$_1$ male mice on either AL or CR diets at different times of day.

Diet	Age	Group	Blood glucose levels (mg/dl)					
			0200 hours	0600 hours	1000 hours	1400 hours	1800 hours	2200 hours
AL	Y	B	–	120	–	132	–	115
		PI	–	104	–	100	–	91
	M	B	96	99	98	105	113	110
		PI	85	48	100	99	102	79
	O	B	86	91	99	100	111	109
CR	Y	B	72	66	74	66	88	74
		PI	86	61	75	56	81	78
	M	B	71	70	52	76	84	74
		PI	72	62	66	41	107	68
	O	B	69	66	75	76	87	70
		PI	66	68	75	56	93	68

AL, ad libitum; CR, calorically restricted; Y, young (4 months); M, middle-aged (12 months); and O, old (>24 months); B, basal; PI, 20 min postinsulin

(during the dark span after normal eating times) as animals aged. This phenomenon has been referred to as a phase delay.

Glucose levels were highest during the time of transition from light to dark, when animals were eating regardless of age or diet (Table 19.3). However, CR mice had much lower glucose levels at any of the ages investigated. Exogenous insulin was not very effective for stimulating additional glucose uptake in CR mice. The only appreciable glucose reduction due to exogenous insulin occurred at the time of day when glucose (and insulin binding) was normally highest (end of the light span/beginning of the dark span). The time dependence for exogenous stimulation of glucose uptake was not as evident in AL mice that had significantly higher blood glucose levels across the full 24-h span. Additionally, in AL mice, insulin produced greater reductions in blood glucose at those times when it was effective.

Recent biochemical studies concerning the effects of CR on specific metabolic processes have supported the conclusion that basic pathways are differentially affected. The activities of enzymes of glycolysis and lipid metabolism have generally been shown to decline, while those associated with gluconeogenesis generally increase or do not change (Koizumi et al. 1987; Feuers et al. 1989). We have been interested in studying specific mechanism(s) through which these modifications in efficiency occur, since a basic understanding might allow for uncovering a mechanism through which CR acts at higher centers. The data presented here serve to specifically describe a mechanism(s) which allows one pathway to act in a more efficient manner, while benefitting another.

In AL-fed rodents, the affinity of the dephosphorylated PK for its major substrate (PEP) just after eating (when glucose levels are highest) is low relative to that seen in CR rodents. Thus, the substrate concentration must be high in order to be metabolized by PK to stimulate ATP production within the glycolytic pathway. The increase in V_{max} seen in aging AL rodents suggests that in order to maintain sufficient glycolytic activity, additional enzyme is synthesized. It is suggested that the relatively high concentration of low-activity enzyme is an age-associated compromise of metabolic function. Rodents on CR diets can achieve a lower K_m. This indicates that far less substrate must be utilized by the glycolytic pathway to produce the required energy source, ATP. This occurs because PK readily binds PEP even when its concentration within the cell is very low and efficiently converts it to pyruvate and ATP in the presence of ADP. Aging does not seem to compromise this mechanism in CR rodents at the same rate that occurs in AL animals.

During the light span when nocturnal rodents are inactive and glucose levels decrease, PK activity decreases as the enzyme is phosphorylated. In AL animals, the K_m decreases, but not to the same level that was found in CR animals. Thus, glycolytic rates remain higher at these times in AL rodents. The inability of aging AL animals to inactivate glycolytic activity in an efficient manner is compromised. The CR animals have the ability to achieve a much higher K_m. This means that glycolytic activity can be regulated downward to minimal levels. Thus, the extremely low levels of glucose seen in CR rodents during the light span are available for brain and muscle function.

Insulin binding was much higher in CR mice at all ages and all times of day. This is an indication that insulin receptor concentration, binding affinity, or both are increased by CR. Alteration in the insulin receptor may play a key role in the increased efficiency of PK regulation realized using CR. Circulating insulin levels have been shown to be higher under conditions of AL than under CR. The inability of high insulin levels to stimulate insulin-mediated processes may be receptor related, and the decline in receptor response which occurs with aging is offset by CR. The current data suggest that changes in pathways of intermediary metabolism which occur in response to CR are multifaceted, and that altered regulation involves modifications in higher centers such as those involved in endocrine control.

It is interesting that the K_m of the PK for PEP under CR conditions could be decreased to far lower levels by insulin than under AL conditions at all ages, and that this ability was maintained across age. It is possible that the different K_m are an indication of age-dependent expression of separate isozymes. Alternatively, it is possible that other isozymes which are responsible for phosphorylation (protein kinases) or dephosphorylation (phosphatase) of PK are differentially expressed with aging and CR offsets this progression. Additionally, similar changes in expression of other genes such as the insulin or insulin receptor genes may be involved. This would result in a redirection of metabolism under CR diets, and based upon current data this redirection would result in an energy savings by one pathway or system for the benefit of another. These events may play a significant role in the extension of life span and in offsetting the time to generation of degenerative disease through caloric restriction.

References

Andres R, Tobin JD (1977) Endocrine Systems. In: Finch CE, Hayflick L (eds) Handbook of the biology of aging. Van Nostrand Reinhold, New York, pp 357–378

Duffy PH, Feuers RJ, Leaky JEA, Nakamura KD, Turturro A, Hart RW (1989) Effect of chronic caloric restriction on physiological variables related to energy metabolism in male Fischer 344 rat. Mech Ageing Dev 48:117–133

Everitt AV (1980) The neuroendocrine system and aging. Gerontology 26:108–119

Feuers RJ, Casciano DA (1985) Insulin and glucagon's reciprocal mediation of a dual regulation mechanism for pyruvate kinase. Biochem Biophys Res Commun 124:651–659

Feuers RJ, Delongchamp RR, Casciano DA, Burkhart JG, Mohrenweiser HW (1980) Assay for mouse tissue enzymes: levels of activity and statistical variation for 29 enzymes of liver or brain. Anal Biochem 101:123–130

Feuers RJ, Duffy PH, Leakey JEA, Turturro A, Mittelstaedt RA, Hart RW (1989) Effect of chronic caloric restriction on hepatic enzymes of intermediary metabolism in the male Fischer 344 rat. Mech Ageing Dev 48:179–189

Finch CE (1977) Neuroendocrine and autonomic aspects of aging. In: Finch CE, Hayflick L (eds) Handbook of the biology of aging. Van Nostrand Reinhold, New York, pp 262–280

Frolkis VV, Bezrukov N, Bogatskaya LN, Verkhratsii NS, Zamostyan VP, Shevchuk VG, Shchegoleva IV (1970) Catecholamines in the metabolic and function regulation in aging. Gerontologia 16:129–140

Harrison DE, Archer JR (1983) Physiological assays for biological age in mice: relationship of collagen, renal function, and longevity. Exp Aging Res 9:245–251

Heuson JC, Legros N (1972) Influence of insulin deprivation on growth of the 7,12-dimethylbenz[a]anthracene-induced mammary carcinoma in rats subjected to alloxan diabetes and food restriction. Cancer Res 32:226–232

Holehan AM, Merry BJ (1986) The experimental manipulation of ageing by diet. Biol Rev 61:329–368

Hunter JD, Feuers RJ, Tsai T, Gronert K, Scheving LE (1990) Circadian variations of radioinsulin-receptor binding in calorically restricted mice and relationships to chronotherapy of human insulin dependent diabetes. Ann Rev Chronopharm (in press) 1991

Johnson BC, Gaijar A, Kubo C, Good RA (1986) Calories versus protein in onset of renal disease in NZB × NZW mice. Proc Natl Acad Sci USA 83:5659–5662

Kmiec Z, Mysliwski A (1983) Age-dependent changes of hormone-stimulated gluconeogenesis in isolated rat hepatocytes. Exp Gerontol 18:173–184

Koizumi A, Weindruch R, Walford RC (1987) Influences of dietary restriction and age on liver enzyme activities and lipid peroxidation in mice. J Nutr 117:361–367

Kritchevsky D, Weber MM, Klurfeld DM (1984) Dietary fat versus caloric content in initiation and promotion of 7,12-dimethylbenz[a]anthracene-induced mammary tumorigenesis in rats. Cancer Res 44:3174–3177

Loyd T (1984) Food restriction increases life span of hypertensive animals. Life Sci 34:401–407

Masoro EJ (1985) Nutrition and ageing – a current assessment. J Nutr 115:842–848

Masoro EJ (1988) Food restriction in rodents: an evaluation of its role in the study of aging. J Gerontol 45:B59–64

McCay CM, Crowell MF, Maynard LA (1935) Effect of retarded growth upon the length of life span and upon the ultimate body size. J Nutr 10:63–79

Oldham JW, Casciano DA, Farr JA (1979) The isolation and primary culture of viable nonproliferating rat hepatocytes. TCA Manual 5:1047–1050

Pollard M, Luckert PH (1985) Tumorigenic effects of direct- and indirect-acting chemical carcinogens in rats on a restricted diet. J Natl Cancer Inst 74:1347–1349

Pollard M, Luckert PH, Pan GY (1984) Inhibition of intestinal tumorigenesis in methylazoxymethanol-treated rats by dietary restriction. Cancer Treat Rev 68:405–408

Ruggeri BA, Klurfeld DM, Kritchevsky D (1987) Biochemical alterations in 7,12-dimethylbenz[a]anthracene-induced mammary tumors from rats subjected to caloric restriction. Biochem Biophys Acta 929:239–246

Sacher GA (1977) Life table modification and life prolongation. In Finch CE, Hayflick L (eds) Handbook of the biology of aging. Van Nostrand Reinhold, New York, pp 582–638

Weindruch R, Walford RL (1988) The retardation of aging and disease by dietary restriction. Thomas, Springfield

Walford RL (1983) Maximum Life Span. Norton, New York

Walford RL, Liu RK, Gerbase-DeLima M, Mathies M, Smith GS (1974) Longterm dietary restriction and immune function in mice: response to sheep red blood cells and to mitogenic agents. Mech Ageing Dev 2:447–454

CHAPTER 20

Effects of Long-Term Caloric Restriction on Hepatic Drug-Metabolizing Enzyme Activities in the Fischer 344 Rat

J.E.A. Leakey[1,4], J.J. Bazare[1], J.R. Harmon[1], R.J. Feuers[2], P.H. Duffy[3], and R.W. Hart[3,4]

Introduction

It is well established that caloric restriction (CR) in rodents can result in an increased maximally achievable life span and a decreased incidence and proliferative rate of neoplasia (Maeda et al. 1985; Masoro 1985, 1988). Although the precise molecular mechanisms underlying these effects have not been satisfactorily elucidated, CR has been associated with a wide range of biochemical and endocrinological alterations that occur in a spectrum of tissues (Weindruch and Walford 1988).

One such biochemical system that is profoundly affected by CR is the group of liver enzymes which are collectively known as drug-metabolizing enzymes (DME). These enzymes are primarily responsible for the activation and detoxication of carcinogens and mutagens, the modulation of toxicity of many other xenobiotics, the biotransformation and metabolic clearance of many endogenous compounds, and the microsomal production of oxygen radicals. Thus DME may influence at least some of the effects of CR upon longevity.

This paper describes the hepatic DME in detail and discusses recent information from the authors' laboratory concerning the effects of CR on DME activities.

Multiple DME Isozymes

In most species DME exist as families of isozymes which exhibit differential regulation by both xenobiotic inducers and hormonal factors (Jakoby 1980; Leakey 1983). The major enzyme families are: the cytochrome P-450 dependent monooxygenases, the uridine diphosphate-(UDP)glucuronyltransferases, the glutathione-S-transferases, the epoxide hydrolases, and the sulfotransferases.

[1]Division of Reproductive and Developmental Toxicology, [2]Division of Genetic Toxicology, and [3]Office of Director, National Center for Toxicological Research, Jefferson, Arkansas 72079, USA
[4]Department of Interdisciplinary Toxicology, University of Arkansas for Medical Sciences, Little Rock, AR 72205, USA

The hepatic cytochrome P-450 dependent monooxygenase system catalyzes the primary route of oxidative metabolism for a wide range of drugs, carcinogens, dietary xenobiotics, and endogenous compounds. Such oxidation produces metabolites exhibiting either increased or decreased toxicity compared to the parent compound (Jakoby 1980; Jefcoate 1983), and may also result in the generation of potentially toxic hydroxyl and superoxide radicals (Ingelman-Sundberg and Hagbjork 1982).

At least 18 cytochrome P-450 isozymes are expressed in rat liver (Gonzalez 1989). These isozymes exhibit overlapping substrate specificities towards many substrates, but several monooxygenase reactions are sufficiently specific for individual isozymes that they may be utilized to infer the presence of those isozymes (Leakey et al. 1989a). Several isozymes are normally present in rat liver as minor constitutive forms, but are readily inducible by xenobiotics. Other isozymes are major constitutive forms in uninduced rat liver. The latter include: cytochromes P-450 IIC11, IIC13, and IIIA2, which are predominant forms in mature male rat liver and cytochromes P-450IIC12 and IIA1, which are predominant forms in mature female rat liver (Gonzalez 1989).

At least eight UDP-glucuronyltransferase isozymes have been identified. These tend to show greater substrate specificity toward endogenous substrates than toward xenobiotics (Burchell and Coughtrie 1989) and are all present as constitutive forms in uninduced rat liver. However, some are also inducible by xenobiotics or hormones (Burchell and Coughtrie 1989).

Rat liver glutathione-S-transferases are a family of dimeric proteins composed of two of at least eight different subunits, such that both homodimers and heterodimers exist (Mannervik and Danielson 1988). These isozymes have broad and overlapping substrate specificities, although activities for selected substrates are higher with certain isozymes. The predominant forms existing in adult rat liver are isozymes, 1:1, 1:2, 2:2, 3:3, 3:4, 4:4, and 8:8 (Mannervik and Danielson 1988).

The hepatic sulfotransferases are also known to exist in multiple forms in the rat and include four arylsulfotransferase isozymes (Sekura et al. 1981) and three steroid sulfotransferase isozymes (Singer 1984).

In many cases several DME isozymes may compete to direct an individual substrate towards different end products. In such cases the metabolic fate of the compound will be influenced by both the K_m and relative concentration of each competing isozyme, even though the compound may never reach saturating concentrations. Small changes in isozyme concentration ratios may therefore cause larger effects in overall hepatic metabolism.

Regulation of Hepatic DME During Aging

The efficiency of drug metabolism and elimination is well known to decrease as a function of old age (Birnbaum 1987; Kitani 1988), and as a consequence many drugs are administered to elderly patients at reduced doses (Schmucker 1979;

Table 20.1. Isozymes exhibiting sexually dimorphic expression in rat liver.

Isozyme	Degree of dimorphic expression	
Predominantly male	(Male)	(Female)
Cytochrome P-450IIA2	100	<5
Cytochrome P-450IIC11	100	<2
Cytochrome P-450IIC13	100	<2
Cytochrome P-450IIIA2	100	<2
UDPGT androgen 17-OH form	100	65–75
GSHT 3:3	100	60–80
Arylsulfotransferase 1 and 2	100	50–70
Predominantly female	(Female)	(Male)
Cytochrome P-450IIA1	100	<2
Cytochrome P-450IIC12	100	<2
UDPGT estrogen 3-OH form	100	60–80
UDPGT bilirubin form	100	60–80
Androgen 5α-reductase	100	4–10
Steroid sulfotransferase I and II	100	<3

Hepatic DME concentrations are expressed as percentages of the values in the predominant sex.
UDPGT, uridine diphosphate-glucuronyltransferase; GSHT, glutathione-S-transferase

Crooks et al. 1983). In the rat, several hepatic DME activities have been shown to decrease with senescence (Kamataki et al. 1985; Galinski et al. 1986; Birnbaum 1987), and these decreases are often associated with a loss of sex-specific hepatic function (Kamataki et al. 1985).

Adult expression of many DME isozymes is sexually dimorphic in rats (Table 20.1; Mulder 1986; Skett 1988), and in several cases such expression has been shown to be regulated by serum growth hormone (Zaphiropouos et al. 1989). The male rat growth hormone secretory pattern is characterized by a pulsatile serum profile, having relatively high pulse amplitudes and low interpulse concentrations, whereas the female serum growth hormone profile exhibits greater pulse frequency, lower pulse amplitude, and interpulse concentrations that are significantly higher than those of males (Eden 1979; Zaphiropouos et al. 1989). It has been shown experimentally that expression of cytochromes P-450IIC11 and IIC13 are directed in male rats by the male growth hormone serum profile, whereas expression of cytochrome P-450IIC12, steroid sulfotransferases (I and II) and testosterone 5α-reductase activity are directed by the female growth hormone serum profile in females (Singer 1984; Zaphiropouos et al. 1989; Gonzalez 1989). Other DME isozymes exhibiting sexual dimorphism (Table 20.1) may also be regulated in part by growth hormone, but other hormones, such as glucocorticoids, thyroid hormones, sex steroids, and insulin may also directly or indirectly influence expression of DME isozymes (Kato 1977; Leakey 1983; Singer 1984; Burchell and Coughtrie 1989; Zaphiropouos et al. 1989).

Sexually dimorphic pituitary growth hormone secretion develops around puberty (Eden 1979) and its pulsatility decays in senescence (Sonntag et al. 1980). Many of the changes in DME expression that occur during senescence can be correlated with changes in growth hormone secretion. For example, expression of

Table 20.2. Effect of aging on microsomal superoxide production in rat liver.

Activity	60 Days ($n = 3$)	22 Months ($n = 4$)
Microsomal cytochrome P-450 (nmol/mg protein)	0.68 ± 0.06	0.52* ± 0.01
Microsomal cytochrome P-450 reductase (nmol min^{-1} mg^{-1} protein)	177.3 ± 20.8	176.6 ± 15.9
Acetyl-cytochrome C reduced per milligram microsomal protein	6.9 ± 3.3	12.3* ± 0.4
Acetyl-cytochrome C reduced per nanomole cytochrome P-450	10.1 ± 4.9	23.6* ± 0.8

Activities (expressed as means ± SEM) are those of male Fischer 344 rats. Acetyl-cytochrome C was prepared by the method of O'Brien (1984) and purified on an *FPLC MonoQ* column (Pharmacia Inc). Reduction of acetyl-cytochrome C refers to the NADPH- and superoxide-dependent microsomal reduction that is inhibited by superoxide dismutase.
*Significantly different from 60-day-old controls on Student's t-test ($p < 0.02$)

cytochromes P-450IIC11 and IIIA2 decreases in aging male rats, but expression of cytochrome P-450IIC12 and testosterone-5α-reductase increases (Kamataki et al. 1985). Thus, the senescent male rat liver appears to be feminized, whereas changes in DME expression are less apparent in aging female rat liver.

Other changes that are not directly dependent upon growth hormone secretion may also occur in aging liver. For instance, we have found that the ability of the microsomal cytochrome P-450 system to generate superoxide greatly increases with age (Table 20.2). Although this is possibly due to age-dependent changes in isozyme composition, reduced metabolic coupling efficiency due to age-related damage of microsomal membrane and protein components may also contribute to this effect.

Long-Term Effects of CR on Hepatic DME

Early work on the effects of CR on hepatic DME activities revealed changes in both monooxygenase activity and in the inducibility of these activities by xenobiotics (Sachan and Das 1982; Weindruch and Walford 1988). Our studies have investigated the effects of long-term CR on a spectrum of hepatic DME activities in aged male Fischer 344 rats (Leakey et al. 1989a, 1989b). As summarized in Table 20.3, the effects of CR under these conditions were isozyme specific and, although several (e.g., groups B and D) appeared to be related to a retardation of the age-dependent hepatic feminization, several isozymes (e.g., groups C and F) were increased by CR but also were either increased or not affected by aging. This suggests that the effects of CR on hepatic DME are more complex than a simple retardation of the aging process.

Further studies have investigated the effects of CR in 9-month-old male rats (Leakey et al., manuscript in preparation). Paradoxically CR appeared to feminize hepatic DME activities at this age (Fig. 20.1). Cytochrome P-450IIC11-dependent

Table 20.3. Effects of long-term CR on hepatic DME activities in aging male rats.

Group aging effect	CR effect	DME isozymes responding
A Decrease	No effect	Epoxide hydrolase, arylsulftransferase (3 and 4) Glutathione-*S*-transferase (1:1)
B Decrease	Increase	Testosterone-16α-hydroxylase (P-450IIC11), Lauric acid hydroxylase (P-450IVA1), N-acetyltransferase, glutathione-*S*-transferase (3.3)
C No effect	Increase	Testosterone-6β-hydroxylase (P-450IIIA2), 4-nitrophenol hydroxylase (P-450IIE1), UDP-glucuronyltransferase (bilirubin form)
D Increase	Decrease	Testosterone-5α-reductase, estrone sulfotransferase
E Increase	No effect	Corticosterone sulfotransferase
F Increase	Increase	Testosterone-7α-hydroxylase (P-450IIA1), γ-glutamyltranspeptidase

Hepatic DME activities which showed significant differences either between young (60 days) and old (22 months) ad libitum fed male rats or between old (22 months) 40% restricted and old ad libitum fed male rats. Data taken from Leakey et al (1989a,b)
CR, caloric restriction; DME, drug metabolizing enzyme; UDP, uridine diphosphate

testosterone-16α-hydroxylase activity was decreased in the restricted rats, whereas testosterone-5α-reductase and corticosterone sulfotransferase activities were increased. Cytochrome P-450IIE1-dependent 4-nitrophenol hydroxylase activity (which is not regulated by growth hormone) was increased by CR at both ages.

Cytochrome P-450IIC11 is a major DME isozyme responsible for the metabolic activation of the hepatocarcinogen aflatoxin B_1 (Shimada et al. 1987). Our data would suggest that CR in 9-month-old-male rats would reduce the risk of hepatocarcinoma due to aflatoxin B_1 exposure. This is in agreement with the work of Pegram et al. (1989) who demonstrated that the rates of both in vitro and in vivo conversion of aflatoxin B_1 into DNA-reactive species are reduced by CR in male rats.

Circadian Effects of CR on Hepatic DME

We have previously shown that CR in rats causes large alterations in circadian profiles of motor activity, body temperature, and respiratory quotient (Duffy et al. 1989). We have also investigated the effects of CR on hepatic DME activities over circadian time points. Although many activities (including testosterone-16α-hydroxylase) did not exhibit circadian variation in either control or restricted rats, CR did alter the circadian rhythms of several enzymes. Furthermore, CR created circadian rhythms in several isozymes which do not normally exhibit circadian variation (Fig. 20.2), and in the case of 4-nitrophenol hydroxylase activity, CR significantly increased the activity only at certain circadian time points.

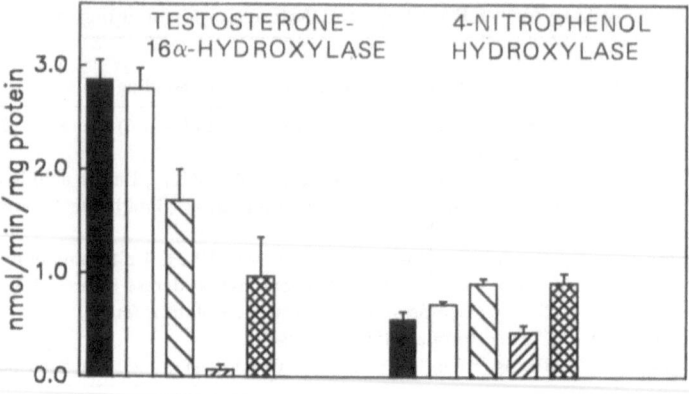

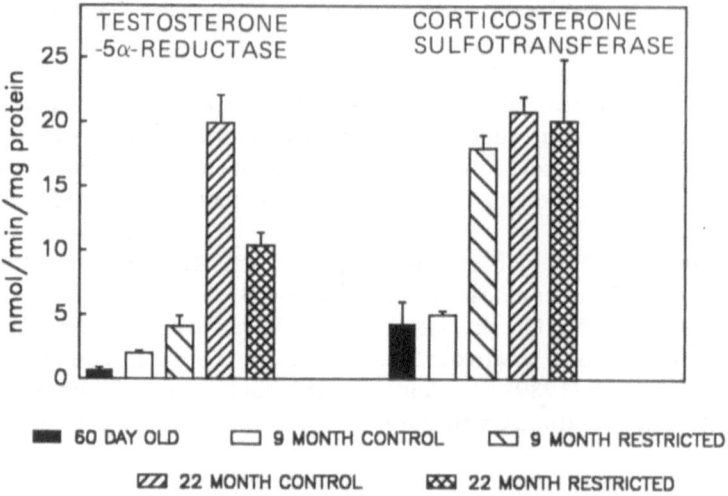

Figure 20.1. Effects of aging and CR on hepatic DME activities in male rats. Male Fischer 344 rats were maintained on a light-dark cycle with lights on from 600 to 1800 hours as described previously (Leakey et al. 1989a). All rats received a standard NIH-31 diet at 1100 hours CST daily and were divided into a control group that was fed ad libitum and a restricted group which received 60% of the amount consumed by the control group, starting at 14 weeks of age. The 22-month-old rats were phase shifted by 12 h 30 days before killing so that the time of feeding of the restricted animals was synchronized to the time within the dark phase when the control rats were most active in feeding (Duffy et al. 1989). The animals that were used for the 9-month time point were not phase shifted, and circadian peak activities (see Fig. 20.2) are displayed in the histogram. The animals were killed and assayed for hepatic DME activities as described by Leakey et al. (1989a, b).

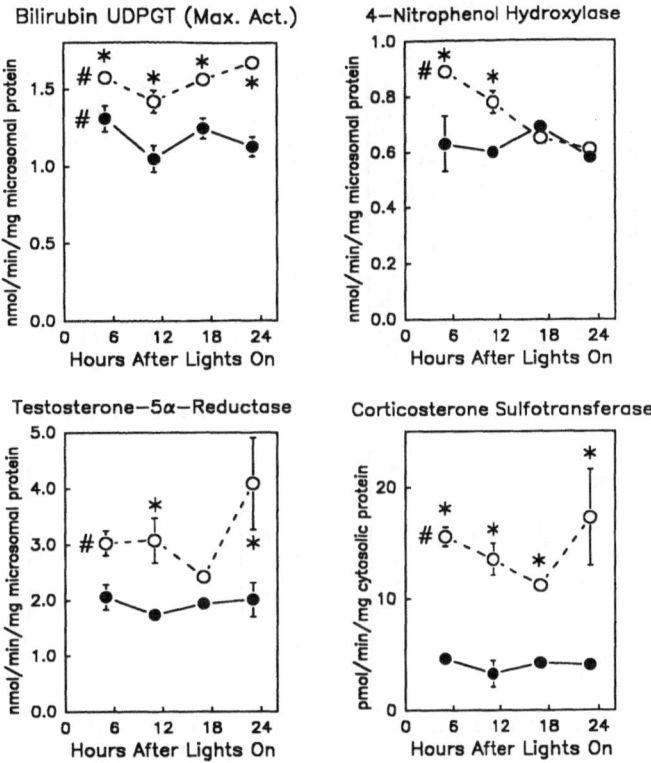

Figure 20.2. Circadian effects of CR on DME activities in 9-month-old male rats. Ad libitum fed control (*solid circles*) and restricted (*open circles*) 9-month-old rats were killed in groups at 6-h intervals beginning 5 h after lights on and assayed for hepatic DME activities. Each experimental group consisted of four to five animals, and activities are given as means $\pm$ SEM. *, Significant differences ($P < 0.05$) between control and restricted activities; #, significant circadian variation when assayed on a SAS GLM program (SAS Institute, Cary, NC)

Such effects may be related to the fact that not only does feeding occur immediately following food allocation in restricted rats, but it is also more rapid, with the total food allocation being consumed over a relatively short time period (Duffy et al. 1989). It follows therefore that circadian factors must be considered when investigating any effects of CR on hepatic enzyme activities in case circadian variations have been created in the activities.

Endocrine Basis for Effects of CR on Hepatic DME

The circadian effects of CR on serum profiles of blood glucose and serum insulin, L-thyroxine, and corticosterone were also investigated (Fig. 20.3). In 9-month-old, day-fed male rats, CR suppressed the night-time peak in serum insulin that

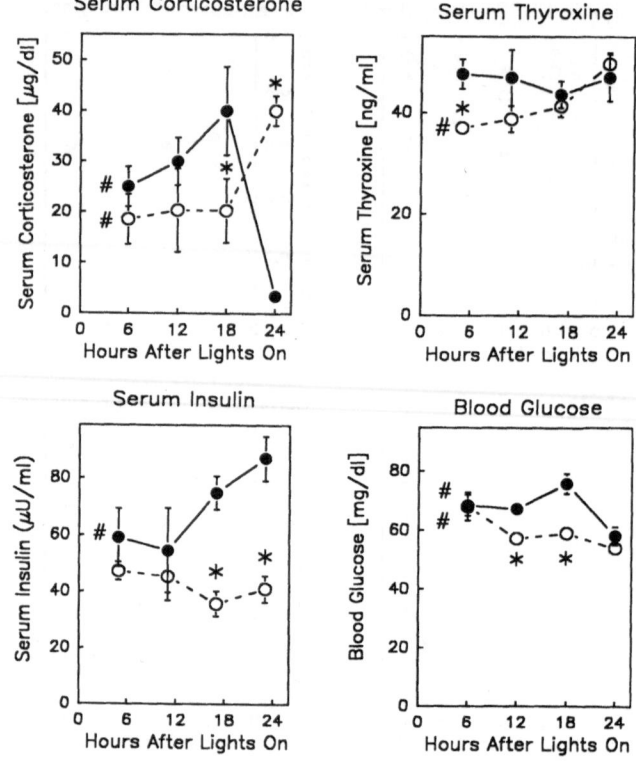

Figure 20.3. Circadian effects of CR on serum hormone profiles in male rats. Trunk blood was collected at killing from the animals used for Fig. 20.2 and assayed for serum hormone concentrations by radioimmunoassay. Blood glucose concentrations were determined using a Lifescan One Touch glucose meter (Johnson and Johnson Co., Mountain View, CA). Symbol notation is the same as for Fig. 20.2.

was observed in the control rats, produced a day-time peak in serum corticosterone, and created a small but significant circadian variation in serum L-thyroxine.

Increased serum corticosterone concentrations may contribute to the altered circadian profiles of hepatic steroid sulfotransferase activity since these isozymes are corticosteroid dependent in rats (Singer 1984). Moreover, increased serum corticosterone may also be responsible for the hepatic feminization effects of CR observed in male rats, since glucocorticoids are known to suppress pituitary growth hormone secretion (Ceda et al. 1987). In addition, decreased insulin concentrations and possibly a decreased insulin/glucagon ratio (Feuers et al., this volume) may contribute to the observed increases in 4-nitrophenol hydroxylase and bilirubin UDP-glucuronyltransferase activities observed in the restricted rats since both these activities are stimulated by low insulin/glucagon ratios (Dong et al. 1988; Leakey et al. 1989a, b).

General Conclusions

CR appears to evoke changes in hepatic DME isozyme expression that may be dependent on age and/or circadian rhythm. These effects are associated, in male rats, with a general early partial feminization of hepatic function followed by a delay in the natural age-dependent feminization which occurs spontaneously during senescence. However, these effects are isozyme specific. Many DME activities remained unchanged by either aging or CR under the conditions of our investigation, while others were increased by CR independently of the hepatic feminization process.

Such changes in hepatic drug metabolizing isozyme profiles may result in altered rates of carcinogen activation and detoxication. They may also alter both the rate of superoxide generation in the endoplasmic reticulum and the transformation and elimination of endogenous metabolites. These alterations may, in turn, influence the aging process in the liver and other tissues.

References

Birnbaum LS 1987) Age-related changes in carcinogen metabolism. J Am Geriatr Soc 35:51–60

Burchell B, Coughtrie WH (1989) UDP-glucuronyltransferases. Pharmacol Ther 43:261–289

Ceda GP, Davis RG, Hoffman AR (1987) Glucocorticoid modulation of growth hormone secretion in vitro. Evidence for a biphasic effect of GH-releasing hormone mediated release. Acta Endocrinol 114:465–469

Crooks J, O'Malley K, Stevenson IH (1983) Pharmacokinetics in the elderly. Clin Pharmacokinet 1:280–296

Dong Z, Hong J, Ma Q, Li D, Bullock J, Gonzales FJ, Geloin HV, Yang CS (1988) Mechanism of induction of cytochrome P-450$_{ac}$ (P-450$_j$) in chemically induced and spontaneously diabetic rats. Arch Biochem Biophys 263:29–35

Duffy PH, Feuers RJ, Leakey JEA, Nakamura KD, Turturro A, Hart RW (1989) Effect of chronic caloric restriction on the physiological variables related to energy metabolism in the male Fischer 344 rat. Mech Ageing Dev 48:117–133

Eden S (1979) Age- and sex-related differences in episodic growth hormone secretion in the rat. Endocrinology 105:555–560

Galinsky RE, Kane RE, Franklin MR (1986) Effect of aging on drug-metabolizing enzymes important in acetaminophen elimination. J Pharmocol Exp Ther 237:107–113

Gonzales FJ (1989) Molecular biology of cytochrome P-450s. Pharmacol Rev 40:244–288

Ingelman-Sundberg M, Hagbjork A-L (1982) On the significance of the cytochrome P-450-dependant hydroxyl radical-mediated oxygenation mechanism. Xenobiotica 12:673–686

Jakoby WB (1980) Enzymic basis of detoxication, vols 1, 2. Academic, New York

Jefcoate CR (1983) Integration of xenobiotic metabolism in carcinogen activation and detoxication. In: Caldwell J, Jakoby WB (eds) Biological basis of detoxication. Academic, New York, pp 32–76

Kamataki T, Maeda K, Shimada M, Kitani K, Nagai T, Kato R (1985) Age-related alteration in the activities of drug-metabolizing enzymes and contents of sex-specific P-450 in liver microsomes from male and female rats. J Pharmacol Exp Ther 233:222–228

Kato R (1977) Drug metabolism under pathological and abnormal physiological states in animals and man. Xenobiotica 7:25–92

Kitani K (1988) Neurohumoral control of liver functions during aging. Gerontology 34:55–63

Leakey JEA (1983) Ontogenesis. In: Caldwell J, Jakoby WB (eds) Biological basis of detoxication. Academic, New York, pp 77–104

Leakey JEA, Cunny HC, Bazare J Jr, Webb PJ, Feuers RJ, Duffy PH, Hart RW (1989a) Effects of aging and caloric restriction on hepatic drug metabolizing enzymes in the Fischer 344 rat. I. The cytochrome P-450 dependent monooxygenase system. Mech Ageing Dev 48:145–155

Leakey JEA, Cunny HC, Bazare J Jr, Webb PJ, Lipscomb JC, Slikker W Jr, Feuers RJ, Duffy PH, Hart RW (1989b) Effects of aging and caloric restriction on hepatic drug metabolizing enzymes in the Fischer 344 rat. II. Effects on conjugating enzymes. Mech Ageing Dev 48:157–166

Maeda H, Gleister CA, Masoro EJ, Murata I, McMahan CA, Yu BP (1985) Nutritional influences on aging of Fischer 344 rats: II Pathology. J Gerontol 40:671–688

Mannervik B, Danielson UH (1988) Glutathione transferases – structure and catalytic activity. CRC Crit Rev Biochem 23:283–337

Masoro EJ (1985) Nutrition and aging – a current assessment. J Nutr 115:842–848

Masoro EJ (1988) Food restriction in rodents: an evaluation of its role in the study of aging. J Gerontol 43:B59–64

Mulder GJ (1986) Sex differences in drug conjugation and their consequences for drug toxicity. Chem Biol Interact 57:1–15

O'Brien PJ (1984) Superoxide production. Methods Enzymol 105:370–378

Pegram RA, Allaben WT, Chou MW (1989) Effect of caloric restriction on aflatoxin B_1-DNA adduct formation and associated factors in Fischer 344 rats. Mech Ageing Dev 48:167–177

Sachan DS, Das SK (1982) Alterations of NADPH-generating and drug metabolizing enzymes by feed restriction in male rats. J Nutr 112:2301–2306

Schmucker DL (1979) Age-related changes in drug disposition. Pharmacol Rev 30:445–456

Sekura RD, Duffel MW, Jakoby WB (1981) Aryl sulfotransferases. Methods Enzymol 77:197–206

Shimada T, Nakamura S, Imaoka S, Funae Y (1987) Genotoxic and mutagenic activation of aflatoxin B_1 by constitutive forms of cytochrome P-450 in rat liver microsomes. Toxicol Appl Pharmacol 91:13–21

Singer SS (1984) Glucocorticoid sulfotransferases in rats and other animal species. Biochem Soc Trans 12:35–38

Skett P (1988) Biochemical basis of sex differences in drug metabolism. Pharmacol Rev 38:269–304

Sonntag WE, Steger RW, Forman LJ, Meites J (1980) Decreased pulsatile release of growth hormone in old male rats. Endocrinology 107:1875–1879

Weindruch R, Walford RL (1988) Retardation of aging and disease by dietary restriction. Thomas, Springfield

Zaphiropouos PG, Mode A, Norstedt G, Gustafsson J-A (1989) Regulation of sexual differentiation in drug and steroid metabolism. Trends Pharmacol Sci 10:149–153

CHAPTER 21

Urinary Biomarkers of Oxidative DNA-Base Damage and Human Caloric Intake*

M.G. Simic[1] and D.S. Bergtold[1]

Introduction

Life span extension is a fascinating goal that has captured human imagination throughout history, prompting much speculation and numerous investigations. Since the early experiments of McCay and coworkers (1935, 1943) with rats, the mechanisms of dietary restriction (DR) and its effects on extension of maximum life span (MLS) and reduction of degenerative diseases has been of great interest (Weindruch and Walford 1988). More recently, reducing the incidence of diseases such as cancer and cardiovascular disorders through nutritional interventions (Ames 1983) has become an economic issue and a driving force in investigations of DR. In contrast to gross observations, however, the progress in mechanistic understanding has been quite modest. Although research on free radical and oxidative processes in biology and medicine (Simic et al. 1988) appears to have growing relevance to our understanding of DR, there has been relatively little interaction between the two areas of study. Nevertheless, novel ideas bridging these fields have been introduced by Harman (1981).

Products of free radical damage to DNA have been known since the 1950s from extensive investigation of the radiation chemistry of model aqueous systems (Scholes 1983; von Sonntag 1987) and also to a limited extent from radiobiological studies of humans and laboratory animals (Pizzarello 1985). One focus of attention has been the reaction products of radiation-generated ·OH radicals with DNA bases. Thymine glycol (Tg) and 5-hydroxymethyluracil (5-hmU) were found to be the major reaction products of ·OH with thymine, whereas 8-hydroxyguanine (8-G-OH) was shown to be the major product of guanine.

It was somewhat surprising, but not completely unexpected, to find Tg and its moiety thymidine glycol (dRTg) as well as 5-hmU in the urine of rat and humans in the absence of radiation treatment (Cathcart et al. 1984). Using mechanisms of radiation chemistry (von Sonntag 1987), the presence of these DNA-base products in the urine was attributed to a complex sequence of reactions, starting

---

*Partial support of this research was provided by the ILSI Risk Science Institute.
[1]National Institute of Standards and Technology, Gaithersburg, MD 20899, USA

with metabolic generation of ·OH radicals, their reaction with cellular DNA, excision of products by repair enzymes, and ultimately ending with the excretion of excised DNA products.[1] Subsequently, 8-G-OH products were also found in the urine (Cundy et al. 1988; Bergtold et al. 1988), as would have been expected if ·OH radicals were the damaging intermediates.

Despite the highly critical reception of metabolic origins of ·OH radicals and their DNA-base products in vivo, the absence of direct proof, and certain confounding factors (for example, 5-hmU is also generated by enzymatic hydroxylation of the thymine methyl group), the presence of hydroxylated DNA base products and their moieties in the urine was a significant discovery, with great potential as a possible measure of integral DNA damage.

New support for the attractive hypotheses that ·OH radicals are generated metabolically and that urinary DNA-base products reflect the extent of DNA damage came from radiation experiments in vivo (Bergtold and Simic 1989; Simic et al. 1989). Both humans and mice produced the same quantity of urinary products for the same number of radiation-generated ·OH radicals (i.e., per normalized radiation energy). These findings indicate that urinary DNA base products may in fact reflect the extent of DNA damage. It also became obvious that the measurements of urinary biomarkers of oxidative DNA-base damage (UBODBD) may be useful in developing short-term in vivo tests of the genotoxicity of foods, food components, diets, and a variety of other agents.

This chapter describes the mechanisms as well as the first observations of the effect of dietary caloric restriction (energy intake) on the reduction of DNA damage in humans, as measured by noninvasive measurements of UBODBD.

Hydroxylated DNA-Base Products as Biomarkers: Background

The use of hydroxylated DNA-base products as biomarkers of DNA damage in vivo is founded on the following principles and radiation-induced processes (Simic and Bergtold 1990; Bergtold et al. 1990; Bergtold and Simic 1989; Simic et al. 1989). Ionizing radiations (e.g., X- and γ-rays) generate ·OH radicals by splitting water, either in aqueous model systems or in organisms. Because the water content is high, the yield of ·OH is considerable in soft tissues. The ·OH radicals generated in the vicinity of DNA react with either DNA bases (major reaction) or deoxyribose (minor reaction). For example, ·OH adds to thymine and guanine at diffusion-controlled rates, $k \approx 10^{10}$ M^{-1} s^{-1} (Buxton et al. 1988), to give free radicals in the first step, which subsequently react with oxygen in a few steps to give stable products. Some of the major reactions involving thymine (T) and guanine (G) are shown in reactions 1 and 2:

---

[1]Low levels of cosmic and background radiation cannot account for even a minute fraction of the levels of DNA base products measured.

$$T \xrightarrow{\cdot OH} \cdot T\text{-}OH \xrightarrow{O_2} \cdot OOT\text{-}OH \to \to T(OH)_2 + P \tag{1a}$$

$$T \xrightarrow{\cdot OH} 5\text{-}U\text{-}CH_2\cdot \xrightarrow{O_2} 5\text{-}U\text{-}CH_2OO\cdot \to \to 5\text{-}U\text{-}CH_2OH \tag{1b}$$

where thymine glycol levels ($T(OH)_2$ or Tg) are about six times higher than 5-hmU, as shown by pulse radiolysis and product analysis (Simic and Jovanovic 1986; Jovanovic and Simic 1986). Small amounts of other products (ΣP) are also formed. Similarly,

$$G \xrightarrow{\cdot OH} \Sigma \cdot G\text{-}OH \xrightarrow{O_2} \to 8\text{-}G\text{-}OH + \Sigma P \tag{2}$$

where only one quarter of $\cdot OH$ radicals gives 8-G-OH, as shown by product analysis (Simic and Jovanovic 1986; Jovanovic and Simic 1989).

Hydroxylated DNA-base products, as well as all other base products, are excised by repair enzymes whose function is to restore the original DNA. To a lesser extent, however, enzymatic repair also generates slightly altered DNA that results from faulty repair (Friedberg 1985). The process is shown in Fig. 21.1. In case of extensive (e.g., lethal) damage, total DNA is eventually lysed:

$$DNA\text{-}P \xrightarrow[lysis]{Ez} \Sigma P + NS \tag{3}$$

Certain compounds, such as antioxidants, may intervene via chemical repair at the free radical level and reduce the DNA damage (Simic 1989). This kind of chemical intervention may either fully restitute the original DNA or may produce a product, DNA-P', which is more amenable to faultless enzymatic repair.

In this investigation dRTg and 8-dRG-OH were selected as biomarkers of oxidative DNA damage since they show a direct correlation with radiation exposure in vivo (Simic et al. 1989). To appear in the urine, these products must first enter the blood and then be removed by the kidneys. The fact that these products have been found in the urine of irradiated humans (Bergtold and Simic 1989; Simic et al. 1989) indicates that these products are either partially metabolized or because of the linear yield with the energy input (Bergtold et al., in preparation) are most likely not metabolized at all.

Measurements of Biomarkers[2]

The general approach for measuring urinary oxidative DNA-base damage biomarkers has been described (Cathcart et al. 1984; Cundy et al. 1988; Ames 1989; Bergtold et al. 1988) since such measurements were first introduced by

---

[2]Certain commercial equipment, instruments, and materials are identified in this work in order to specify adequately the experimental procedure. Such identification does not imply recommendation or endorsement by the National Institute of Standards and Technology, or does it imply that the material or equipment identified is necessarily the best available for these purposes.

$$\text{DNA} \xrightarrow{\cdot OH} \text{DNA-R}\cdot \rightarrow \text{DNA-P} \xrightarrow{Ez} \text{DNA (or DNA')} + \Sigma P + NS$$

$$\begin{array}{c} \text{Chemical} \\ \text{Repair} \end{array} \quad \text{DNA-P'} \xrightarrow{Ez} \text{DNA (or DNA'')} + \Sigma P + NS$$

Figure 21.1. Chemical and enzymatic repair of damaged DNA by ·OH radicals. Excision enzymes and glycosylases (*Ez*) eliminate products (ΣP) and parts of DNA, which appear as nuclear subcomponents (*NS*, DNA bases, nucleosides, nucleotides, and oligonucleotides)

Ames (Cathcart et al. 1984). In Ames's studies the background levels of DNA-base products were measured by high-performance liquid chromatography (HPLC) coupled with either UV or electrochemical detection. Since the levels of the compounds measured were low and the samples were complex, gas chromatography/mass spectrometry/selected ion monitoring (GC/MS/SIM) techniques for the measurement of DNA-base products (Dizdaroglu and Bergtold 1986) were introduced for measuring these products in urine (Bergtold et al. 1988) because they are more sensitive and capable of reliable detection of marker compounds in biological matrices. Multiple ions, characteristic of each specific compound, were monitored at the appropriate retention times to verify the identities of the compounds. The best quantitative measurements were made when internal standards labeled with stable isotopes were used for quantifying DNA products.

Generation of Free Radicals In Vivo

Direct proof of free radical processes in vivo is extremely difficult because these highly reactive species are short lived (Simic et al. 1988). For example, superoxide radical, $\cdot O_2^-$, has been found to be generated by xanthine (or hypoxanthine)/xanthine oxidase in cells and biosystems (McCord and Fridovich 1968), by NADPH oxidase in polymorphonuclear leukocytes (PMN) (Babior 1988; Klebanoff and Clark 1985), and by mitochondria (Forman and Boveris 1982; Nohl and Jordan 1986; Richter et al. 1988).

In the presence of superoxide dismutase (SOD), H_2O_2 is generated (McCord and Fridovich 1968):

$$2\cdot O_2^- + 2H^+ \xrightarrow{SOD} H_2O_2 + O_2 \tag{4}$$

Iron (II) and some complexes of Fe(II) decompose hydrogen peroxide with concomitant generation of ·OH radicals (Haber and Weiss 1932):

$$H_2O_2 + Fe(II) \rightarrow \cdot OH + OH^- + Fe(III) \tag{5}$$

For example, formation of ·OH radicals in muscle tissue by organic solvents (Karam and Simic 1989) and alcohol (Simic et al. 1989) has been demonstrated

Table 21.1. Specific daily yield (nmol/kg per day) of dRTg and 8-dRG-OH in human urine for high-calorie (2100 ± 200 kcal/day) and low-calorie (1100 ± 100 kcal/day) diets.

Urinary biomarkers	High-calorie intake	Low-calorie intake
dRTg	0.25 ± 0.05	0.11 ± 0.01
8-dRG-OH	0.34 ± 0.06	0.11 ± 0.03
Body weight (kg)	82.0 ± 0.5	78.5 ± 0.5

Measurements were taken after 10 days of constant calorie intake and are averages of four high/low cycles over a period of 1 year.
dRTg, thymidine glycol; 8-dRG-OH, 8-hydroxydeoxyguanosine

via formation of o-Tyr, a biomarker of ·OH radical reaction with phenylalanine (Karam and Simic 1988).

Caloric Restriction and Urinary Biomarkers

Reduction of caloric intake of the same type of diet has been shown to result in reduced levels of UBODBD (Simic and Bergtold 1990). The levels of dRTg and 8-dRG-OH in human urine for both high-calorie ($\approx$2000 kcal/day) and low-calorie ($\approx$1000 kcal/day) intake are shown in Table 21.1. They represent an average of four sets of measurements over a 1-year period for the same person under similar conditions during the high- or low-intake portion of each cycle (body weight, calorie intake, type of food, and level of physical activity).

On the basis of this newly obtained evidence for the direct effect of the dietary calorie intake on the extent of oxidative damage to DNA, as indicated by the levels of UBODBD in Table 21.1, and previous suggestions and observations on the effects of caloric restriction on the survival of rodents (McCay et al. 1935; 1943); on the role of oxidative processes in carcinogenesis and aging (Harman 1981; Totter 1980; Ames 1983; Cutler 1984; Cerutti 1985); on the effect of altered bases on the fidelity of DNA polymerase (Kuchino et al. 1987); on urinary biomarkers (Cathcart et al. 1984; Ames et al. 1985; Bergtold et al. 1988; Simic et al. 1989; Ames 1989); on general biological and biochemical parameters (Weindruch and Walford 1988); and on the formation of oxidative DNA base products by cancer-promoter-activated neutrophils (Frenkel 1989), a unified oxidative DNA damage theory of caloric intake, endogenous carcinogenesis, and aging is proposed. This is a more specific extension of a unified theory of aging (Sohal and Allen 1985). Several lines of evidence support this hypothesis.

Of all endogenous processes that may lead to significant DNA damage (methylation, deamination, depurination), oxidative processes appear to be the most significant (Totter 1980; Ames 1983), and they correlate fairly well with specific metabolic rate (SMR; Ames 1989). The rate of aging directly correlates to SMR, implicating endogenous metabolic processes as an inherent factor in aging and carcinogenesis (Cutler 1984). The age-specific cancer rate is inversely correlated

Table 21.2. Correlations between SMR, MLS, cancer onset, the yield of dRTg in urine under normal conditions, and increase induced by ionizing radiations (X- and γ-rays).

Species	SMR (cal g^{-1} day^{-1})	MLS (years)	Cancer onset[a] (years)	dRTg (nmol kg^{-1} day^{-1})	ΔdRTg (nmol day^{-1} J^{-1})
Mouse	180	3.5	1	7.2 ± 1	3.0 ± 0.6
Human	25	100	50	0.29 ± 0.1	3.1 ± 0.8

SMR, specific metabolic rate; MLS, maximum life span; dRTg, thymidine glycol
[a]2% probability.

with MLS, while the cumulative cancer risk increases at about the fifth power of age (Ames et al. 1985). Superoxide radical, $\cdot O_2^-$, generated by xanthine/xanthine oxidase is an efficient transforming agent in JB6 mouse lymphoma cells (Seed et al. 1987). Oxidative products (clastogenic factors) in cells exposed to oxidative stress may activate oncogenes and cause cell proliferation (Crawford and Cerutti 1988). PMNs may be stimulated by tumor promoters to generate $\cdot O_2^-$ and H_2O_2 (Goldstein et al. 1981) and to induce damage to DNA, as evident from the formation of Tg and 5-hmU (Frenkel 1989). Hydroxy radicals ($\cdot$OH) generated in vivo produce proportional levels of normalized (per unit weight) UBODBD in both mice and humans (Simic et al. 1989), providing plausible evidence for the hypothesis of metabolic generation of $\cdot$OH radicals (Cathcart et al. 1984).

Finally, higher caloric intake in humans generates higher levels of UBODBD (Table 21.1), i.e., higher levels of oxidative species, which in turn induce higher levels of DNA damage. Since MLS is inversely correlated with UBODBD (Table 21.2), longevity would depend on the rate of DNA damage. Organisms with higher SMR or, as suggested here, higher energy intake per unit body weight, have shorter lives. Another consequence of greater energy intake may be onset of cancer at an earlier age (Weindruch et al. 1986).

The effect of type of food on cancer induction, aging, and MLS is not fully resolved. The role of dietary fat in carcinogenesis (Hubbard and Erikson 1987; Birt 1990) and metastasis (Erickson and Hubbard 1990) has generated great interest. Further investigations of the effect of dietary fat in equicaloric diets on UBODBD may provide useful insights into this important issue.

Conclusions

The use of urinary biomarkers of oxidative DNA-base damage (UBODBD) is proposed as a noninvasive approach to assessing the potential genetic damage induced by foods, food components, and other compounds and agents, as well as a method for optimization of diets. Replacement of integral factors (for example, MLS or onset of cancer) by intensity factors, such as UBODBD, allows faster, more economical assessment of deleterious processes in vivo and earlier prognostication. Finally, with this method, the assessment of individual risk versus statistical population risk becomes feasible.

References

Ames BN (1983) Dietary carcinogens and anticarcinogens. Oxygen radicals and degenerative diseases. Science 221:1256–1264

Ames BN (1989) Mutagenesis and carcinogenesis: endogenous and exogenous factors. Environ Mol Mutagen 14(16):66–77

Ames BN, Saul RL, Schwiers E, Adelman R, Cathcart R (1985) Oxidative DNA damage as related to cancer and aging. In: Sohal RS, Birnbaum LS, Cutler RG (eds) Molecular biology of aging: gene stability and gene expression. Raven, New York, pp 137–144

Babior BM (1988) The respiratory burst oxidase. In: Simic MG, Taylor KA, Ward JF, von Sonntag C (eds) Oxygen radicals in biology and medicine. Plenum, New York, pp 815–821

Bergtold DS, Simic MG (1989) Background levels of DNA damage. Free Radic Res Commun 6:195–198

Bergtold DS, Simic MG, Alessio H, Cutler RG (1988) Urine biomarkers for oxidative DNA damage. In: Simic MG, Taylor KA, Ward JF, von Sonntag C (eds) Oxygen radicals in biology and medicine. Plenum, New York, pp 483–490

Bergtold DS, Berg CD, Simic MG (1990) Urinary biomarkers in radiation therapy of cancer. In: Emerit I, Packer L, Auclair C (eds) Antioxidants in therapy and preventive medicine. Plenum, New York

Birt DF (1990) The influence of dietary fat on carcinogenesis. Nutr Rev 48:1–5

Buxton GV, Greenstock CL, Helman WP, Ross AB (1988) Critical review of rate constants for reactions of hydrated electrons, hydrogen atoms and hydroxyl radicals. Phys Chem Ref Data 17:513–886

Cathcart R, Schwiers E, Saul RL, Ames BN (1984) Thymine glycol and thymidine glycol in human and rat urine: a possible assay for oxidative DNA damage. Proc Natl Acad Sci USA 81:5633–5637

Cerutti PA (1985) Prooxidant stress and tumor promotion. Science 227:375–381

Cundy KC, Kohen R, Ames BN (1988) Determination of 8-hydroxydeoxyguanosine in human urine: a possible assay for in vivo oxidative DNA damage. In: Simic MG, Taylor KA, Ward JF, von Sonntag C (eds) Oxygen radicals in biology and medicine. Plenum, New York, pp 479–482

Crawford D, Cerutti P (1988) Expression of oxidant stress-related genes in tumor promotion of mouse epidermal cells JBG. In: Cerutti P, Nygaard OF, Simic MG (eds) Anticarcinogenesis and radiation protection. Plenum, New York, pp 183–190

Cutler RG (1984) Antioxidants, aging, and longevity. In: Pryor WA (ed) Free radicals in biology, vol 6. Academic, New York, pp 371–428

Dizdaroglu M, Bergtold DS (1986) Characterization of free radical-induced base damage in DNA at biologically relevant levels. Anal Biochem 156:182–188

Erickson KL, Hubbard NE (1990) Dietary fat and tumor metastasis. Nutr Rev 48:6–14

Forman HJ, Boveris A (1982) Superoxide radical and hydrogen peroxide in mitochondria. In: WA Pryor (ed) Free radicals in biology, vol. 5. Academic, New York, pp 65–90

Frenkel K (1989) Oxidation of DNA bases by tumor promoter-activated processes. Environ Health Perspect 81:45–54

Friedberg EC (1985) DNA repair. Freeman, New York

Goldstein BD, Witz G, Amoruso M, Stone DS, Troll W (1981) Stimulation of human polymorphonuclear leukocyte superoxide anion radical production by tumor promoters. Cancer Lett 11:257–262

Haber F, Weiss J (1932) Uber die Katalyse des Hydroperoxydes. Naturwissenschaften 20:948–950

Harman D (1981) The aging process. Proc Natl Acad Sci USA 78:7124

Hubbard NE, Erickson KL (1987) Enhancement of metastasis from a transplantable mouse mammary tumor by dietary linoleic acid. Cancer Res 47:6171–6175

Jovanovic SV, Simic MG (1986) Mechanisms of OH radical reaction with thymine and uracil derivatives. J Am Chem Soc 108:5968–5972

Jovanovic SV, Simic MG (1989) The DNA-guanyl radical: kinetics and mechanisms of generation and repair. Biochim Biophys Acta 1008:39–44

Karam LR, Simic MG (1988) Detection of irradiated meats: a use of hydroxyl radical biomarkers. Anal Chem 60:1117A–1119A

Karam LR, Simic MG (1989) Mechanisms of free radical chemistry and biochemistry of benzene. Environ Health Perspect 82:185–190

Klebanoff SJ, Clark RA (1985) The neutrophil: function and clinical disorder. North Holland, New York

Kuchino Y, Mori F, Kasai H, Inoue H, Iwai S, Miura K, Ohtsuka E, Nishimura S (1987) Misreading of DNA templates containing 8-hydroxydeoxyguanosine at the modified base and at adjacent residues. Nature 327:77–79

McCay CM, Crowell MF, Maynard LA (1935) The effect of retarded growth upon the length of the life span and upon the ultimate body size. J Nutr 10:63

McCay CM, Sperling G, Barnes LL (1943) Growth, aging, chronic diseases and life span in rats. Arch Biochem Biophys 2:469

McCord JM, Fridovich I (1968) The reduction of cytochrome c by milk xanthine oxidase. J Biol Chem 243:5753

Nohl H, Jordan W (1986) The mitochondrial site of superoxide formation. Biochem Biophys Res Commun 138:533–539

Pizzarello DJ (ed) (1985) Radiation biology. CRC, Boca Raton

Richter C, Park JW, Ames BN (1988) Normal oxidative damage to mitochondrial and nuclear DNA is extensive. Proc Natl Acad Sci USA 85:6465–6467

Scholes G (1983) Radiation effects on DNA. Br J Radiol 56:221–231

Seed JL, Nakamura Y, Colburn NH (1987) Implication of superoxide radical anion in promotion of neoplastic transformation in mouse JB6 cells by TPA. In: Cerutti P, Nygaard OF, Simic MG (eds) Anticarcinogenesis and radiation protection. Plenum, New York, pp 175–182

Simic MG (1989) Mechanisms of inhibition of free-radical processes in mutagenesis and carcinogenesis. Mutat Res 202:377–386

Simic MG, Bergtold DS (1990) New approaches in biological dosimetry: urine biomarkers. In: Ricks RC, Fry SA (eds) The medical basis for radiation accident preparedness. Elsevier, New York, pp 489–498

Simic MG, Jovanovic SV (1986) Free radical mechanisms of DNA base damage. In: Simic MG, Grossman L, Upton AC (eds) Mechanisms of DNA damage and repair. Plenum, New York, pp 39–50

Simic MG, Taylor KA, Ward JF, von Sonntag C (eds) (1988) Oxygen radicals in biology and medicine. Plenum, New York

Simic MG, Karam LR, Bergtold DS (1989) Generation of oxy radicals in biosystems. Mutat Res 214:3–12

Sohal RS, Allen RG (1985) Relationship between metabolic rate, free radicals, differentiation and aging: a unified theory. In: Woodhead AD, Blackett, Hollaender A (eds) Molecular biology of aging. Plenum, New York, pp 75–104

Totter JR (1980) Spontaneous cancer and its possible relationship to oxygen metabolism. Proc Natl Acad Sci USA 77:1763–1767

von Sonntag C (1987) The chemical basis of radiation biology. Taylor and Francis, New York

Weindruch R, Walford RL (1988) The retardation of aging and disease by dietary restriction. Thomas, Springfield.

Weindruch R, Walford RL, Fliegel S, Guthrie D (1986) The retardation of aging by dietary restriction: longevity, cancer, immunity and lifetime energy intake. J Nutr 116:641

Part V
Models for Dietary Restriction Research

CHAPTER 22

Models for Investigating How Dietary Restriction Retards Aging: The Adaptation Hypothesis*

R.L. Walford[1]

Introduction

The studies reported by McCay et al. (1935) that dietary restriction (DR) initiated in early youth of rats extends both average and maximum life spans have been extended by numerous investigators (for comprehensive review see Weindruch and Walford 1988). Figure 22.1 schematizes the field of DR from both historical and conceptual standpoints. While there is a rough correlation between the time line and the "events" listed to the right of the time line in Fig. 22.1, I have made exceptions in order to improve the conceptual groupings [for example, Sacher (1977), Williams et al. (1987), and Hibbs and Walford (1989) all contributed to survival curve analysis, but at quite different periods]. From 1935 until the early 1970s essentially all studies of DR concerned its effect(s) on survival and/or diseases, especially tumors. From about 1972 on, and beginning with assessment of the effects of DR on the immune system (Walford et al. 1972, 1974), a major preoccupation of investigators has been with documenting the physiologic and "biomarker" effects of DR. Since the mid 1980s, the main emphasis has been the search for mechanisms. However, important studies are still ongoing in all these areas.

While the vast majority of DR studies have used rats and mice, and nearly all ideas about mechanisms have been limited to dealing with observations in these rodents, the age-retarding effects of DR are clearly not merely a "rodent phenomenon." DR extends life span in protozoa, rotifers, daphnia, *Drosophila*, nematodes, spiders, mollusks, and fish, as well as rodents (reviewed in Weindruch and Walford 1988). Preliminary biomarker data suggest a positive effect of DR on primates (Roth 1990). Regrettably most theoreticians have wholly neglected the nonrodent data. Therefore I will first consider how the nonrodent data impinge upon various ideas about how DR works. I will next discuss items listed in Fig. 22.1 which I believe need emphasis in the ordering of our (or at least

*This work was supported by NIA grant AG-00424.
[1]Department of Pathology, UCLA School of Medicine, Los Angeles, CA 90024, USA

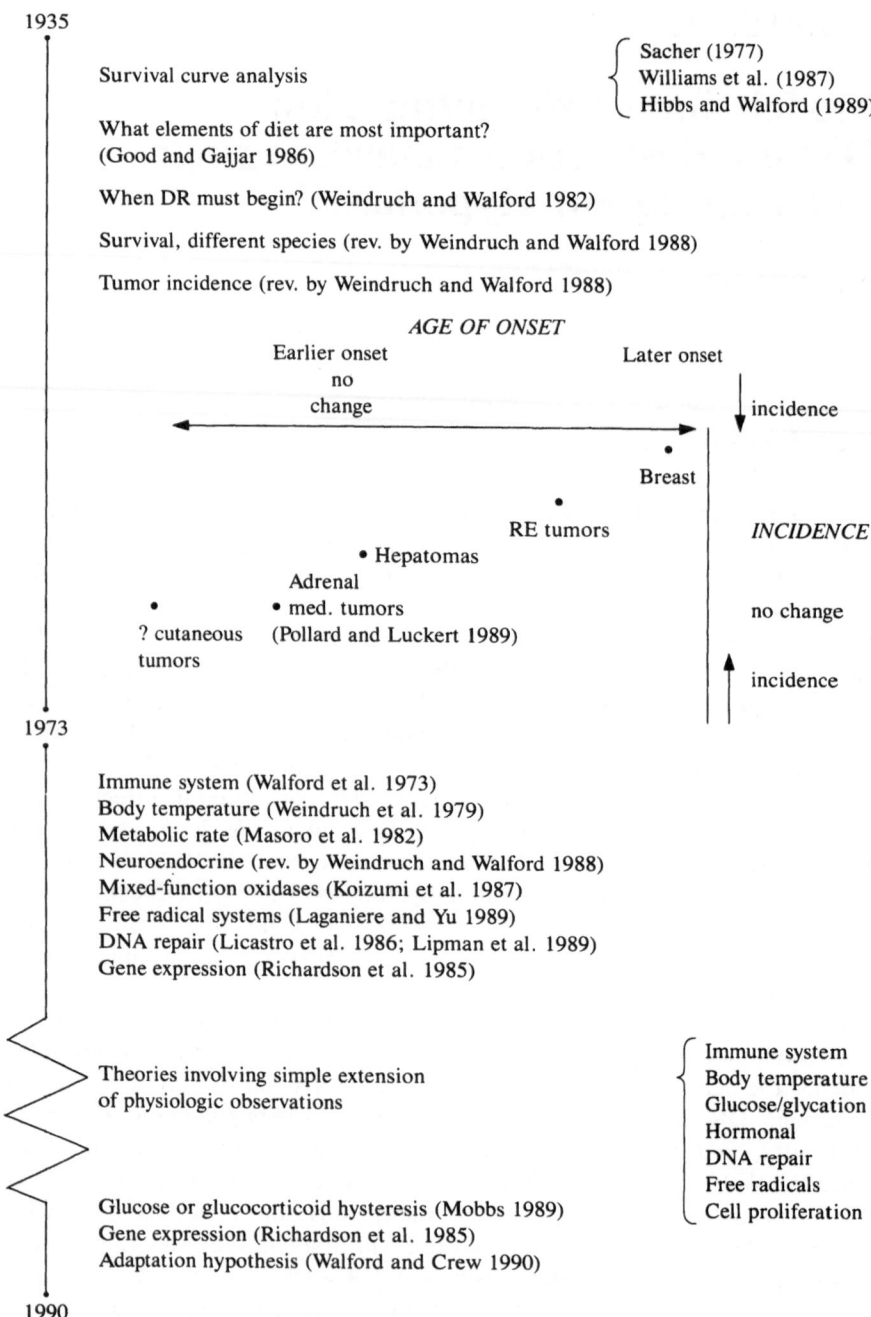

Figure 22.1. Schematization of dietary restriction (DR) research from historical and conceptual standpoints

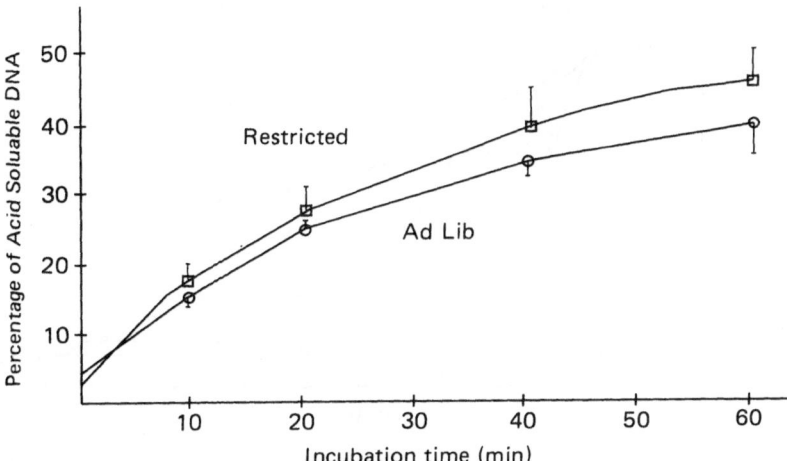

Figure 22.2. Micrococcal nuclease digestibility of liver cell nuclei isolated from 7- to 8-month-old DR (*squares*) and ad libitum-fed (*circles*) mice (R.L. Walford and C.E. Castro, unpublished observations)

my) thoughts about mechanisms. Finally, I will discuss the "Adaptation Hypothesis" recently set forth by myself and Mark Crew (Walford and Crew 1990).

Dietary Restriction Effects on Life Span

The DR effect on life span is phylogenetically independent. So far as tested, and with rare and special-case exceptions (Weindruch and Walford 1988), DR extends life span across the entire animal kingdom. There is no basis in experimental observation for the idle speculation that long-lived vertebrates, such as humans, have somehow already "used up" any extension that might be derived from caloric restriction. In fact a considerable amount of inferential evidence points in the opposite direction (Weindruch and Walford 1988).

Whether the life span extension caused by DR is mediated by similar mechanisms in various species is unknown. Evidence might be obtained by measuring different but potentially common biomarkers of aging in the different species, including, for example, DNA repair, the free radical scavenging system, the time of onset of reproductive senescence, and parameters such as the micrococcal nuclease digestibility of nuclei from DR compared to control animals (Fig. 22.2). Lacking these comparative studies, I shall assume for the purposes of this essay that similar mechanisms obtain.

DR in rodents is associated with a considerable drop in average internal body temperature, as first noted in my laboratory (Weindruch et al. 1979; Cheney et al. 1983) and confirmed in Hart's (Duffy et al. 1989). Body temperature lowering

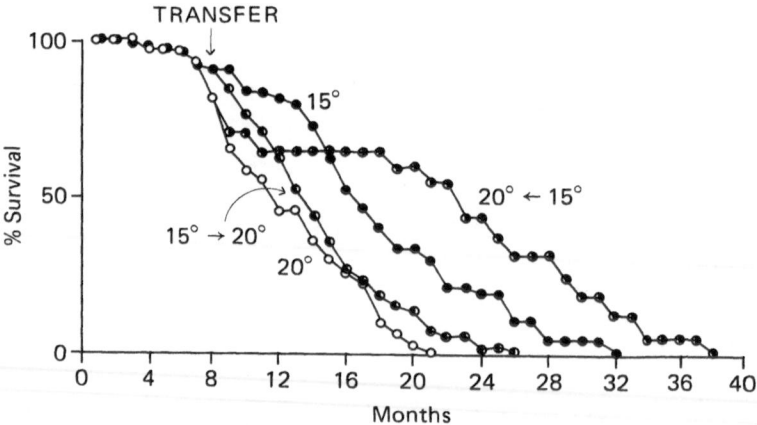

Figure 22.3. Survival curves of annual fish at different temperatures. Fish were retained throughout life at 15 °C or 20 °C, or transferred from one of these temperatures to the other at 8 months of age (Liu and Walford 1975)

has therefore been proposed as the mechanism underlying the effects of DR (Duffy et al. 1989). Indeed, lowering of core body temperature does in fact retard aging quite markedly in poikilothermic animals (Liu and Walford 1972), as illustrated in Fig. 22.3. But DR has been shown to retard aging in advanced poikilothermic vertebrates such as fish (Comfort 1963; Walford 1983), and in fish the mechanism cannot be lowering of body temperature since poikilothermic body temperature equals that of the environment, and test and control animals would have identical temperatures. Unless the mechanism is different between fish and rodent, these observations rule out temperature lowering as a primary mechanism.

The rate of "basal" (unstimulated) cell proliferation for mitotic cellular populations is reduced by DR, and this has been proposed as a primary mechanism (Johnson et al. 1986; Weindruch and Walford 1988). Does DR extend life span in invertebrates if limited to the life period when all or virtually all cells are post-mitotic? The answer may be "no" but the situation is not entirely clear from published literature (Weindruch and Walford 1988). But here clearly cross-species modelling should be rewarding.

It has been fashionable to document that an older DR animal tests, for a certain parameter, to resemble a younger animal, and then to suggest that this parameter relates to the basic DR mechanism(s). This has been done for the immune response, protein synthesis, DNA repair, free radical systems, and others. But it can be equally argued that the old DR animal is kept young by some entirely different mechanism, but of course tests younger by what may be regarded as biomarker parameters.

There is a type of DR change, however, which is not subject to the above criticism. Blood glucose will serve as illustrative example. It does not show a

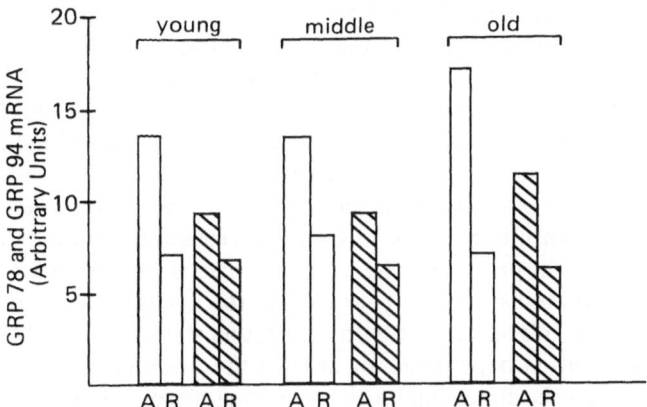

Figure 22.4. Levels of mRNA for glucose regulatory proteins GRP78 (*open bars*) and GRP94 (*hatched bars*) in ad libitum-fed (*A*) and DR mice (*R*) (adapted from Spindler et al. 1990)

consistent age-related alteration (Leiter et al. 1988). However, DR leads to a consistently lowered blood glucose (Masoro et al. 1989; Koizumi et al. 1989). This lowering is not, therefore, simply a reversion to or maintenance at a more youthful level. The DR-induced low blood glucose level has been interpreted as supporting a glycation hypothesis of aging (Masoro et al. 1989) as well as a hysteresis hypothesis (Mobbs 1989). The downregulation in DR mice of expression of genes for the glucose regulatory proteins GRP78 and GRP94 (Spindler et al. 1990), as shown in Fig. 22.4, might also fit a glycation hypothesis.

An effect on gene expression has been suggested as a potential primary mechanism for the effect of DR (Cheney et al. 1980; Lindell 1982; Richardson et al. 1985). We have extended this idea, linking it with an evolutionary approach first suggested in another form by Totter (1985), to arrive at the following adaptation hypothesis (Walford and Crew 1990). The extension of life span and other effects which accompany "undernutrition without malnutrition" represent an adaptive response to diminished food availability and are brought about by a redirection of available energy from growth, "basal" cell proliferation and turnover, reproduction, and possibly what may be loosely subsumed under the term developmental rate, towards maintenance and repair, thereby increasing the animal's chance of surviving the period of low food availability. This adaptive response, this alteration in the physiologic status of the animal, is, we have proposed, mediated by the induction or modulation of a limited number of specific transacting factors which bind to shared sequences on the various genes whose expression leads to the numerous global effects noted in the DR animal. (Or taking the black-box approach, I suggest that one might also view the induced changes as selection for altered coupling constants in physiologic reactions (Hibbs and Walford 1989).) So far, studies in my laboratory (M. Crew and

R.L. Walford, unpublished observations) using gel shift assays have suggested qualitative changes in at least one transacting factor, NF-1, in the DR animal. Whereas collaborative studies with Spindler's laboratory at the mRNA level have indicated a downregulation of glucose regulatory proteins GRP78 and GRP94 (Spindler et al. 1990a), no alterations in mRNA for enhancer-binding protein, C-jun, IGF1 (insulin-like growth factor 1), RNA polymerase II elongation factor S-2, CCAAT, or SP-1 (stimulatory protein-1) have been found in DR animals but mRNA for hepatic insulin receptor mRNA is increased (Spindler et al. 1990b). Further studies are in progress. The view has also been expressed that DNA-binding nuclear proteins may in part dictate the configuration of chromatin (Sevall 1983). This might connect with the results shown in Fig. 22.2.

I find the adaptation hypothesis, or at least the orientation or point of view behind that hypothesis, appealing for the following reasons: It allows for the concept that up- or downregulation of a relatively few agents (transacting factors in this instance) can exert the enormous number of effects documented for DR. By way of contrasting example, I find it difficult to believe that a change in level of glycation (or of blood glucose levels) can delay reproduction, alter tumor incidence and time of onset, influence DNA repair, alter levels of free radical scavengers, cause lymphopenia but with an enhanced immune response when challenged (Weindruch and Walford 1988), plus the numerous other alterations associated with DR. This conceptual difficulty applies to all categories I have labeled "theories involving extension of physiologic observations" in Fig. 22.1. A redirection of energy also fits well with the finding that the metabolic rate per lean body mass is not diminished by DR (McCarter et al. 1989).

Additionally, the adaptation hypothesis emphasizes that many of the changes in the DR animal may not relate *directly* to extending survival, i.e., to repair and maintenance, but simply to saving energy from some systems for use by other systems. Many hormonal changes due to DR, for example (reviewed in Weindruch and Walford 1988), instead of having a direct cause-and-effect relationship to life extension mechanisms, may reflect simply a shift in reproductive potential, or the shutting off of "basal" cell proliferation to save energy. The multitudinous phenomena of DR may thus be divided by an adaptive hypothesis into groups of phenomena which seem to have a common basis. If we could, by a mechanism that did not depend upon caloric deprivation, upregulate the postulated transacting factor(s), it would not necessarily be required to downregulate growth or reproduction. If our aim is to find the mechanism whereby DR extends life span, and to apply that to humans, the adaptive hypothesis predicts that not all the effects of DR itself would follow, except presumably the extension of life span.

As a corollary to the above, and noting that most, but by no means all, age-related parameters are affected by DR (Walford et al. 1987), and that a recent study showed a substantial differential effect on enzymes within the same (P-450) system (Leakey et al. 1989), I suggest that DR may not be quite the all-encompassing age-retardation syndrome it is generally treated as being, but may represent a broad, but not universal, "reverse progeroid" type of syndrome. This would accord with the fact that Gompertz plots of DR populations sometimes show an alteration in slope, sometimes a shift in intercept (Harris et al. 1990; Hibbs and Walford 1989). This feature can be accommodated within the adaptation hypothesis. One prediction therefore is that disease types and causes of death in DR animals may, if carefully enough scrutinized, be somewhat different from those in controls.

References

Cheney KE, Liu RK, Smith GS, Leung RE, Mickey MR, Walford RL (1980) Survival and disease patterns in C57BL/6J mice subjected to undernutrition. Exp Gerontol 15: 237-258

Cheney KE, Liu RK, Smith GS, Meredith PJ, Mickey MR, Walford RL (1983) The effect of dietary restriction of varying duration on survival, tumor patterns, immune function and body temperature in B10C3F$_1$ female mice. J Gerontol 38:420-430

Comfort A (1963) Effect of delayed and resumed growth on the longevity of a fish (lebistas reticulatus, Peters) in captivity. Gerontologia 8:150-156

Duffy PH, Feuers RJ, Leakey JA, Nakamura KD, Turturro A, Hart RW (1989) Effect of chronic caloric restriction on physiological variables related to energy metabolism in the male Fischer 344 rat. Mech Ageing Dev 48:117-133

Good RA, Gajjar AJ (1986) Diet, immunity, and longevity. In: Hutchinson ML, Munro HN (eds) Nutrition and aging. Academic, New York, pp 235-249

Harris SB, Weindruch R, Smith GS, Mickey MR, Walford RL (1990) Dietary restriction alone and in combination with oral ethoxyquin/2-mercaptoethylamine. J Gerontology 45:B141-147

Hibbs AR, Walford RL (1989) A mathematical model of physiological processes and its application to the study of aging. Mech Ageing Dev 50:193-214

Johnson BC, Gajjar A, Kubo C, Good RA (1986) Calories versus protein in onset of renal disease in NZBxNZW mice. Proc Natl Acad Sci USA 83:5659-5672

Koizumi A, Weindruch R, Walford RL (1987) Influence of dietary restriction and age on liver enzyme activities and lipid peroxidation in mice. J Nutr 117:361-367

Koizumi A, Wada Y, Tsukada M, Hasegawa J, Walford RL (1989) Low blood glucose levels and small islets of Langerhans in the pancreases of calorie restricted mice. Age 12:93-96

Laganiere S, Byung PY (1989) Effect of chronic food restriction in aging rats. II. liver cytosolic antioxidants and related enzymes. Mech Ageing Dev 48:221-230

Leakey JE, Cunny HC, Bazara J, Webb PI, Feuers RJ, Duffy PH, Hart RW (1989) Effects of aging and caloric restriction on hepatic drug metabolizing enzymes in the Fischer 344 rat. The cytochrome p-450 dependent monooxygenase system. Mech Ageing Dev 45:145-155

Leiter EH, Premdas F, Harrison DE, Lipson LG (1988) Aging and glucose homeostasis in C57BL/6J male mice. Fed Proc 2:2807-2811

Licastro F, Weindruch R, Walford RL (1986) Dietary restriction retards the age-related decline of DNA repair capacity in mouse splenocytes. In: Facchini A, Haaijman JJ, Labo G (eds) Immunoregulation and aging. Eurage, Rijswijk, pp 53-61

Lindell TJ (1982) Molecular aspects of dietary modulation of transcription and enhanced longevity. Life Sci 31:625-629

Lipman JM, Turturro A, Hart RW (1989) The influence of dietary restriction on DNA repair in rodents: A preliminary study. Mech Ageing Dev 48:135-143

236   R.L. Walford

Liu RK, Walford RL (1972) The effect of lowered body temperature on lifespan and immune and non-immune processes. Gerontologia 18:363–388

Liu RK, Walford RL (1975) Mid-life temperature-transfer effects on lifespan of annual fish. J Gerontol 30:129–131

Masoro EJ, Yu BP, Bertrand HA (1982) Action of food restriction in delaying the aging process. Proc Natl Acad Sci USA 79:4239–4243

Masoro EJ, Katz MS, McMahan CA (1989) Evidence for the glycation hypothesis of aging from the food-restricted rodent model. J Gerontol 44:B20–22

McCarter RJ, Herlihy JT, McGee JR (1989) Metabolic rate and aging: effects of food restriction and thyroid hormone on minimal oxygen consumption in rats. Aging 1:71–76

McCay CM, Crowell ME, Maynard LA (1935) The effect of retarded growth upon the length of the life span and upon the ultimate body size. J Nutr 10:63–80

Mobbs CV (1989) Neurotoxic effects of estrogen, glucose, and glucocorticoids: neurohormonal hysteresis and its pathological consequences during aging. Rev Biol Res Aging (in press)

Pollard M, Luckert PH (1989) Spontaneous diseases in aging lobund-wistar rats. In: Snyder DL (ed) Dietary restriction and aging. Liss, New York, p 51

Richardson A, Roberts MS, Rutherford MS (1985) Aging and gene expression. In: Rev Biol Res Aging 2:395–419

Roth G, cited in Dec 1989/Jan 1990 issue of Longevity p 26

Sacher GA (1977) Life table modification and life prolongation. In: Finch CE, Hayflick L (eds) Handbook of the biology of aging. Van Nostrand Reinhold, New York, pp 582–638

Sevall JS (1983) DNA-binding nonhistone proteins. In: Honilica LS (ed) Chromosomal nonhistone proteins. CRC, Boca Raton, pp 25–46

Spindler SR, Crew MD, Mote PL, Grizzle JM, Walford RL (1990) Dietary energy restriction in mice reduces hepatic expression of glucose regulated proteins 78 and 94 (GRP78 and GRP94) mRNA. J Nutr 120:1412–1417, 1990

Spindler SR, Grizzle JM, Walford RL, and Mote PL (1990b), Aging and restriction of dietary calories increases hepatic insulin receptor mRNA, and aging increases the level of hepatic glucocorticoid receptor mRNA in female C3B10RF$_1$ mice. J Gerontol (in press)

Totter JR (1985) Food restriction, ionizing radiation, and natural selection. Mech Ageing Dev 30:261–271

Walford RL (1983) Maximum life span. Norton, New York

Walford RL, Crew MD (1990) How dietary restriction retards aging: an integrative hypothesis. Growth, Dev Aging 53:139–141

Walford RL, Harris SB, Weindruch R (1987) Dietary restriction and aging: historical phases, mechanisms, current directions. J Nutr 117:1650–1654

Walford RL, Liu RK, Mathies M, Lipps L, and Konen T (1972) Influence of calorie restriction on immune function, relevance for an immunologic theory of aging. Proc. 9th Internat. Congr. Nutrition 1:374–381 (Karger, Basel)

Walford RL, Liu RK, Gerbase-DeLima M, Mathies M, Smith GS (1972, 1974) Longterm dietary restriction and immune function in mice: response to sheep red blood cells and to mitogenic agents. Mech Ageing and Dev 2:447–454

Weindruch R, Walford RL (1988) The retardation of aging and disease by dietary restriction. Thomas, Springfield

Weindruch R, Kristie JA, Cheney K, Walford RL (1979) The influence of controlled dietary restriction on immunologic function and aging. Fed Proc 38:2007–2016

Weindruch R, Walford RL (1982) Dietary restriction of mice beginning at 1 year of age: effect on life-span and spontaneous cancer incidence. Science 215:1415–1418

Williams JR, Spencer PS, Stahl SM, Borzelleca JF, Nichols W, Pfitzer E, Yunis EJ, Carchman R, Opishinski JW, Walford RL (1987) Interactions of aging and environmental agents: the toxicological perspective. In: Baker SR, Rogul M (eds) Environmental toxicity and the aging process. Liss, New York, pp 81–135

CHAPTER 23

Aging, Dietary Restriction, and Cancer in a Germ-Free Animal Model

D.L. Snyder[1,2], B.S. Wostmann[1], and M. Pollard[1]

Introduction

Freedom from infectious disease and from tumors induced by microbial carcinogens or by viruses has made germ-free (GF) animals a valuable tool in cancer and aging research. Gordon et al. (1966) suggested that protection from the deleterious effects of microbes on the aging host would permit a better view of the endogenous aging process. His research showed that GF mice lived at least 8 months longer than conventional (CV) mice (24 vs. 16). Research by Morris Pollard at the University of Notre Dame's Lobund Laboratory showed that GF rats outlived CV rats (Pollard 1971) and spontaneously developed liver tumors (Pollard and Luckert 1979) and prostate adenocarcinomas (Pollard 1973). Young GF rats and mice have lower resting oxygen consumption, cardiac output, and reduced heart size (Wostmann 1975), and adult GF rats have reduced body size (Snyder and Wostmann 1987) when compared to CV rats. The longer life span and reduced body size of GF rats are traits similar to those reported for long-lived diet-restricted (DR) rats. Pollard and Wostmann (1985) examined the effect of DR on cancer development and aging in GF Lobund-Wistar (L-W) rats. Ten GF L-W rats restricted to approximately 70% of daily food intake lived well beyond any ad libitum-fed CV or GF L-W rats and showed few of the tumors common to old L-W rats.

The results of Pollard and Wostmann's early study initiated the much larger Lobund Aging Study at the University of Notre Dame. Designed to determine what pathologies would develop over the life span of ad libitum and DR CV and GF L-W rats, the study included investigators outside of the university who had interests in aging and its associated pathologies and biochemical changes. The present article is intended only to highlight the findings of the Lobund Aging Study. An earlier account of this study was published in the proceedings of the

[1]Lobund Laboratory, University of Notre Dame, Notre Dame, IN 46556, USA
[2]Current address: Center for Gerontological Research, The Medical College of Pennsylvania, 3300 Henry Avenue, Philadelphia, PA 19129, USA

symposium on the effects of dietary restriction on aging and disease in germ-free and conventional Lobund-Wistar rats (Snyder 1989).

Methods

Germ-free L-W rats were first propagated at the University of Notre Dame in 1958. The CV breeding colony was derived from the GF colony, and at regular intervals GF males and females are added to the CV colony to maintain close genetic proximity. Only male rats were used in this study. The GF rats were housed in plastic or steel isolators and were maintained using routine gnotobiotic procedures (Subcommittee on Standards for Gnotobiotics 1970). The CV rats were housed in plastic isolators to limit exposure to the local environment. CV isolators were opened only to introduce feed and water and to clean the cages. Ad libitum-fed rats (GF-F and CV-F) were housed four to a cage in commercial plastic boxes. DR rats (GF-R and CV-R) were taken from their respective breeding colonies at 6 weeks of age and housed individually in plastic cages. All rats were fed steam-sterilized natural ingredient diet L-485, our colony diet since 1968 (Kellogg and Wostmann 1969). The DR rats were never allowed more than 12 g of diet per day. This method of feeding becomes restrictive at about 8 weeks of age and results in a 30% reduction in feed intake in adult rats (Snyder and Wostmann 1987). A complete description of the housing and diet is provided in Snyder et al. (1990).

Rats were killed if they were in a moribund condition due to a palpable tumor or had severe weight loss over several months. Healthy rats were selected for killing from 3 to 30 months of age for measurement of physiological parameters related to aging. After an overnight fast, each rat was anesthetized with halothane and blood was removed from the exposed heart with a needle and syringe. Blood was allowed to clot for 30 min and then centrifuged. The serum was frozen at $-70°C$ in individual aliquots for later analysis. Individual tissues were quickly removed, weighed, and frozen in liquid nitrogen or preserved according to the protocol of each investigator receiving tissue samples. Rats were examined for tumors and tissue samples were fixed for histological examination.

Results and Discussion

Each of the four experimental groups had a characteristically different pattern of growth. Up to 4 months of age the GF-F rats grew at the same rate as the CV-F rats, but after 5 months the GF-F rats had a slightly slower rate. By 18 months the average weight of CV-F rats was just over 480 grams, but the GF-F rats averaged just over 450 grams. Since the cecum of GF rats is 20 grams larger than that of CV rats, the corrected difference in body weights between GF-F and CV-F rats may be considered closer to 50 grams. The CV-R and GF-R rats grew at a similar rate, but adult GF-R rats tended to be slightly heavier. Further details on the growth of GF and CV L-W rats are available in Snyder and Wostmann (1987).

One hundred CV-F, 88 CV-R, 96 GF-F, and 127 GF-R rats were used to determine the survival characteristics of the male L-W rat. The length of life of the four experimental groups closely followed their adult body weights. The median length of life of the heavier CV-F and GF-F rats were 31.0 and 33.6 months, respectively, while the lighter CV-R and GF-R rats were 38.6 and 37.8 months, respectively. The 10th percentile and maximum survival ages were CV-F: 36.3/38.9, GF-F: 39.0/40.5, CV-R: 43.9/50.9, and GF-R: 44.8/47.4. Statistical analysis showed that the survival of each of the DR groups was different from that of the corresponding ad libitum-fed groups, that the survival of the GF-F rats was different from that of the CV-F rats, and that the survival of the GF-R and CV-R rats were equivalent (Snyder et al. 1990). The small increase in life span between GF-F and CV-F rats may be due partly to the lack of prostatitis in GF rats and a natural DR state as indicated by a smaller adult body size. GF status was without additional benefit when food intake and final body weights were similar.

Most tumors do not appear in L-W rats until after 30 months of age. The majority of these tumors develop spontaneously in the liver, the adrenal medulla, and the mammary glands. Although the frequencies of these pathologies were lower in the CV-R, GF-F, and GF-R rats than in the CV-F rats, the major effect of DR and GF life was to delay the age of occurrence of these tumors. Hepatomas were found in 58% of moribund CV-F rats at a mean age of occurrence of 32.5 months. The corresponding figures for hepatomas in the other three experimental groups were CV-R: 34% and 41.1, GF-F: 47% and 37.3, and GF-R: 42% and 38.4. Prostate adenocarcinomas were found in 22% of moribund CV-F rats at a mean age of occurrence of 28.1 months. The corresponding figures for prostate adenocarcinomas in the other three experimental groups were CV-R: 7% and 40.0, GF-F: 7% and 22.0, and GF-R: 6% and 37.0. DR also reduced the incidence of prostatitis in CV rats and delayed the mean age of occurrence from 23.3 to 41.0 months.

It should be noted that nephrosis was not a common finding in any of the healthy or moribund L-W rats examined in this study. When it occurred, it was usually considered mild and was never the cause of death. The use of soy protein rather than casein protein along with a natural resistance to kidney disease seems to have prevented nephrosis in L-W rats. A complete description of the pathological finding of the Lobund Aging Study has been reported in Pollard et al. (1989) and Snyder et al. (1990).

Healthy and moribund L-W rats were examined for cardiomyopathy and pituitary adenomas by outside investigators. Cornwell and Thomas (1989) reported that the occurrence of cardiac fibrosis was age related and most extensive after 30 months of age. DR significantly reduced the amount of fibrosis at all ages. Sano et al. (1989) reported that gonadotroph nodules were common in ad libitum-fed L-W rats older than 26 months, but the development of this age-associated lesion was delayed by DR and GF status parallel to prolongation of life span.

Thus a restriction of only 30% in food intake caused a considerable delay in the development of a broad array of pathologies common to L-W rats and extended the life span by approximately 8 months in CV rats and 4 months in GF rats.

A number of age-associated physiological changes were seen in the rats examined during the Lobund Aging Study. Serum levels of thyroid stimulating hormone (TSH), thyroxine (T4), triiodothyronine (T3), prolactin (PRL), and testosterone (T) have been reported for the 4 experimental groups from 6 to 30 months of age (Snyder et al. 1988). TSH and T4 levels were slightly higher in the GF-F and GF-R rats, both hormones declined gradually from the youngest to the oldest rats examined, and DR had no effect on either hormone. T3 levels were also unaffected by DR and did not change over time in CV rats. Within the GF animals, DR had no effect, young rats had considerably higher T3 levels, and old rats had considerably lower T3 levels than CV rats of the same age. PRL levels were generally similar between CV and GF rats, but DR prevented the age-associated rise in PRL levels seen in ad libitum-fed rats. T levels showed a gradual decline from adult to old age in all rats, but DR dramatically raised T levels at all ages. There was no reduction in testes size in DR rats when compared to the ad libitum-fed rats. These results indicated that a reduction in thyroid hormones and in reproductive function was not necessary to promote an extended life span in DR rats.

Serum insulin, blood glucose, serum cholesterol, and serum triglycerides in 6–30 month old rats were reported by Snyder and Towne (1989). Ad libitum-fed rats had slightly higher serum insulin levels, but there were no differences due to age or GF status. Serum glucose levels were similar in all rats except for a tendency to high levels in the oldest CV-F and GF-F rats. Serum cholesterol and triglyceride levels were higher in CV rats, reduced by DR and increased with age. A recently completed study at Lobund Laboratory using only CV L-W rats has shown that DR will prevent the age-associated increase in glycosylated hemoglobin levels. In this study serum glucose levels were similar in CV-F and CV-R rats from 6 to 30 months, but while CV-F rats showed a steady increase in glycosylated hemoglobin after 18 months of age, there was no increase with age in CV-R rats.

Although DR had very little effect on growth-regulating hormones such as insulin, T4, and T3, DR did significantly reduce serum levels of somatomedin-D/insulin-like growth factor-I (SMD-C) in the L-W rats in this study (Prewitt and D'Ercole 1989). SMD-C is a growth-hormone-dependent anabolic peptide that has a high correlation with nitrogen metabolism and body weight. Serum SMD-C levels were highest in the CV-F rats from 5 to 18 months of age but fell to levels equal to the DR rats at 30 months of age. GF status also reduced serum SMD-C levels but not to the extent seen in DR rats.

Age-associated changes in the bone were observed by two investigators. Nishimoto et al. (1990) found declines in serum bone γ-carboxyglutamic acid (Gla) protein, and matrix concentrations of calcium, magnesium, and bone Gla protein in all four experimental groups as they aged. The decline in these parameters, which indicates a loss of bone mineral and a reduction in bone mineralization with age, was not altered by DR or GF status. Rosen et al. (1989) showed a significant loss of alveolar bone with age in all four experimental groups that was not altered by DR or GF status. These results are contradictory to those of Kalu et al. (1988) who reported that DR appeared to stop an age-associated decrease in bone density in Fischer 344 rats. It is possible that nephrosis caused by a casein diet and then corrected through DR contributed to the effects seen in the Fischer

344 rats. DR may have very little direct effect on age-associated changes in bone metabolism when nephrosis is absent as in the L-W rats.

Several investigators examined the hypothesis that DR alters the accumulation of oxidative and free radical-initiated damage that may be responsible for the degenerative changes associated with aging. Chen and Lowry (1989) found that the liver activity of three enzymes which are part of the antioxidative defense system was enhanced by DR in the CV L-W rats. Superoxide dismutase (responsible for removing superoxide radicals), catalase, and selenium-dependent glutathione peroxidase (both decompose cellular hydrogen peroxide) were enhanced in 30-month-old CV-R rats when compared to 30-month-old CV-F rats. GF status, however, tended to lower the activity of superoxide dismutase and catalase when compared to CV rats. Selenium-dependent glutathione peroxidase activity, however, was enhanced in both GF-F and GF-R rats. The lower metabolic rate of GF rats may have reduced the production of free radicals and therefore lowered the activity of the enzymes responsible for removing free radicals. Lang et al. (1989) measured whole blood levels of glutathione in 6- to 40-month-old rats. Glutathione serves as the substrate in the enzymatic reaction that decomposes hydrogen peroxide. DR increased blood glutathione levels by 20%–100% at all ages. After a peak at 18 months blood glutathione declined with age, but the decline was slower in DR rats. Starke-Reed (1989) looked for the accumulation of oxidized proteins in the testes of aging L-W rats. There was a large increase in oxidized protein between 6 and 18 months of age in the CV-F and CV-R rats, but at 30 months the amount remained high in the CV-F rats and declined in the CV-R rats. The levels of oxidized proteins were lower in both groups of GF rats when compared to CV rats. Because of its susceptibility to oxidation, glutamine synthetase activity was also measured in the testes. The CV-F rats showed a decline in activity of 40% between 6 and 30 months, but there was no change in the other three groups. This study also suggests that GF status imparts a protection against free radical damage that is similar to that provided by DR. From the three studies mentioned above it would appear that DR does enhance the cellular mechanisms which destroy free radicals once they are produced. Within the GF-F rats such protection may also be present, but it is not additive with DR.

Other investigators in the Lobund Aging Study have examined the influence of DR on immune function, body composition, gastrointestinal function, muscle enzymes, and microsomal monooxygenases. The complete reports of these and other studies will soon be published. Hopefully the integration of all the findings from the Lobund Aging Study will help in the search for a universal hypothesis for the action of DR on the aging process.

Conclusion

Our study confirms the ability of DR to extend life span in the rat by showing that rats which are virtually free of kidney disease (due to consumption of a casein-free diet) and free of infectious disease (from living in a GF environment) still receive considerable benefit from reduced dietary intake. It is also apparent from the Lobund Aging Study that GF status, even with its reduced metabolic rate, has a relatively small effect on life span when compared to DR.

The GF animal has been suggested as an ideal model for aging research because of the freedom from microbial interference during aging, but the cost of maintaining GF animals for extended periods is prohibitive. Our study has shown that conventional L-W rats derived from a clean breeding colony, housed in minimal barrier isolation, and fed a natural ingredient diet will live almost as long as GF L-W rats. Freedom from kidney disease and the presence of many age-associated pathologies rather than one overriding cause of death may make the L-W rat an ideal animal for aging research.

References

Chen LH, Lowry SR (1989) Cellular antioxidant defense system. In: Snyder DL (ed) Dietary restriction and aging. Liss, New York, pp 247–256

Cornwell GG, Thomas BP (1989) Cardiac fibrosis in the aged germfree and conventional Lobund-Wistar rat. In: Snyder DL (ed) Dietary restriction and aging. Liss, New York, pp 69–74

Gordon HA, Bruckner-Kardoss E, Wostmann BS (1966) Aging in germ-free mice: life tables and lesions observed at natural death. J Gerontol 21:380–387

Kalu DN, Masoro EJ, Yu BP, Hardin RR, Hollis BW (1988) Modulation of the age-related hyperparathyroidism and senile bone loss in Fischer rats by soy protein and food restriction. Endocrinology 122:1847–1854

Kellogg TF, Wostmann BS (1969) Stock diet for colony production of germfree rats and mice. Lab Anim Care 19:812–814

Lang CA, Wu W, Chen T, Mills BJ (1989) Blood glutathione: a biochemical index of life span enhancement in the diet restricted Lobund-Wistar rat. In: Snyder DL (ed) Dietary restriction and aging. Liss, New York, pp 241–246

Nishimoto SK, Padilla SM, Snyder DL (1990) The effect of food restriction and germfree environment on age-related changes in bone matrix. J Gerontol Biol Sci 45:B164–168

Pollard M (1971) Senescence in germfree rats. Gerontologia 17:333–338

Pollard M (1973) Spontaneous prostate adenocarcinomas in aged germfree Wistar rats. J Natl Cancer Inst 51:1235–1241

Pollard M, Luckert PH (1979) Spontaneous liver tumors in aged germfree Wistar rats. Lab Anim Sci 29:74–77

Pollard M, Wostmann BS (1985) Aging in germfree rats: the relationship to the environment, diseases of endogenous origin, and to dietary modification. In: Archibald J, Ditchfield J, Rowsell HC (eds) The contribution of laboratory animal science to the welfare of man and animals. Fischer, New York, pp 181–186

Pollard M, Luckert P, Snyder DL (1989) Prevention of prostate cancer and liver tumors in L-W rats by moderate dietary restriction. Cancer 64:686–690

Prewitt TE, D'Ercole AJ (1989) Modest dietary restriction and serum somatomedin-C/insulin like growth factor-I in young, mature and old rats. In: Snyder DL (ed) Dietary restriction and aging. Liss, New York, pp 157–162

Rosen S, Strayer M, Glocker W, Marquard J, Beck FM (1989) Effects of aging, diet restriction and microflora on oral health in humans and animals. In: Snyder DL (ed) Dietary restriction and aging. Liss, New York, pp 75–85

Sano T, Kovacs K, Stefaneanu L, Asa S, Snyder DL (1989) Spontaneous pituitary gonado-troph nodules in aging male Lobund-Wistar rats. Lab Invest 61:343-349

Snyder DL (ed) (1989) Dietary restriction and aging. Liss, New York

Snyder DL, Towne B (1989) The effect of dietary restriction on serum hormone and blood chemistry changes in aging Lobund-Wistar rats. In: Snyder DL (ed) Dietary restriction and aging. Liss, New York, pp 135–146

Snyder DL, Wostmann BS (1987) Growth rate of male germfree Wistar rats fed ad libitum or restricted natural ingredient diet. Lab Anim Sci 37:320–325

Snyder DL, Wostmann BS, Pollard M (1988) Serum hormones in diet-restricted gnotobi-otic and conventional Lobund-Wistar rats. J Gerontol Biol Sci 43:B168–173

Snyder DL, Pollard M, Wostmann BS, Luckert P (1990) Life span, morphology, and pathology of diet-restricted germ-free and conventional Lobund-Wistar rats. J Gerontol Biol Sci 45:B52–58

Starke-Reed PE (1989) The role of oxidative modification in cellular protein turnover and aging. In: Snyder DL (ed) Dietary restriction and aging. Liss, New York, pp 269–276

Subcommittee on Standards for Gnotobiotics, Committee on Standards, Institute of Laboratory Animal Resources, National Resource Council (1970) Gnotobiotes, stan-dards and guide lines for breeding, care and management of laboratory animals. National Academy of Sciences, Washington

Wostmann BS (1975) Nutrition and metabolism of the germfree mammal. World Rev Nutr Diet 22:40–92

CHAPTER 24

Chronic Caloric Restriction in Old Female Mice: Changes in the Circadian Rhythms of Physiological and Behavioral Variables

P.H. Duffy[1], R.J. Feuers[1], J.E.A. Leakey[1], and R.W. Hart[1]

Introduction

One of the most frequently reported observations in the gerontological literature is the significant increase in life span (Sacher 1977; McCay et al. 1935) and the prevention or delayed onset of various types of chronic diseases (Walford et al. 1974; Maeda et al. 1985) associated with chronic caloric restriction (CR). CR has also been shown to be effective in inhibiting various types of spontaneous tumors (Sarkar et al. 1985) and chemically induced lesions (Ruggeri et al. 1987; Kritchevsky et al. 1984) in rats and mice. However, little is known about the primary mechanisms by which CR interacts with environmental factors to alter the timing of the biological clock. One hypothesis that may account for the slowing down of age-related physiological and disease processes is that the synchronization of organisms to their surroundings is controlled or modified by qualitative or quantitative changes in nutritional parameters and that changes in 24-h (circadian) rhythms that are a direct result of CR may ultimately alter primary mechanisms of aging. Several studies in which total calories were restricted to 76% of ad libitum (AL) by limiting access to food to a few hours daily (single meal feeding) (Nelson et al. 1975; Philippens et al. 1977) or when intake was decreased by reducing the total weight of the food pellets fed to the caloric restricted group to 60% of the AL level (Duffy et al. 1989a; McCarter et al. 1985) have reported significant alterations in the circadian rhythms of various physiological and biochemical parameters. However, in other studies, no significant difference in longevity was found in CR mice fed a single meal during the early light phase, a single meal during the dark phase, or fed multiple meals (six times) during the dark phase (Nelson and Halberg 1986a). While CR in itself was found to prolong life, the added imposition of frequent photoperiod schedule shifting had no statistically significant effect on mean survival time, regardless of whether the meal schedule reinforced or opposed shifts in the photoperiod (Nelson and

[1]DHHS/PHS/Food and Drug Administration, National Center for Toxicological Research, Jefferson, AR 72079, USA

Halberg 1986b). This may mean that varying the dietary regimen may have no effect on longevity if the total amount of food consumed is maintained at a constant restricted calorie level.

CR was found to significantly alter physiological parameters that relate to total energy metabolism (Duffy et al. 1989a), as well as intermediary metabolism (Feuers et al. 1989) and cyt-P450 metabolism (Leakey et al. 1989). Considered in their entirety, the results of some other studies that relate to genetic information systems may imply that CR-induced changes in the circadian timing of physiological variables at the whole animal level such as deep body temperature, motor activity, and oxygen metabolism (Duffy et al. 1989a) precede and are directly proportional to specific changes at the molecular level such as proto-oncogene expression (Nakamura et al. 1989) and DNA repair (Lipman et al. 1989), thereby suggesting a cause and effect relationship between these variables.

Chronic CR decreased the mesor (circadian rhythm-adjusted mean value) and increased the amplitude of the daily body temperature rhythm (Nelson et al. 1975; Duffy et al. 1989a) and respiratory quotient and activity (Duffy et al. 1989a). Significant decreases in metabolism and temperature were found to occur during the transition period between carbohydrate metabolism and fatty acid metabolism in CR rats and increased motor activity, with no significant increase in metabolic rate, suggesting increased metabolic efficiency in CR rats (Duffy et al. 1989a).

The effects of CR on the circadian regulation of physiological and behavioral variables have been studied in male rats (Duffy et al. 1989a). However, few if any studies have monitored the effects of chronic CR on these parameters in female mice. Therefore, the present study was undertaken to monitor behavioral endpoints such as food and water consumption, number of feeding and drinking episodes, and gross motor activity and physiological endpoints such as oxygen and carbon dioxide metabolism, respiratory quotient, and body temperature in female B6C3F$_1$ hybrid mice that were AL fed or fed a CR diet. The specific aims of this study were to:

1. Determine possible effects of CR in female mice
2. Determine whether physiological and behavioral performances in CR mice are altered by the frequency and duration of feeding as well as the amount of food consumed
3. Explore mechanisms by which CR may affect variables that modulate energy output and energy efficiency, such as metabolism and thermoregulation
4. Determine primary mechanisms by which CR decreases the rate of aging and the susceptibility to chronic diseases
5. Determine whether the effects of CR on homeostatic regulation at the whole animal level are related and correlated to specific changes at the cellular and molecular level
6. Study the interaction between CR and environmental synchronizers such as photoperiod and feeding regimen, related to specific changes in circadian (24-h) rhythms

7. Develop an optimum CR feeding regimen to be used to properly synchronize behavior among the various test groups
8. Determine the importance of using chronobiological testing techniques in nutrition and gerontological research

Materials and Methods

Animal Husbandry and Feeding Regimens

The work presented here was the first in a series of studies to develop physiological and behavioral markers to be used to investigate primary mechanisms by which chronic CR modifies disease, toxicity, and the aging process. The CR and AL test animals used in this study were obtained from a National Institute on Aging (NIA) animal colony and the feeding and environmental regimen were established before the principal investigators in the present study became involved in the project. Any changes in environmental and animal husbandry factors at the time of the experiment would have required the monitoring of variables under different experimental conditions than were used by other NIA investigators, thereby invalidating the comparison between these studies. The timing of experimental factors and the availability of CR animals precluded the evaluation of these animals under different experimental regimens.

The female B6C3F$_1$ mice used in this study were raised at the National Center for Toxicological Research (NCTR) in a specific-pathogen-free (SPF) environment at 23°C and were entrained to a 12/12, light/dark cycle with lights on from 0600 to 1800 hours daily. These hybrid mice were the result of an F$_1$ cross between C57BL6 and C3H mice. All animals were fed AL and were maintained on this regimen from the time the animals were weaned until they were entered into the experiment. The test animals were then divided into a control group which was maintained on AL feeding and a CR group which received 61%–67% of the AL food ration starting at 14 weeks of age and continuing for the duration of the experiment. The CR ration was dispensed daily in precision pellets of exact weight to ensure that each mouse received exactly the same number of calories. Previous studies have shown that this level of restriction causes maximum life extension (McCay et al. 1935; Maeda et la. 1985). The calorie content of the CR ration was based on the food consumption in the AL animals. Food was left in the hopper until eaten so that CR and control animals were not forced to eat at a specific circadian stage, thereby allowing the test animals to express spontaneous feeding behavior. AL mice were given more food than they consumed in a daily interval so that food was available to them at all times. All mice were fed a standard NIH-31 diet at 1000 hours daily. The experimental protocol was initiated when the test animals were 28 months of age, at which time the subjects were transferred from the SPF barrier to the testing area, where they were maintained under similar environmental conditions. Test cages were placed in an acoustically sealed incubator located in a separate isolated room, and food was dispensed through a special feeder tube so that the mice could not see or hear laboratory personnel. Therefore, inadvertent visual or auditory stimuli could not affect the results of the experiment.

Physiological and Behavioral Testing Procedures

The equipment and testing procedures used in this study were similar to those used previously (Duffy et al. 1987). Briefly, food and water consumption as well as feeding and total motor activity were continuously monitored with a system that measured the force applied to a load cell. Drinking activity was monitored by a digital pulse-counting device triggered when the mouse made contact with the drinking tube. Deep body temperature and motor activity were measured by a transmitter capsule that was surgically implanted in the peritoneal cavity of the mouse 60 days before the animals were tested, and data were collected on a computerized data acquisition system (Data Sciences, Inc., Roseville, Minnesota). Oxygen consumption, carbon dioxide production, and respiratory quotient (RQ) were measured with an open-circuit flow system equipped with a microprocessor-controlled flow and absolute pressure regulating device. Animals were placed individually in lexan polycarbonate metabolism cages, and sample flow from several chambers was multiplexed to an Applied Electrochemistry model S3A oxygen analyzer (Ametek, Inc., Pittsburgh, Pennsylvania) and a model 864 infrared carbon dioxide analyzer (Beckman Instruments, Inc., Fullerton, California). A model 1200 microcomputer system (ADAC Corp., Woburn, Massachusetts) was used to provide process control, calibration, and collection in the metabolism monitoring system.

Ten AL-fed animals and ten CR animals (28 months of age) were placed individually in test cages (11 in $\times$ 7 in $\times$ 7.5 in) for 2 weeks prior to the initiation of the experiment. Both test groups were fed 5 h after the onset of the light phase of the photoperiod cycle. All other environmental factors were identical to those described previously. Data were collected for all parameters at 3-min intervals over a 10-day testing period. The photoperiod was then altered by 12 h and the animals were allowed to phase shift for a 30-day period so that food was presented to restricted mice 5 h after the onset of the dark phase of the photoperiod cycle. Food consumption in the AL group was at its maximum level at this time of day. All physiological and behavioral variables were subsequently measured continuously for another 10-day period. This additional procedure was utilized to synchronize the feeding regimen so that restricted mice were eating at the same time of day as when control mice ate most freely. The set of restricted mice which had been tested under light-fed (LF) conditions were tested under dark-fed (DF) conditions.

Conventional statistical methods (means and standard errors; Student's t-test) and the cosinor statistical method (Halberg et al. 1972) were used to analyze the data. Using the least squares method, the time of day when maximum values occur (acrophase) and the extent of the daily change (amplitude = ½ the fitted curve peak-through difference) were calculated for individual animals (single cosinor) and test groups (population-mean cosinor) (Halberg et al. 1972). CR significantly altered the waveform of the circadian rhythms for some variables so that cosinor analysis did not always accurately define the results. Therefore, terms such as range, mean values, etc., are given to aid in interpreting the data.

Results

The results from the present study were divided into two categories and the results compared among AL, LF, and DF mice. Basic statistical data relating to physiological variables and behavioral variables are given in Tables 24.1 and 24.2, respectively. Values for CR animals are expressed as percentages of AL values and the results of multiple t-test analysis delineate significant differences among the various test groups.

Physiological Variables

Various physiological parameters that are related to total energy output were compared among the AL, LF, and DF test groups. Body temperature was monitored in all three test groups, but metabolic variables were not measured in DF mice. Average body temperature per day and the range in body temperature per day were significantly higher in AL mice than in either of the CR groups. However, no significant difference was found between LF and DF for these variables. The rates of whole body oxygen utilization and carbon dioxide protection were significantly reduced in DF mice as a result of CR. When the oxygen and carbon dioxide respiratory gases were calculated per gram of lean body mass (LBM), no significant difference was found between AL and DF animals. Qualitative changes in metabolic pathways were determined by calculating the ratio of carbon dioxide protection to oxygen utilization, which we identified as RQ. Exposure to chronic CR caused a significant decrease in average daily RQ as well as a twofold increase in daily RQ range.

Behavioral Variables

Most of the behavioral variables that were related to food and water consumption monitored in this study were found to be significantly altered by CR in both of the feeding regimens (LF and DF). Both groups were given the same amount of food, but the amount of food spilled and not eaten was higher in the DF group than in the LF group. This resulted in slightly higher food consumption in the LF group (LF = 67% of AL; DF = 61% of AL), but no significant difference in total water consumption was found among the three groups. Additionally, the amount of food consumed per gram of LBM was significantly lower in both of the CR groups than AL, while water consumption per gram LBM was greater in LF than AL. However, food consumption and water consumption per gram LBM were greater in LF than DF. Average food and water consumption per episode and the

Table 24.1. Results of t-test analysis and means and standard errors for various physiological measures in female B6C3F$_1$ mice.

Measurement	AL group (Mean ± SE)	LF-restricted group (Mean ± SE)	(% AL)	DF-restricted group (Mean ± SE)	(% AL)	Significance levels for various comparisons AL vs. LF	AL vs. DF	LF vs. DF
Average body temperature/day (group average) (°C)	36.61 ± 0.06	34.92 ± 0.34	95.38	35.14 ± 0.17	95.98	A***	B***	—
Average body temperature/h (max-min) (24th range) (°C)	37.40 – 35.45 (1.94)	38.88 – 28.41 (10.47)	539.69	38.77 – 31.91 (6.86)	353.61	A***	B***	C*
Average O$_2$ consumption/day (g LBM) (ml g^{-1} h^{-1})	3.43 ± 0.12			3.29 ± 0.09	95.92	—	—	—
Average CO$_2$ consumption/day (g LBM) (ml g^{-1} h^{-1})	3.17 ± 0.12			2.98 ± 0.13	94.01	—	B*	—
Average RQ/day	0.92 ± 0.01			0.89 ± 0.02	96.74	—	—	—
RQ variation/day (max-min) (24-h range)	0.98 – 0.86 (0.12)			1.01 ± 0.74 (0.25)	208.47	—	B***	—

AL, ad libitum-fed; LF, restricted group fed during light period; DF, restricted group fed during dark period; LBM, lean body mass; A, AL × LF restricted comparison (significant effect); B, AL × DF restricted comparison (significant effect); C, LF restricted × DF restricted comparison (significant effect)
*$p < .05$
***$p < .001$

Table 24.2. Results of *t*-test analysis and means and standard errors for various behavioral measures in female B6C3F₁ mice.

Measurement	AL group (Mean ± SE)	LF-restricted group (Mean ± SE)	(% AL)	DF-restricted group (Mean ± SE)	(% AL)	Significance levels for various comparisons AL vs. LF	AL vs. DF	LF vs. DF
Total food consumption/day (g)	4.87 ± 0.16	3.27 ± 0.04	67.15	2.97 ± 0.02	61.13	A***	B***	C***
Food consumption (g⁻¹ LBM) (g)	0.20 ± 0.00	0.16 ± 0.00	82.14	0.15 ± 0.00	74.49	A***	B***	C***
Total water consumption/day (g)	3.71 ± 0.18	3.91 ± 0.17	105.37	3.60 ± 0.12	97.03	—	—	—
Water consumption (g⁻¹ LBM) (g)	0.15 ± 0.01	0.19 ± 0.01	128.00	0.18 ± 0.01	118.00	A**	—	C***
Average food consumption/episode (g)	0.35 ± 0.03	1.43 ± 0.07	413.04	1.12 ± 0.10	325.22	A***	B***	—
Average water consumption/episode (g)	0.33 ± 0.02	0.91 ± 0.15	277.61	0.44 ± 0.03	136.20	A*	B**	C*
Number feeding episodes/day	14.12 ± 1.12	2.29 ± 0.10	16.23	2.65 ± 0.18	18.78	A***	B***	—
Number drinking episodes/day	11.38 ± 0.86	4.32 ± 0.52	37.95	8.10 ± 0.69	71.18	A***	B**	C**
Average time active/feeding episode (min)	17.85 ± 3.29	67.60 ± 3.05	378.81	63.61 ± 5.25	356.42	A***	B***	—
Average time active/drinking episode (min)	7.11 ± 1.36	33.32 ± 6.42	468.50	12.26 ± 1.70	172.47	A**	B*	C**
Total time feeding/day (min)	251.97 ± 24.87	154.94 ± 5.76	61.49	168.62 ± 7.57	66.92	A**	B**	—
Total time feeding/day (g⁻¹ food) (min)	51.79 ± 7.11	47.40 ± 1.52	91.52	56.70 ± 3.54	109.47	—	—	C*
Total time drinking/day (min)	80.93 ± 7.48	143.89 ± 4.86	177.80	99.34 ± 11.13	122.76	A***	—	C**
Total time drinking/day (g⁻¹ water) (min)	21.82 ± 2.02	37.14 ± 1.96	170.22	27.60 ± 3.10	126.50	A**	—	C*
Total food consumption/total water consumption × 100 (%)	133.40 ± 4.30	84.40 ± 0.04	63.27	82.63 ± 2.30	61.94	A***	B***	—
Average activity/day (pulse/h)	9.65 ± 1.98	15.06 ± 2.74	156.06	14.13 ± 2.28	146.42	A***	B**	—

AL, ad libitum-fed; LF, restricted group fed during light period; DF, restricted group fed during dark period; LBM, lean body mass; A, AL × LF restricted comparison (significant effect); B, AL × DF restricted comparison (significant effect); C, LF restricted × DF restricted comparison (significant effect)
*p<.05
**p<.01
***p<.001

average time spent eating and drinking per episode were lower in AL than LF or DF. Average water consumption alone per episode and average time spent drinking per episode were greater in LF and DF. Conversely, the total number of feeding and drinking episodes per day and the ratio of food consumption to water consumption per day were higher in AL than LF or DF, and the number of drinking episodes per day was greater in DF than LF. Comparisons were also made among the test groups related to the time spent consuming a given substrate corrected for the amount of substrate consumed. The total time spent in drinking per day per gram of water consumption was higher in LF than AL. However, the total time feeding per day per gram of food consumption was greater in DF than LF, whereas the total time drinking per day per gram of water consumption was greater in LF than DF. Indigenous ambulatory behavior was also altered by chronic CR. Spontaneous total motor activity was significantly increased in LF- and DF-restricted groups compared to AL controls, and motor activity was higher in LF mice than DF mice.

Circadian Analysis

The timing and amplitude of the daily fluctuation in a variety of physiological and behavioral parameters were also altered by chronic CR in a time-dependent fashion. This is to say that CR caused dramatic effects in these measures at some stages of the circadian cycle and few, if any, effects at other stages. Circadian profiles for food and water consumption in female mice are illustrated for AL animals in Fig. 24.1A and for DF and LF animals in Fig. 24.1B. All values are expressed in hours after lights on (HALO). As was seen in a previous rat study (Duffy et al. 1989a), food and water were consumed by control mice at a significant rate throughout the dark phase of the circadian cycle, but food and water intake subsided to a low level shortly after the onset of the light phase. AL mice seemed to anticipate the dark period by sharply increasing their consumption rate before the lights were turned off so that maximum food intake occurred shortly after the onset of light phase. The waveform for food and water consumption in the two CR groups was distorted significantly compared to AL values. Unlike the AL mice that consumed their food and water at high rates over a 12-h time interval, CR mice started feeding immediately after food was presented to them and consumed nearly all of their ration during the next 4 h. Therefore, food and water consumption in LF mice was shifted out of phase with DF and AL mice.

Variations in the feeding regimen such as restricting the caloric intake and changing the time of day that food was presented to female mice were found to significantly modify the circadian rhythms for oxygen consumption and motor activity. Daily patterns for these variables, expressed in HALO, are represented for AL mice in Fig. 24.2A and for LF and DF mice in Fig. 24.2B. Although there is a high correlation between motor activity and oxygen utilization in AL mice, activity started to rise before the onset of lights off (anticipatory behavior), whereas motor activity in AL mice did not increase significantly until after lights

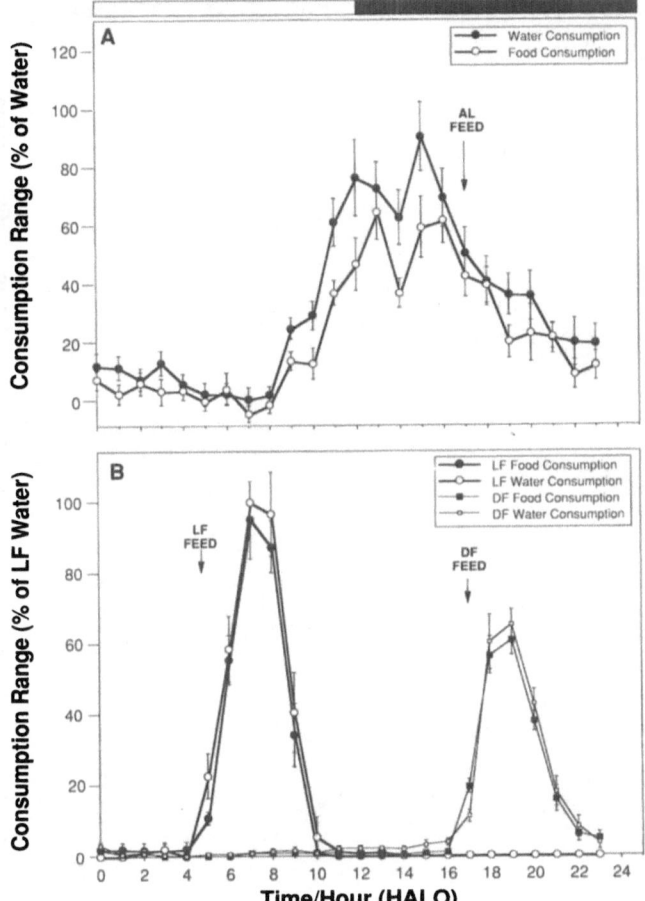

Figure 24.1A,B. Circadian rhythms for food and water consumption in old female B6C3F₁ mice are expressed as hourly mean values ± standard error. Data are given for ad libitum-fed mice (*AL*) in **A** and for light-fed (*LF*) and dark-fed (*DF*) caloric-restricted mice in **B**. All values are converted to percentage of daily range. Feeding time is denoted by (→ *FEED*) and *bars* above the graphs denote periods of light and dark

off. Both parameters reached their maximum levels within 5 h after the commencement of the nighttime phase and were maintained at significant levels throughout this period. Large differences in the timing of circadian rhythms for activity were expressed by LF and DF mice, since maximal activity occurred just before food was presented to the various groups. Therefore, motor activity in DF was fairly well synchronized with AL mice but was shifted 12 h out of phase with LF mice. As seen in the food consumption variable, activity in CR was compressed into a short 4-h interval. In contrast to AL mice, activity seemed to

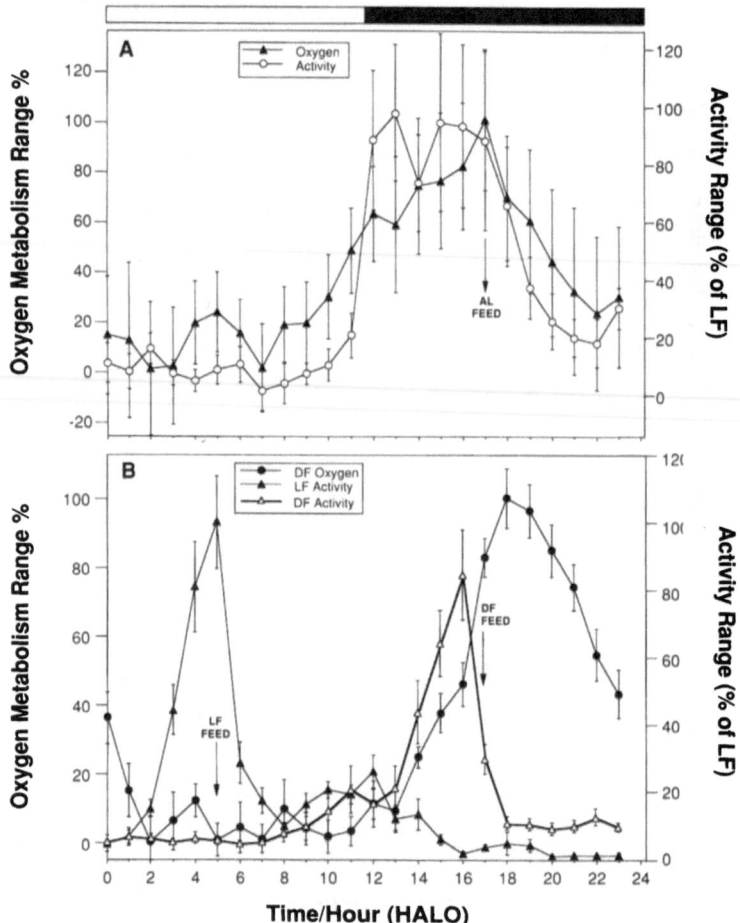

Figure 24.2. Circadian rhythms for oxygen metabolism and total motor activity in old female B6C3F$_1$ mice are expressed as hourly mean values ± standard error. Ad libitum-fed (*AL*) mice are represented in **A** and light-fed (*LF*) and dark-fed (*DF*) mice are represented in **B**. All values are converted to percentage of daily range. Feeding time is denoted by (→ *FEED*) and *bars* above the graphs denote periods of light and dark

precede oxygen metabolism in DF mice and dropped to a low level just before oxygen reached maximum level, indicating that DF mice were only active for a short time during the first half of the dark phase, whereas the rhythm for oxygen consumption in DF mice was not compressed to the same degree as activity.

RQ and body temperature were found to be highly synchronized with each other as well as with oxygen metabolism and activity in all three test groups, and the circadian timing of these variables was highly dependent on and correlated to the

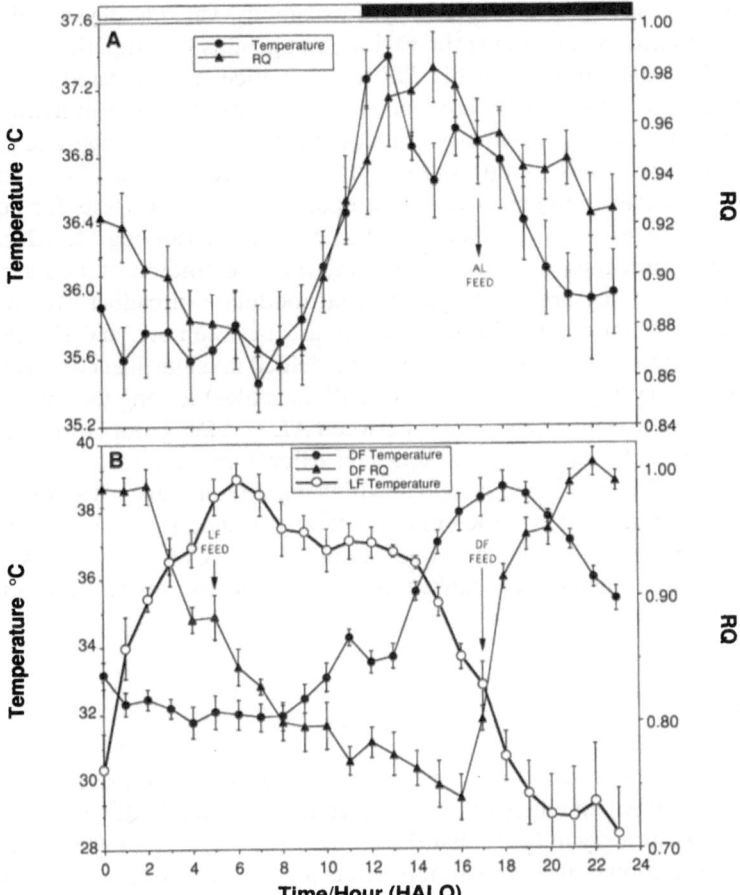

Figure 24.3. Circadian rhythms for body temperature and respiratory quotient (*RQ*) in old female B6C3F$_1$ mice are given as hourly mean ± standard error. Data for ad libitum-fed (*AL*) mice are represented in **A** and data for light-fed restricted (*LF*) and dark-fed restricted (*DF*) mice are represented in **B**. Feeding time is denoted by (→ *FEED*) and *bars* above the graphs denote periods of light and dark

feeding regimen. Daily patterns for body temperature and RQ are illustrated for AL mice in Fig. 24.3A and for LF and DF mice in Fig. 24.3B. Body temperature and RQ undergo low-amplitude fluctuations in AL animals and high-amplitude variations in LF and DF mice, so that at the circadian stage preceding the presentation of food, RQ and temperature values dropped well below the AL readings, and at the circadian stage following feeding, RQ values in DF mice exceeded those for AL mice. As seen in other variables, body temperature waveforms in the CR groups were similar but were out of phase in proportion to the respective feeding times.

The overall timing of circadian rhythms among various physiological and behavioral measures and among the various test groups was significantly altered by CR. Values for acrophase in AL, LF, and DF mice are compared in Fig. 24.4 for all measures and the results are given relative to two timing regimens, HALO and hours after feeding acrophase. Circadian acrophase and amplitude parameters for the various test groups and variables are given in Table 24.3, together with 95% confidence intervals and a comparison of various tests for equality. Nearly all of the respective acrophase and amplitude tests among AL, LF, and DF mice were significantly different, indicating that the amount of food given and the timing of feeding are major factors that modulate circadian rhythms. The exceptions were the comparisons of acrophase between AL and DF mice for water consumption and total activity, which yielded no significant differences. A comparison of average acrophase (across all variables) among the three groups indicated that the timing was similar between AL and DF groups (3 h apart) and that acrophase in LF mice were shifted 9 h earlier than AL mice and 12 h earlier than DF mice. The relative sequence in which parameters reached acrophase was similar between LF and DF mice in that activity peaked first, followed by food and water consumption and then body temperature. Conversely, in AL mice, food and water consumption peaked first, followed by temperature and then motor activity.

Discussion

The results of the present study together with those from previous studies in rats (Duffy et al. 1989a) and mice (Nelson and Halberg 1986a; Duffy et al. 1989b, to be published) strongly indicate that the amount of food consumed as well as the timing and duration of feeding strongly affect a wide variety of physiological and behavioral measures at the whole animal level. CR seemed to modify feeding behavior so that the eating of multiple meals in short episodes, which is the normal profile for AL, was replaced by a compressed feeding regimen characterized by the consumption of only a few meals over a longer time interval as was found in LF and DF mice. These results largely explain why the effects of reduced caloric intake on temperature and metabolic measures reported here are very similar to those found in previous studies in which CR was obtained by permitting animals to feed ad libitum for only a few hours (single ad libitum meal feeding) (Nelson et al. 1975; Philippens et al. 1977) rather than reducing the amount of food presented without limiting the duration of feeding, as was used in this study. Although both of these regimens produce compressed meal feeding and reduced caloric consumption, the dispensing of precise restricted amounts of food to animals in single cages is the optimal procedure for CR studies because it insures that each animal receives the same amount of food, thereby producing similar adaptations in physiological and behavioral performance.

The level of water consumption in female CR mice was not reduced to the same degree as food intake, as was previously reported in male rats (Duffy et al.

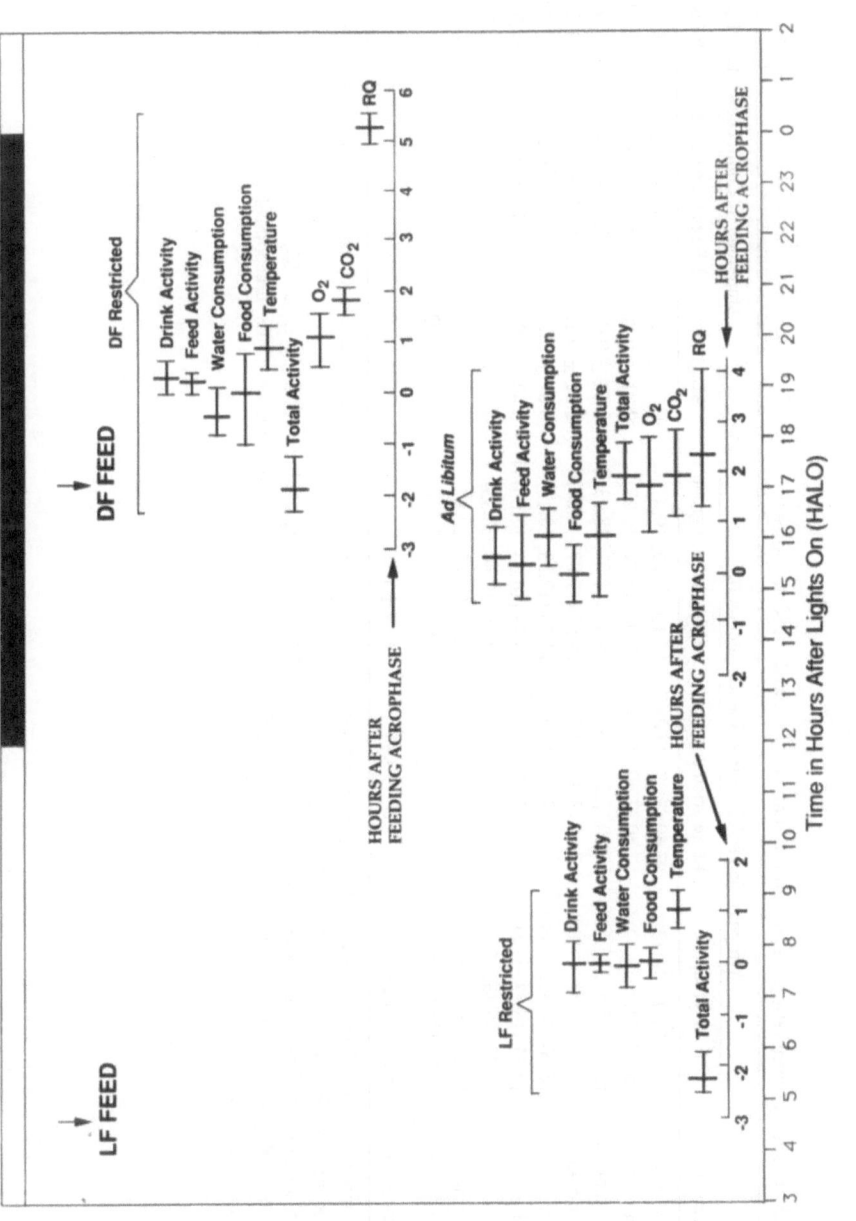

Figure 24.4. The relative timing (acrophase ± 95% confidence interval) of various physiological and behavioral measures are represented for ad libitum-fed, light-fed (*LF*), and dark-fed (*DF*) old female B6C3F₁ mice. *Bars* above the graphs signify periods of light and dark and (→ *FEED*) denotes feeding time. Time is expressed as hours after lights on (HALO) and hours after feeding acrophase to illustrate the degree of synchronization to the photoperiod cycle and/or feeding cycle. *RQ*, respiratory quotient

Table 24.3. Results of population mean cosinor analysis* in female B6C3F₁ mice.

Measurement	Acrophase in HALO + Confidence Interval (h)			Amplitude + Confidence Interval (h)		
	AL group	LF-restricted group	DF-restricted group	AL group	LF-restricted group	DF-restricted group
Drinking activity	15.6 (15.07, 16.13)	7.60 (7.07, 8.00)	19.10 (18.87, 19.47)	0.38 ± 0.11	0.48 ± 0.13	0.50 ± 0.11 A
Feeding activity	15.47 (14.8, 16.40)	7.60 (7.40, 7.73)	19.06 (18.87, 19.27)	0.01 ± 0.00	0.01 ± 0.00	0.01 ± 0.00
Water consumption	16.00 (15.47, 16.53)	7.50 (7.20, 7.80)	18.40 (18.07, 18.93) N	0.68 ± 0.08	2.32 ± 0.54	5.94 ± 3.25
Food consumption	15.27 (14.73, 15.80)	7.60 (7.33, 7.87)	18.87 (17.87, 19.60)	0.85 ± 0.12	1.88 ± 0.04	1.23 ± 0.25
Temperature	16.00 (14.86, 16.66)	8.60 (8.27, 9.00)	19.73 (19.33, 20.13)	0.87 ± 0.24	5.21 ± 1.21	3.30 ± 0.68
Total activity	17.27 (16.73, 17.80)	5.33 (5.07, 5.87)	16.93 (16.53, 17.60) N	8.49 ± 1.90	23.78 ± 9.87	24.74 ± 9.36
Oxygen	17.00 (16.13, 17.93)		19.93 (19.40, 20.33)	0.07 ± 0.01		0.17 ± 0.04
Carbon dioxide	17.20 (16.4, 18.07)		20.67 (20.47, 20.87)	0.10 ± 0.02		0.19 ± 0.02
Respiratory quotient	17.60 (16.60, 19.27)		24.06 (23.80, 0.33)	0.05 ± 0.02		0.13 ± 0.02
Group mean ± SE	16.37 ± 0.29	7.37 ± 0.44	19.64 ± 0.65			

AL, ad libitum-fed group; LF, restricted group fed during light period; DF, restricted group fed during dark period; A, results of tests for equality—comparison of amplitude between ad LF × DF (not yielding significant different; $p > .05$)

N, results of tests for equality—comparisons of acrophase between AL × DF (not yielding significant difference; $p > .05$)

* All tests for equality among various groups for acrophase and amplitude yielded significant differences ($p < .02$); exceptions N, A, and AL × LF (amplitude for drinking activity; $p < .05$)

1989a). This result may suggest that a high level of water consumption must be maintained to regulate a proper osmotic balance as well as reducing the appetites of CR mice by partially filling their stomachs with liquid, thereby satiating their hunger. However, unlike CR rats that reduce water consumption to a small but significant extent compared with AL animals (Duffy et al. 1989a), in CR mice, no significant decrease in water consumption was found.

Average body temperature was found to be reduced, and the daily range in body temperature and metabolic variables was found to be increased by CR in mice as well as rats (Duffy et al. 1989a). This increase in amplitude is probably due to the combined effects of compressed meal feeding and reduced caloric intake. However, the magnitude of these changes was found to be greater in small rodents such as mice than in larger rodents such as rats, suggesting that the effects of CR are species specific and that the relative body mass of an organism may be an important factor in modulating CR-induced changes in these variables.

The results of the present study suggest that there are several factors that may relate to basic mechanisms by which CR downregulates body temperature and metabolism at certain circadian stages:

1. Restricted meal feeding in CR mice causes exogenous energy supplies as well as an endogenous glycogen reserve to be rapidly metabolized in CR animals to meet immediate energy requirements. Therefore, decreased metabolic output and lower body temperature may be required during intervals when metabolic substrates are nearly exhausted.
2. CR may limit the transport of essential amino acids, such as L-tryptophan, to the brain during periods of low RQ (fatty acid metabolism) (Scott and Potter 1970; Leveille 1970). Limited neurotransmitter production (serotonin, etc.) could lead to decreased temperature and metabolic output prior to feeding (Li and Anderson 1987).
3. CR mice have smaller deposits of adipose tissue, which may be important in insulating these animals against excessive heat loss, thereby increasing the rate of thermal conductance and lowering body temperature.
4. CR seems to promote greater metabolic efficiency so that more work is done relative to the amount of available energy supplies.

Both of the restricted groups in this study expressed higher activity levels than the AL groups without increasing their metabolic rate. Similar results of increased energy conservation in CR rats were reported in studies related to intermediary metabolism. CR seemed to promote glucose metabolism under conditions of limited energy production and expenditure, thereby increasing metabolic efficiency and decreased energy storage (Feuers et al. 1989).

Significant alterations in metabolic pathways were associated with chronic CR, as indicated by rapid fluctuations in RQ over the circadian interval. Fatty acids seemed to the primary metabolic substrate for several hours before feeding (RQ=.75), while carbohydrates were the predominant energy source in postabsorptive CR mice (RQ=1.00). Periods of low RQ were associated with reduced temperature and metabolic output, suggesting that torpor may be triggered when carbohydrate reserves are depleted and fatty acids are the only source of energy. These findings would also indicate that qualitative changes in energy storage must have occurred to generate the supplies of fat required during periods of torpor. Scott and Potter (1970) and Leveille (1970) reported that the predominant form of energy storage in CR rodents was lipid synthesis, indicating that increased fatty acid production at one circadian stage (before eating) may have been necessary to furnish the high levels of fat (low RQ) that were required at another circadian stage (after eating). These results account for the high-amplitude oxygen metabolism, body temperature, and RQ rhythms found in the present study. Nelson et al. (1975) and Duffy et al. (1989a) reported similar findings in mice and rats, respectively.

There are several possible mechanisms by which reduced metabolism and body temperature may affect maximum achievable life span. Molecular factors which could affect survival, such as DNA repair (Lipman et al. 1987), DNA damage (Lindahl and Nyberg 1972), and proto-oncogene expression (Nakamura et al. 1989), were found to be decreased by CR and were highly correlated to body temperature, while immunological response was reported to be increased by lowering body temperature (Liu and Walford 1972).

CR studies provide valuable insight into the relative importance of environmental synchronizers such as feeding and photoperiod. The fact that LF mice consumed their ration immediately after it was presented, even though it was during the middle of the light phase of the circadian cycle, may indicate that the feeding synchronizer was dominant over the photoperiod synchronizer in CR mice. Unlike a previous study in old rats (Duffy et al. 1989a), where the activity component was split into two components, one corresponding to the presentation of food and the other corresponding to the onset of the dark phase, in mice, all motor activity was shifted to synchronize with the onset of feeding. These results suggest that the relative ability of various environmental factors to reset or synchronize the biological clock may be different in mice than in rats.

The sequence in which parameters react to environmental stimuli may indicate possible cause-and-effect relationships among these measures that may be related to primary mechanisms by which CR modulates basic physiological and behavioral processes. In the present study the upswing in motor activity expressed by CR mice during the torpid period prior to feeding may have been necessary to increase body temperature to a normal level during a period of fat metabolism when temperature and metabolic homeostasis was inoperative. However, the fact that spontaneous motor activity dropped to a low level with no reduction in temperature or metabolism immediately after the animals ingested exogenous carbohydrates suggests that normal endogenous metabolic regulation was reestablished and motor activity was no longer required to modulate metabolic and temperature output. Since the onset of food consumption, followed by increased metabolism and temperature in AL mice, occurred during the dark phase prior to the initial surge in spontaneous activity, normal homeostatic temperature regulation seemed to be maintained without the aid of motor activity.

It is interesting to note that the physiological performance of CR mice during torpor closely emulates that seen in extremely long-lived species that express torpor without CR (Duffy et al. 1987). Characteristics such as low body temperature and metabolism (decreased homeothermic regulation), accumulation of lipid stores prior to torpor, metabolism of fatty acids during torpor (low RQ), the expression of motor activity and the switch to carbohydrate metabolism during arousal, and the correlation between low food supplies and hypothermia are all common to hibernating animals (Lyman 1982) and CR animals (Duffy et al. 1989a). These data strongly suggest that the torpor associated with CR is a form of minihibernation, which is triggered by an alternative form of regulation resulting from a low-calorie diet. As in true hibernation, CR-induced torpor is a form of controlled hypothermia regulation rather than a lack of regulation. The fact that CR animals and hibernating rodents have increased life spans compared to rodents that do not express these traits leads to speculation that short- or long-term torpor may slow the rate of aging.

A previous study found that CR increased life span to the same extent in all feeding regimens regardless of whether food was presented during the light phase, the dark phase, or at six intervals during the dark phase of the circadian cycle (Nelson et al. 1975). However, this does not diminish the importance of using chronobiological testing techniques in nutritional, gerontological, and toxicological research. If feeding does not occur in control and CR animals at the same circadian stage, variables will not be in the same phase of their rhythm even though they are sampled at the same clock time. This factor could cause investigators to misinterpret circadian variability for specific effects resulting from CR. Since the presentation of food was found to be the dominant environmental synchronizer in the study, it is imperative that synchronized nocturnal feeding be used in all future CR experiments, particularly when data can only be sampled at one circadian stage or only a few stages. The results of this study cast serious doubt on the conclusions drawn in previous experiments that used daytime feeding regimens if chronobiological variables were not accounted for. However, the results of the present study provide a useful tool that can be used in reevaluating existing CR studies in which the animals were not properly synchronized. The timing of acrophase for most variables was found to be similar among the three test groups when data were corrected for feeding time. Maximum food consumption (acrophase) occurred approximately 3.3 h after the onset of lights off in AL animals. Therefore, if data from CR and AL mice are normalized to hours after feeding acrophase, results from the various test groups can be accurately compared at the same circadian stages, thereby eliminating a large component of chronobiological variation.

References

Duffy PH, Feuers RJ, Hart RW (1987) Effect of age and torpor on the circadian rhythms of body temperature, activity, and body weight in the mouse (Peromyscus leucopus). In: Pauly J, Scheving L (eds) Advances in chronobiology, Part B. Liss, New York, pp 111–120
Duffy PH, Feuers RJ, Leakey JA, Turturro A, Hart RW (1989a) Effect of chronic caloric restriction on physiological variables related to energy metabolism in the male Fischer 344 rat. Mech Ageing Dev 48:117–133

Duffy PH, Feuers RJ, Hart RW (1989b) Effect of chronic caloric restriction on the circadian regulation of physiological and behavioral variables in old male $B_6C_3F_1$ mice. Chronobiol Int (submitted)

Feuers RJ, Duffy PH, Leakey JA, Turturro A, Mittelstaedt RA, Hart RW (1989) Effect of chronic caloric restriction on hepatic enzymes of intermediary metabolism in the male Fischer 344 rat. Mech Ageing Dev 48:179–189

Halberg F, Johnson EA, Nelson W, Runge W, Sothern RB (1972) Autorhythmometry procedures for physiological self-measurements and their analysis. Physiol Teacher 1:1–11

Kritchevsky D, Weber MM, Klurfeld DM (1984) Dietary fat versus caloric content in initiation and promotion of 7,12-dimethylbenz(a)-anthracene-induced mammary tumorigenesis in rats. Cancer Res 44:3174–3177

Leakey JEA, Cunny HC, Bazare J, Webb PJ, Feuers RJ, Duffy PH, Hart RW (1989) Effects of aging and caloric restriction on hepatic drug metabolizing enzymes in the Fischer 344 rat. I. The cytochrome P-450 dependent monoxygenase system. Mech Ageing Dev 48:145–155

Leveille GA (1970) Adipose tissue metabolism: influence of periodicity of eating and diet composition. Fed Proc 29:1294–1301

Li ETS, Anderson GH (1987) Dietary carbohydrate and the nervous system. Nutr Rev 7:1329–1339

Lindahl T, Nyberg B (1972) Rate of depurination of native deoxyribonucleic acid. Biochemistry 11:3610–3618

Lipman JM, Turturro A, Hart RW (1989) The influence of dietary restriction on DNA repair in rodents: a preliminary study. Mech Ageing Dev 48:135–143

Liu RK, Walford RL (1972) The effect of lowered body temperature on life span and immune and non-immune processes. Gerontologia 18:363–388

Lyman CP (1982) Who is who among the hibernators. In: Lyman W, Willis JS, Malan A, Wang LCH (eds) Hibernation and torpor in mammals and birds. Academic, New York

Maeda H, Gleiser CA, Masoro EJ, Murata I, McMahan CA, Yu BP (1985) Nutritional influences on aging of Fischer 344 rats. II. Pathology. J Gerontol 40:671–688

McCarter R, Masoro EJ, Yu BP (1985) Does food restriction retard aging by reducing the metabolic rate? Am J Physiol 248:E488–E490

McCay CM, Crowell MF, Maynard LA (1935) The effect of retarded growth upon the length of life span and upon the ultimate body size. J Nutr 10:63–79

Nakamura KD, Duffy PH, Turturro A, Hart RW (1989) The effect of dietary restriction on MYC proto-oncogene expression in mice: a preliminary study. Mech Ageing Dev 48:199–205

Nelson W, Halberg F (1986a) Meal-timing, circadian rhythms and life span of mice. J Nutr 116:2244–2253

Nelson W, Halberg F (1986b) Schedule-shifts, circadian rhythms and life span of freely-feeding and meal-fed mice. Physiol Behav 38:781–788

Nelson W, Scheving LE, Halberg F (1975) Circadian rhythms in mice fed a single daily meal at different stages of lighting regimen. J Nutr 105:171–184

Philippens KMH, Mayersbach HV, Scheving LE (1977) Effects of the scheduling of meal-feeding at different phases of the circadian system in rats. J Nutr 107:176–193

Ruggeri BA, Klurfeld DM, Kritchevsky D (1987) Biochemical alterations in 7,12-dimethylbenz[a]anthracene-induced mammary tumors from rats subjected to caloric restriction. Biochim Biophys Acta 929:239–246

Sacher GA (1977) Life table modification and life prolongation. In: Finch C, Hayflick L (eds) Handbook of the biology of aging. Van Nostrand Reinhold, New York

Sarkar NH, Fernandes G, Telang NT, Kourides IA, Good RA (1982) Low-calorie diet prevents the development of mammary tumors in C3H mice and reduces circulating prolactin level, murine mammary tumor virus expression, and proliferation of mammary alveolar cells. Proc Natl Acad Sci USA 79:7758–7762

Scott DF, Potter VR (1970) Metabolic oscillation in lipid metabolism in rats on controlled feeding schedules. Fed Proc 29:1553–1559

Walford RL, Liu RK, Gerbase-Delima M, Mathies M, Smith GS (1974) Long term dietary restriction and immune function in mice: response to sheet red blood cells and to mitogenic agents. Mech Ageing Dev 2:447–454

CHAPTER 25

Genes Regulate Effects of Dietary Restriction*

D.E. Harrison[1], J.R. Archer[1], and B. Kent[2]

Introduction

Knowledge of how food restriction retards aging processes might suggest clinically useful treatments in man. This nutritional manipulation extends life spans of short-lived rodents, and is most effective if begun early in life (McCay et al. 1939; Ross 1976; Barrows and Kokkonen 1977; Young 1979; Yu et al. 1985; Weindruch et al. 1986). It is the only treatment that consistently extends mammalian life spans beyond the longest known for the species tested without food restriction.

Food restriction might extend longevity by (1) retarding immune dysfunction with age or (2) reducing long-term metabolic rates. Food restriction preserves immune responses in normal aging mice (Walford et al. 1974; Weindruch and Suffin 1980; Weindruch et al. 1982) and retards immunopathological changes in autoimmunity-prone mice (Fernandes et al. 1978). T-cell precursor frequencies remain near young levels in old, food-restricted C57/B16J (B6) × CBA/HT6CaJ (CBA) F_1 hybrids (Miller and Harrison 1985). These data fit the idea that immune dysfunctions are important in causing aging and suggest that beneficial effects of food restriction result from retarding immune dysfunction with age. Higher levels of immune function obviously would contribute to extended longevity by increasing resistance to infectious agents and to certain tumors. In addition, autoimmunity may increase with age as normal immune responses decline, and autoimmune reactions may cause many of the deleterious effects of aging (Walford 1969).

However, other sets of data contradict the idea that immune dysfunctions are important in aging. Some immune responses are reduced in food-restricted mice (Christadoss et al. 1984) and longevity is extended in genetically obese C57BL/6J (B6)-*ob/ob* females without improving their immune responses (Harrison et al. 1984). B6-*ob/ob* mice have impaired cellular immunity (Meade and

*This work was supported by grants AG06232 and AG00594 from the National Institute on Aging and by grant DK25687 from the National Institute on Diabetes and Kidney Disease. These institutes are not responsible for its content, nor do they necessarily represent the official views of those agencies.
[1]The Jackson Laboratory, Bar Harbor, ME 04609, USA and [2]Mt. Sinai Medical Center, One Gustave Levy Pl, New York, NY 10029, USA

Sheena 1979) expressed when mutants are sensitized in vivo (Nichols et al. 1978). While there may be an immunosuppression caused by the elevated insulin and corticosteroid levels of *ob/ob* mice, levels of these hormones were elevated at young ages when the splenic immune responses were normal. The immune responses per spleen then showed accelerated rates of declines with age (Harrison et al. 1984) that food restriction failed to alter.

Food restriction may also retard aging by reducing metabolic rates. Rubner (1908) calculated that domestic animals of a wide range of sizes use over their lifetimes a similar number of calories per gram body mass. Pearl (1928) proposed the "rate of living" hypothesis: the higher the metabolic rate per gram body mass, the shorter the life span of a species. Thus, food restriction might extend longevity by retarding metabolic rates.

Why reducing metabolic rates may reduce aging rates is explained by current theories that aging is caused by free radical damage (Harman 1981), mostly from toxic oxygen radicals, so the rate of O_2 utilization should affect rates of aging. This is supported by the finding of Sacher and Duffy (1979) that longevities in 26 mouse strains and F_1 hybrids were negatively correlated with average O_2 consumption. Sacher's (1977) calculations using data published by Ross (1969) showed that rats fed different diets had a wide range of longevities, but all groups had similar lifetime calorie intakes per gram. In precise studies of metabolic rates over 3–4 weeks, young Sprague-Dawley rats adapted to a restricted diet by reducing their O_2 consumption (Forsum et al. 1981), and the degree of reduction in Wistar rats was proportional to the reduction in body weight (Hill et al. 1985). However, 3–4 weeks does not represent long-term food restriction.

In the only study that measured effects of food restriction over a long time, F344 rats consumed equal amounts of O_2 per gram lean body mass per day whether food was restricted or fed ad libitum (McCarter et al. 1985). This contradicts the metabolic hypothesis and fits lifelong food restriction studies, in which Masoro et al. (1982) and Weindruch et al. (1986) found that rats and mice limited to one half to two thirds of normal caloric consumption were much smaller and longer lived, and so consumed more calories per gram during their lives than did ad libitum-fed controls. In our initial results using mice of four genotypes fed high- and low-calorie diets over about 18 months, oxygen consumption per mouse was slightly higher for those fed low-calorie diets (food restricted) in three of the four genotypes. The type of housing was very important in B6D2F$_1$ mice, with significantly higher oxygen consumption per mouse when they were caged singly.

These results do not necessarily contradict the idea that an important cause of aging is free radical damage. Increased O_2 utilization might not be harmful if protective mechanisms against free radicals (such as antioxidants) were also increased proportionally. While protective measures are vital to test, unfortunately, there are too many antioxidants and enzymes with potential protective action to measure practically. The critical point is the rate of oxidative damage to the organism. Total lipid peroxidation reactions in tissue homogenates may be measured by methods that correlate with maximum life span potentials of the species (Cutler 1985). Direct measures of oxidative damage to long-lasting molecules such as DNA (Cathcart et al. 1984) should also be used to determine whether food restriction reduces such damage.

B6-*ob/ob* mice are important because they greatly increase weight gain and adiposity when fed the same amount as normal mice (Coleman 1985). Thus they appear to use different metabolic pathways than normal mice, perhaps producing and burning fat rather than glycogen. Food restriction extends longevities of *ob/ob* mice (Lane and Dickie 1958; Harrison et al. 1984; Harrison and Archer 1987); the latter two studies showed that it retards their collagen aging and aging in several other biological systems, although in the 1984 study food-restricted obese mice rapidly lost immune responses with age.

Materials and Methods

This paper reviews many studies that have already been reported. References will be given and their methods summarized. Methods will be more detailed when they have not been published elsewhere.

Mice

We studied mice of the following genotypes: C57BL/6J (B6), CBA/CaH-T6J (CBAT6), BALB/cByJ (BALB), DBA/2J (D2), WB/ReJ (WB), CBA/HT6CaJ (CBAT6), AEJ/GnRkJ, LP/J, Mol/J, RF/J, SM/J, Spretus/J, the F_1 hybrids B6D2F$_1$, WBCBAT6F$_1$, B6CBAT6F$_1$, BALBD2F$_1$ and WBB6F$_1$, and genetically obese B6-*ob/ob* mice (congenic with the B6 line). All were produced and maintained at The Jackson Laboratory, which is fully accredited by the American Association for Accreditation of Laboratory Animal Care. All mice were virgins and were introduced into the aging colony at weaning (4 weeks of age, retired breeders were 8 months old). When introduced, they were fed by one of the regimens detailed below and were housed in filter-hooded, double-sided plastic cages in isolated rooms under positive pressure with high-efficiency particulate air (HEPA) filtered at room temperature ($22 \pm 2\,°C$), lit from 7 am to 7 pm. Mice were housed four per side with others of the same genotype, except where noted that individuals were housed singly. Details of the animal health assessment procedure for over 30 microorganisms have been published (Harrison and Archer 1983). No significant incidences of pathogenic organisms were detected, and only mice appearing healthy to an experienced observer were used. In many cases, the tests did not harm the subjects, so subsequent longevities were determined after testing.

Feeding Protocols

All mice received chlorinated water ad libitum, acidified to prevent growth of *Pseudomonas*, and were fed as follows:

Protocol I. In the earlier food restriction protocol used for much of the data in this paper (Harrison et al. 1984; Miller and Harrison 1985; Harrison and Archer 1987), all mice are fed a grain-based, pelleted, pasteurized diet (96WA, Emory

Table 25.1. Protocol IIA, AIN 76-based diets.

Ingredient	Percentage weight	
	Ad libitum-fed mouse diet	Restricted mouse diet
Casein	18.0	25.56
DL-Methionine	0.27	0.382
Sucrose	27.42	23.21
Corn starch	32.26	27.49
Dextrose (monohydrate)	5.0	4.24
Fiber (cellulose type)	5.0	5.0
Corn oil	5.0	4.14
AIN 76 mineral mix	5.25	7.43
AIN 76 vitamin mix	1.50	2.12
Choline bitartrate	0.30	0.424

Diets used in food restriction protocol IIA. Food-restricted mice are given 72% the weight of diets eaten by ad libitum-fed animals. Amounts of most nutrients are constant, but caloric intake is reduced to 68% by reducing the amount of carbohydrate and oil.

Morse, Guilford, CT), containing 22% protein, 7% fat, 50% nitrogen-free extract (mostly carbohydrates), and 357 kcal/100 g. Restricted mice receive their rations each day in a single feeding between 12:00 and 2:00 pm, 6 days a week, with double rations on the 6th day, and none on the 7th. Ad libitum-fed mice have access to food at all times. These feeding regimens are continued throughout life. Restricted mice are given two thirds of the amount of food consumed by ad libitum-fed mice, with one or more food pellets each when fed. There were no unusual losses from fighting among food-restricted mice, and the variations in body weights of cage-mates were similar in fed and restricted groups, suggesting that the dominant mouse did not consume an unusually large share of food in the restricted cages.

Automated Feeding System Used in Protocols IIA and IIB. Mice are fed a measured amount of powdered food every day using an automated system consisting of long troughs into which the food is placed each week, with an evenly spaced string of eight sliding dividers, with each space between dividers holding a 1-day supply of food. A timing mechanism activates a small motor to slide the dividers along one space at the designated feeding time each day. Between the last two dividers is a hole in the trough through which the food falls into the mouse cage. To use the feeders, the trough is filled to excess in a larger container, and all but the correct amount of food is removed, using scrapers that ride on the trough sides to remove food a measured distance towards the bottom. The sliding dividers are pressed firmly into the powdered food and reattached to the motor and timer. Each trough is long enough to extend over several cages, with several sets of sliding dividers, so that refilling the troughs is efficient.

Protocol IIA. Total amounts of all nutrients except carbohydrates and fats are the same among animals fed ad libitum and restricted, and the fiber percentage is the same. The diet is semichemically defined, based on the AIN 76 diet, as described by Bieri et al. (1977), but with the following alterations. Protein is reduced as shown in Table 25.1, and starches are changed from 50% sucrose plus 15% corn starch to 27.4% sucrose, 32.3% corn starch, and 5% dextrose to

Table 25.2. Protocol IIB, chronic food restriction diets.

Ingredient	Percentage weight	
	High energy	Low energy
Casein	8.0	9.76
Soybean meal	12.5	15.25
Alfalfa meal	5.0	6.1
Wheat	20.0	24.4
Oats	17.0	20.73
Corn starch	20.07	15.91
Corn silk	11.0	0.00
Calcium phosphate	3.0	3.66
Calcium carbonate	0.80	0.98
Sodium chloride	0.50	0.61
Safflower oil	1.0	1.2
Mineral mix	0.77	0.94
Vitamin mix	0.30	0.35

Diets used in food restriction protocol IIB. Amounts of nutrients fed are constant, except that 11% corn oil and an extra 7% cornstarch are given in the high-energy diet. Using automatic feeding systems, food-restricted mice are given the low-energy diet at 82% of the weight of the high-energy diet given to control animals; their caloric intake is about 70% of that in controls. In the study of early weight gain (Fig. 25.5), two similar diets are added: pelleted high-energy diet, and pelleted low-energy diet with an added 18% non-nutritive oat bran so that nutrient percentages are the same in both (except corn oil and cornstarch); these diets are fed ad libitum.

increase amounts of complex starches. The resulting powder can still be pelleted. The vitamin and mineral mix supplements are those recommended by the AIN 76 committee (Bieri et al. 1977) but amounts are increased by a factor of 1.5 to insure against potential deficiencies developing with age. In addition, vitamin K is increased 10-fold in the vitamin mix. This diet, detailed in Table 25.1, is purchased from TEKLAD and was designed with the help of Dr. R. J. Rose.

Protocol IIB. This diet is based on natural products, and was specially designed by Dr. J. Knapka for lifelong feeding of high- and low-energy diets. Total amounts of all components except corn starch and corn oil are the same in the high- and low-energy diets, as shown in Table 25.2. The low energy diet is used in two forms. One is fed at 82% of the high-energy diet using the automated feeding system to provide measured amounts of each daily. The other contains an additional 18% oat bran and is fed ad libitum.

Measurements in Different Biological Systems

Food (given in food restriction experiments) is the amount eaten recorded as grams per day per mouse and measured for at least 1 week during the period when the physiological systems were being tested. Weight is body weight in grams measured when the other tests were performed. Tail length is measured in centimeters from the root of the tail where it joins the anus to the tip of the tail. Changes with age in different biological systems can be used to evaluate treatments to alter aging rates (Harrison and Archer 1988b), in this case dietary restriction.

Tail collagen aging is given as the denaturation rate—the number of minutes required for a single tendon fiber from the middle of the tail to be denatured in 7 M urea at 45°C, so that it is broken by a 2-g weight (Harrison et al. 1978). Urine-concentrating ability is measured as the osmolality in mosmol/kg of urine from a mouse deprived of water for 48 h. Urine samples are held in capillary tubes to prevent evaporation, food remains present, and mice are injected with 1.0 ml saline intraperitoneally at time zero (Harrison and Archer 1983).

Thymus-dependent immune responses were measured using spleen cells, as detailed previously (Astle and Harrison 1984; Harrison et al. 1984). Formation of direct plaques in response to sheep red blood cells (SRBC) was measured in vivo by a modified Jerne Plaque assay. Proliferation in response to the mitogen phytohemagglutinin (PHA) in vitro was measured by tritiated thymidine uptake. Since the spleen must be removed, these measures can only be done once, and may affect subsequent longevity.

Limiting dilution assays were done using blood cells, as detailed previously (Miller 1984; Miller and Harrison 1985). Helper and killer T-cell frequencies were calculated from titration curves in which doses of 5, 10, 20, or 40 whole blood leukocyte responder cells were added to each microtiter well, which contained appropriate media, supplements, and filler cells. Each well was scored as positive if its supernatant supported T-lymphocyte proliferation (for helpers) or if its cells caused target cell lysis (for killers) at least three standard deviations above the control values for wells containing no responder cells.

Tight wire refers to a test measuring the time in seconds that a mouse can hold itself on a wire suspended above foam padding. Timing is begun after the mouse grips the wire with both front paws and one hind paw. There is no exit from the wire, but mice are removed and testing ends if they cling for 240 s in a single trial. Otherwise the maximum score of five trials is recorded (Campbell 1982). Correlations between tight-wire clinging times and body weights among mice of the same age and genotype showed that only a small portion of the variance could be explained by increased weight causing decreased clinging times. For example, in B6 males at 500–614 days of age, correlation coefficients (r) between clinging times and body weights for ad libitum-fed and food-restricted $+/+$ mice and for food-restricted ob/ob mice were all negative, -0.266, -0.159, and -0.409, but only explained 7%, 3%, and 17%, respectively, of the variances in clinging times (Harrison and Archer 1987).

Open field is a measure of the activity displayed during 5 min in an 80-cm, square, well-lit, open-topped box with 80-cm sides, measured as the number of 15-cm-sided squares that the mouse crosses (Sprott and Eleftheriou 1974). Hair growth is a test measuring the fraction of a 2-cm square, shaved area with 2-cm sides centered on the back near the tail into which hair has begun to regrow after 25 days. The subjective element in estimating the area is minimized by defining regrowth as the first appearance of hair, and dividing the shaved area to be scored into 8 equal portions with a transparent screen. Excellent repeatability was found when 22 mice were randomly recaged and retested an hour after the initial trial; 20 were scored exactly the same way both times, and scores of the other 2 varied by only one of the 8 screen divisions (Harrison and Archer 1987).

Wound healing is a test measuring the number of days required until the wound made in the tail in removing a portion of one of the dorsal tail tendon bundles for the tail collagen test feels smooth to the touch when running the index finger along the tail. Tails were checked 7 days after the wounds were made and every 2–3 days thereafter. Since this test is subjective, all animals in an experiment were scored by the same technician. In a blind test of repeatability, the correlation coefficient between successive runs was 0.88 with 16 mice. In the same experiment, one that was designed to evaluate the technician, wounds were made at different times over 3 weeks, but presented as if all had been made simultaneously. Expected healing times were 27.2, 34.5, 41.2, and 34.8 days for four groups of mice; reported times were 24.5, 36.5, 40.0, and 35.6, respectively, demonstrating good objectivity (Harrison and Archer 1987).

Hemoglobin is tested as the concentration of hemoglobin in blood, measured by taking blood from the retroorbital sinus heparinized in a 70-ml microhematocrit capillary tube (Fisher), mixing 10 ml blood with 2.50 ml Drabkin's reagent (1.0 g sodium bicarbonate, 0.05 g potassium cyanide, 0.20 g potassium ferricyanide dissolved in 1000 ml water) and measuring absorption 5–60 min later using a spectrophotometer (Helena Digispec) at a wavelength of 540 nm.

Metabolic rates were measured by indirect calorimetry on individual mice for 24-h periods using an adaptation of the open circuit technique for oxygen consumption used by Brooks and White (1978), Forsum et al. (1981), Hill et al. (1985), and McCarter et al. (1985). An accurately measured airflow through the cage was dried with calcium sulfate and monitored by a Beckman oxygen electrode. The electrode signal was processed through a MacLab interface (Precision Instruments) on an Apple Macintosh Plus microcomputer to measure the amount of oxygen used.

Statistical significances of differences between groups were tested using the Student-Newman-Keuls multiple range test or a multifactor ANOVA (using the Statview SE program by Abacus Concepts, Berkeley, CA). Correlation coefficients (r) were calculated for linear correlations using the same program.

Results and Discussion

Food Restriction Effects on Obese and Normal B6 Females

Harrison et al. (1984) studied the effects of lifelong food restriction beginning at 4 weeks of age in genetically obese (*ob/ob*) and normal B6 females (Table 25.3). Food restriction greatly reduced body mass in obese mice, but they retained very high percentages of body fat, nearly 50%. Kidney function, measured as urine concentrating ability, was not impaired by obesity. Immune responses were

Table 25.3. Effects of food restriction and obesity in B6 females.

Measure	B6 – +/+ females		B6 – *ob/ob* females	
	Fed ad libitum	Restricted	Fed ad libitum	Restricted
Food (g/day)	3.0	2.0	4.2	2.0
Weight (g)	30 ± 1	20 ± 2	59 ± 5	28 ± 2
Fat (% body weight)	22 ± 6	13 ± 3	67 ± 5	48 ± 1
Tail collagen (denaturation; min)	46.5 ± 2.2	33.3 ± 0.8	77.0 ± 8.9	33.7 ± 0.9
Urine concentration (mosmol)	2600 ± 140	3000 ± 180	3400 ± 120	3400 ± 80
Immune responses[a]				
SRBC	6 ± 3	16 ± 8	2 ± 2	3 ± 2
PHA	124 ± 31	127 ± 28	33 ± 18	21 ± 15
Spleen weight (mg)	150 ± 8	78 ± 4	75 ± 17	36 ± 3

Values are mean ± SEM. Mice in the first three lines were 180–360 days old with ages matched in all groups, and 4–8 of each type tested. Tail collagen data are given for 488-day-old animals (4–6 per group), urine concentrations for 329- to 364-day-old mice (13–17 per group), and immune responses for groups averaging 550–560 days of age (8–14 per group). Ad libitum-fed 100-day-old *ob/ob* controls had the following values for SRBC and PHA responses, and spleen weights with 21 mice: 31 ± 5 and 201 ± 41, and 105 ± 8. Values for 21 ad libitum-fed 100-day-old +/+ controls were: 27 ± 5 and 254 ± 18, and 112 ± 3. Most of these data were presented in Harrison et al. (1984), and food restriction was done using protocol I.
[a]Immune responses are given per spleen as number of direct anti-SRBC plaques × 10^4 in vivo and cpm tritiated thymidine uptake × 10^5 in vitro. SRBC, sheep red blood cells; PHA, phytohemagglutinin

normal in young obese mice (see legend to Table 25.3), but were less than 10% of normal in 550-day-old obese mice. Food restriction failed to mitigate this rapid immunological aging (Table 25.3). Denaturation rates of tail collagen were accelerated in obese mice, but this accelerated aging process was greatly retarded by food restriction to exactly the same rate of collagen aging as found in food-restricted normal mice (Table 25.3).

Food restriction extended longevities in the same pattern by which it retarded collagen aging: much more in obese mice than in normal mice, so that food-restricted obese and normal mice had similar longevities (Fig. 25.1). These results suggest that reduced food consumption and reduced collagen aging, but not reduced adiposity or improved immune functions, are essential in extended longevity of *ob/ob* mice. These data do not support the suggestions that food restriction extends longevity by reducing fat or improving immune responses and suggest that food restriction acts by retarding collagen aging.

Food Restriction Effects on B6CBAF₁ Males and Obese and Normal B6 Males

A similar food restriction experiment compared B6-*ob/ob* and +/+ males and B6CBAF₁ males (Harrison and Archer 1987). Food-restricted F1 hybrids behaved as expected, having results significantly more like those of younger mice

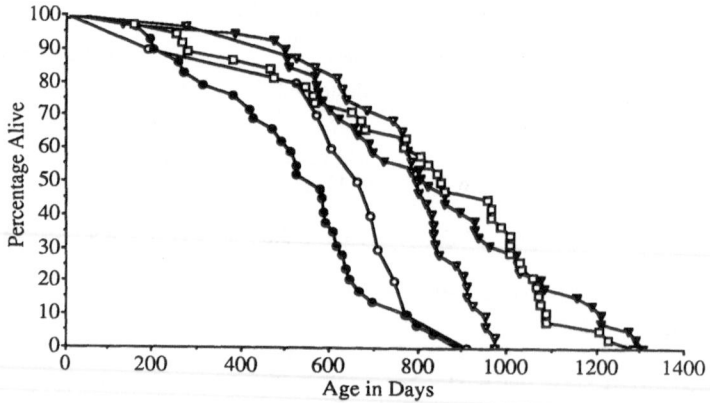

Figure 25.1. Food restriction by protocol I. Longevity plots as percent remaining alive versus age in days for B6 females. Data for ad libitum-fed *ob/ob* mice that were then restricted from 8 months of age plotted as *open circles* (*n*=10), ad libitum-fed *ob/ob* mice as *solid circles* (*n*=29), ad libitum-fed +/+ mice as *open triangles* (*n*=30), food-restricted *ob/ob* mice as *solid triangles* (*n*=39), and food-restricted +/+ mice as *open squares* (*n*=38). Some of these data are in Table 25.6 and in Harrison et al. (1984)

in most systems (tail collagen, urine concentration, hair regrowth, tight wire, and open field activity), but the opposite results in wound healing (Table 25.4). Longevities were extended from a mean of 974 days (already quite long lived) to 1180 days, and the longest life span was 1742 days, a new record for *Mus* (Fig. 25.2).

B6-*ob/ob* males benefitted greatly from food restriction in all measures, as females had (Table 25.3), and their longevities were greatly extended (Fig. 25.3). However, B6 +/+ males showed a dramatic difference in response to the same food restriction treatment, with longevities significantly reduced from a mean of 858 days to 593 days by food restriction (Fig. 25.3). Nevertheless, restricted B6 +/+ males had test results significantly more like those of younger mice than ad libitum-fed B6 +/+ males in tail collagen, tight wire, and open field activity. Hair regrowth, however, was much slower in the restricted B6 males and was the only measure that predicted the reduction in longevity; wound healing was poorer, as it had been in F₁ males (Table 25.5).

Taken alone, the experiments with F₁ hybrids would have led to oversimplified conclusions. Longevities were extended by food restriction, and values were more like those of young animals in five physiological systems. If this had been the only group studied, it would have seemed reasonable to suggest that all five systems aged more slowly and could be used to evaluate antiaging treatments in any mouse genotype. However, food restriction reduced longevities in B6 +/+ males while maintaining three of the same five systems at levels characteristic of younger mice. Thus the life-extending effects of a particular food restriction regimen may depend on the genotypes of the animals treated and the composition of

Table 25.4. Effects of food restriction on B6CBAF$_1$ males.

Measure	Food restricted		Ad libitum fed	
	Adult	Aged	Adult	Aged
Food (g/day)	2.0	2.1	3.3	3.1
Weight (g)	24.9 ± 0.3*	24.8 ± 0.5*	39.1 ± 0.7	40.0 ± 0.9
Tail length (cm)	10.4 ± 0.04*	10.9 ± 0.03	10.8 ± 0.04	11.0 ± 0.04
Tight-wire clinging(s)	233 ± 4*	199 ± 10*	178 ± 11	77 ± 10
Open field (no. of squares crossed)	105 ± 7*	87 ± 4*	49 ± 4	58 ± 4
Tail collagen (denaturation; min)	33.5 ± 1.4*	95.9 ± 7.4*	40.6 ± 1.2	205 ± 7
Urine concentration (mosmol)	3929 ± 40*	3453 ± 98*	3156 ± 68	2831 ± 84
Hair growth[a]	69.4 ± 6.3	67.2 ± 6.8*	53.3 ± 5.2	36.3 ± 6.9
Wound healing[b]	19.4 ± 0.7	32.6 ± 3.6	16.2 ± 0.5*	20.0 ± 1.3*

Values are mean ± SEM, with 30–36 mice per group; measures are defined in text. Adults were 294–389 days old, and aged mice 775–804 days old when tested, with ages matched in the two dietary groups. These data were presented in Harrison and Archer (1987), and food restriction was done using protocol I. *Values significantly different from others of the same age on the same line with $p < .05$ by the Student-Newman-Keuls multiple range test.
[a]Percent of shaved area regrowth in 25 days.
[b]Number of days until healed.

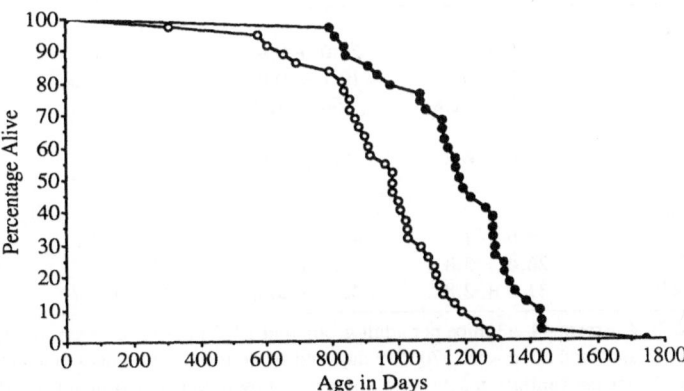

Figure 25.2. Food restriction by protocol I. Longevity plots as percent remaining alive versus age in days for B6CBAF$_1$ males. Data for ad libitum-fed mice are plotted as *open circles* ($n=35$) and food-restricted mice as *solid circles* ($n=34$). Some of these data are in Table 25.6 and in Harrison and Archer (1987)

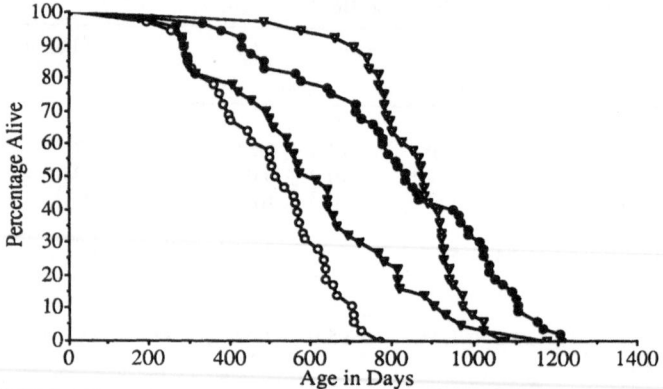

Figure 25.3. Food restriction by protocol I. Longevity plots as percent remaining alive versus age in days for B6 males. Data for ad libitum-fed *ob/ob* mice are plotted as *open circles* (n=36), ad libitum-fed +/+ mice as *solid circles* (n=47), food-restricted *ob/ob* mice as *open triangles* (n=36), and food-restricted +/+ mice as *solid triangles* (n=37). Some of these data are given in Table 25.6 and in Harrison and Archer (1987)

Table 25.5. Effects of food restriction on B6 males.

Measure	Food restricted		Ad libitum fed	
	Adult	Aged	Adult	Aged
Food (g/day)	1.8	2.1	3.0	2.8
Weight (g)	22.5 ± 0.8*	24.0 ± 0.9*	33.3 ± 0.4	35.6 ± 0.6
Tail length (cm)	9.7 ± 0.06*	10.1 ± 0.04	10.0 ± 0.04	10.1 ± 0.06
Tight-wire clinging(s)	149 ± 11.8*	60 ± 12.9*	77 ± 8.4	35 ± 6.2
Open field (no. of squares crossed)	116 ± 6.2	111 ± 10.3*	109 ± 5.2	83 ± 6.1
Tail collagen (denaturation; min)	25.6 ± 1.3	46.8 ± 4.3*	27.8 ± 1.3	60.3 ± 2.5
Hair growth[a]	26.4 ± 3.8	18.3 ± 5.8	68.9 ± 3.7*	40.0 ± 6.6*
Wound healing[b]	31.6 ± 2.8	42.7 ± 3.8	20.1 ± 0.7*	20.9 ± 0.9*

Same as Table 25.4, except 44–53 mice per adult group and 17–24 per aged group. Adult mice were 260–373 days old and aged mice were 570–679 days old when tested, with ages matched in the two dietary groups. Urine concentration data were not affected by food restriction in B6 males, so were not included.

*Values significantly different from others of the same age on the same line with $p < .05$ by the Student-Newman-Keuls multiple range test.
[a]Percent of shaved area regrowth in 25 days.
[b]Number of days until healed.

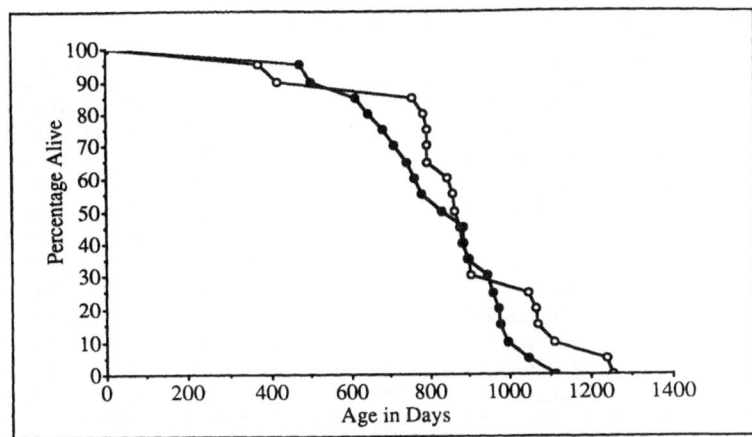

Figure 25.4. Food restriction by protocol IIA. Longevity plots as percent remaining alive versus age in days for normal B6 males fed semichemically defined diets with all nutrients the same in both groups except carbohydrates and fats. Data for food-restricted mice are plotted as *open circles* (*n*=20) and for ad libitum-fed mice as *solid circles* (*n*=20). These data are summarized in Table 25.6

the particular diets and may not always correlate with effects in different biological systems. This illustrates how particular biological systems may age at different rates in different individuals.

Why were B6 males damaged by exactly the same food restriction treatment that benefitted females and obese mice of the same inbred strain and B6CBAF$_1$ hybrids? The most likely possibility is that they required higher levels than the other groups of an essential nutrient that is not present in adequate amounts when they are restricted to about two thirds of their normal food intake. Therefore B6 males were studied using food restriction protocol II with a defined diet (Table 25.1), so that amounts of protein and all micronutrients, and the percentage of fiber, are held constant in fed and restricted groups. When calories were reduced by reducing only carbohydrates and fats, B6 males were not harmed by food restriction, although they benefitted only slightly (Fig. 25.4). There apparently was an unexpectedly specific requirement for some micronutrient in B6 males. This illustrates that an antiaging treatment that is successful in some individuals may be harmful in other individuals, even if they are related.

Comparisons of Longevity Patterns

Longevities of normal and obese B6 female mice food restricted using protocol I are shown in Fig. 25.1. Death rates increased more slowly with age in the food-restricted groups than in the groups fed ad libitum. This caused maximum longevities in restricted +/+ mice to be about 30% greater than those of fed +/+

Table 25.6. Effects of food restriction on longevities.

Genotype	Gender	Treatment	Median	Mean	SE	n	Range max. 10%
B6 +/+	M	Ad libitum fed[a]	774	818	40	20	1046–1115
B6 +/+	M	Restricted[a]	855	878	50	20	1239–1250
B6 +/+	M	Ad libitum fed	878	858	19	45	994–1172
B6 +/+	M	Restricted	591	593	40	48	967–1145
B6-*ob/ob*	M	Ad libitum fed	515	490	21	53	700–770
B6-*ob/ob*	M	Restricted	821	817	36	48	1102–1210
B6CBAF$_1$	M	Ad libitum fed	985	951	35	35	1188–1296
B6CBAF$_1$	M	Restricted	1191	1185	36	34	1432–1742[b]
B6 +/+	F	Ad libitum fed	799	771	28	32	954–976
B6 +/+	F	Restricted	850	810	48	38	1089–1287
B6-*ob/ob*	F	Ad libitum fed	552	526	36	29	776–893
B6-*ob/ob*	F	Restricted	814	823	46	39	1209–1307

[a]These mice were food restricted using protocol IIA. All other mice were food restricted using protocol I. In both cases, calories fed restricted groups were 66%–70% of those fed ad libitum. All groups not separated by a space were studied over the same time period on the same diet, and are, from top to bottom, the same mice in Figs. 25.4, 25.2, 25.3, and 25.1, respectively, with more mice added to those in Fig. 25.1.
[b]The longest-lived mouse was fed ad libitum after 1541 days of age.

mice, although the former began dying at similar ages. This pattern of longevity curve suggests that food restriction did not benefit general health and reduce early mortality, but may have retarded some types of basic aging processes and increased maximum longevities.

Obese mice food restricted from weaning began dying about 300 days later than fed obese mice; however, food restriction late in life had little benefit (Fig. 25.1). Maximum longevities were increased over 400 days by food restriction, suggesting benefits to both general health and basic aging processes. The longevity summaries at the bottom of Table 25.6 illustrate these results.

Figure 25.2 compares longevities of F$_1$ hybrid mice food restricted using protocol I. Death rates increase with age in parallel, but the increase starts later in the food-restricted group. Thus the median, mean, and maximum longevities for restricted mice are about 200 days longer than those for fed groups, suggesting that some basic aging processes may have been retarded.

Longevities of normal and obese B6 male groups food restricted using protocol I are shown in Fig. 25.3. Death rates increased more slowly with age in the food-restricted groups than in the groups fed ad libitum. This caused maximum longevities in restricted +/+ mice to be almost as long as those of fed +/+ mice, although the former began dying at a much younger age. The decline in mean and median longevities as a result of food restriction in +/+ mice was over 260 days, while the decline in maximum longevity was only 27 days. This suggests that food restriction had most of its deleterious effects on health, reducing mean longevities, rather than on fundamental aging processes which would have reduced maximum longevities. Food-restricted obese mice began dying much later than fed obese mice,

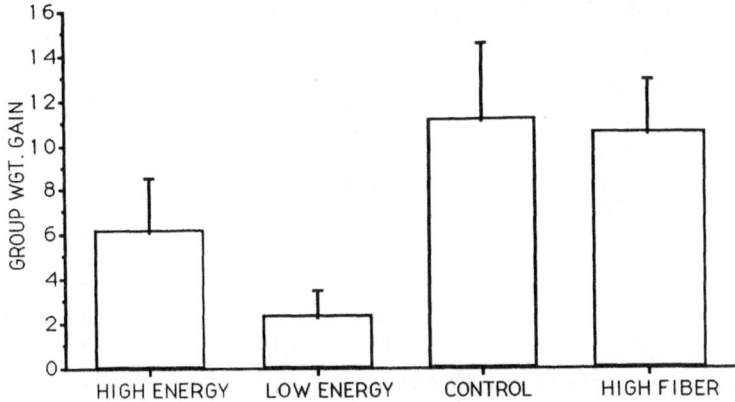

Figure 25.5. Weight gains using the automatic feeding system and allowing ad libitum feeding. Four-week-old mice were fed for 4 weeks using protocol IIB. Groups were fed *HIGH-ENERGY* diets, or *LOW-ENERGY* (food restricted with caloric contents about 70% of that in HIGH-ENERGY) diets using daily automatic feeding at 11:00PM every night. The *CONTROL* group received pelleted high-energy diet ad libitum, and the *HIGH-FIBER* group received pelleted low-energy diet with an added 18% nonnutritive oat bran ad libitum. Diets are detailed in Table 25.2. Results are summarized for 56 mice in each group, using 8 mice caged 4 per side of each of the B6, D2, B6D2F$_1$, WBCBAT6F$_1$, B6CBAT6F$_1$, BALBD2F$_1$, and WBB6F$_1$ types. *Bar heights* give means, while *error bars* are SDs

giving increases in mean and maximum longevities of about 300 days; increases in maximum longevities were even greater, over 400 days, as expected if death rates increased with age more slowly in the restricted *ob/ob* group.

Figure 25.4 compares longevities of restricted and ad libitum-fed B6 males given a defined diet (protocol IIA) so that calories were reduced two thirds by reducing only carbohydrates and fat. Death rates increase with age in parallel, but the increase may start later in the food-restricted group. This is a preliminary experiment, and longevities may have been reduced because the mice at about 800 days of age had to reside for several weeks in an area with temperatures lower than normal. During this period more than half the mice in the restricted group died; possibly the death rate was increased as a result of the changed environment. Nevertheless, food restriction clearly was not harmful to longevities.

Figure 25.5 gives initial weight gains testing the diets in Protocol IIB, Table 25.2. The low-energy diet is used in two forms. One is fed at 82% of the high-energy diet using the automated feeding system to provide measured amounts of each daily (LOW ENERGY in Fig. 25.5). The other contains an additional 18% oat bran (HIGH FIBER in Fig. 25.5) and is fed ad libitum to test whether mice consume the same amounts of this and of the high-energy diet fed ad libitum (CONTROL in Fig. 25.5), avoiding the need for automated feeding systems. Results were similar in all seven genotypes tested, so they are pooled. When food was supplied ad libitum, mice ate more of the high-fiber

Table 25.7. Effects of genotype on collagen aging.

Genotype	Denaturation times (min)	(n)
AEJ/GnRkJ	71.6 ± 15.4	7
BALB/cByJ	56.3 ± 5.5	5
C57Bl/6J	38.0 ± 6.4	8
DBA/2J	70.8 ± 17.4	8
LP/J	68.2 ± 21.9	5
Mol/J	57.2 ± 25.3	7
RF/J	62.7 ± 21.1	7
SM/J	65.0 ± 16.1	11
Spretus/J	61.4 ± 24.3	9

Data given as mean ± SD. All mice were caged singly and given the same amount of food using protocol IIA, 3 g of high-energy diet per day) from 4 weeks of age; tail tendons were tested at 73–76 weeks of age.

diet, showing very high weight gains similar to those given the high-energy control diet ad libitum.

Since experiments are better controlled when amounts of most nutrients are the same, as using our protocol II diets, it is interesting to compare longevities of ad libitum-fed B6 males given the defined diet (protocol IIA – Fig. 25.4 and line 1 of Table 25.6) or a Jackson Laboratory standard, grain-based diet (protocol I – Fig. 25.3 and line 3 of Table 25.6). Longevities appear similar with means of 818 and 858 days, respectively, suggesting that the defined diet does not cause important deleterious effects in this genotype. Nevertheless life spans may be slightly less using the protocol IIA diet, and Dr. Knapka has pointed out that the AIN 76 diets were never intended for lifelong studies. Therefore we have collaborated with him to evaluate the diets in Protocol IIB. These also have the advantage that amounts of most nutrients are the same in the restricted (low-energy) and control (high-energy) diets.

Defining Genetic Effects on Aging Apart from Effects on Food Consumption

Genotype affects food consumption, and as the foregoing material and everything else presented in this conference shows, reduced food consumption is well known to increase longevities. Thus there is a danger that genetic studies will inadvertently study genes that alter voluntary feeding, which in turn affects aging. This would be a waste, since effects of food restriction are directly studied by comparing mice of the same genotype consuming different amounts of food. This problem can be avoided if mice of different genotypes are meal fed, all receiving the same amount of food, as is done using automatic feeding systems. Collagen aging is evaluated in Table 25.7. B6 mice appear to have younger collagen than other mice of the same age of eight different strains. This effect cannot result from lower voluntary food consumption.

Table 25.8. Effects of food restriction on precursor T-cell frequencies.

Age	Diet	Helpers	Killers
7–10 months	Ad libitum	105 ± 43(4)	55 ± 22(4)
32–35 months	Ad libitum	16 ± 7(6)	10 ± 4(6)
32–35 months	Restricted	56 ± 19(5)	45 ± 11(6)

Frequencies are given as mean numbers of helper or killer precursor cells per 10^3 peripheral blood leukocytes determined by limiting dilution assays, ± standard deviation, with the number of individual mice tested given in parentheses. These data are from Miller and Harrison (1985), and mice were food restricted using protocol I.

Does Food Restriction Act by Retarding Collagen Aging?

The relationships between retarding collagen aging and extending longevity are not consistent. The tail tendon collagen test (Tcd) shows excellent correlations with chronological age, yet it does not correlate as expected with subsequent longevity of individuals or strain longevity (Harrison et al. 1978; Ingram et al. 1982; Harrison and Archer 1988b), although collagen aging seemed slower in a long-lived mouse species (Harrison et al. 1978). The agreement between reduced tail tendon collagen aging rates and increased longevities in restricted B6 female +/+ and *ob/ob* mice suggests that collagen aging is important to longevity in these animals (Table 25.3, Fig. 25.1). Collagen aging seems to be generally retarded by food restriction (Table 25.3–25.5), which generally extends longevities, yet collagen aging was also retarded in B6 +/+ males whose longevities were reduced by food restriction (Table 25.5, Fig. 25.3).

Does Food Restriction Act by Retarding Aging of Immune Responses?

Although food restriction extended longevities in female obese mice without retarding immune aging, it seemed to retard immune aging in the SRBC response of normal mice (Table 25.3). Food restriction also retarded changes with age in frequencies of T-helper and -killer cell precursors using limiting dilution assays (Miller 1984; Miller and Harrison 1985). Even in extremely old (32–35 months) B6CBAF1 mice that were restricted to 70% of the normal food intake, T-precursor-cell frequencies in the circulation were almost at young levels (Table 25.8). This supports the idea that some of the benefits of food restriction result from retarding immune aging.

Does Food Restriction Act by Reducing Metabolic Rates?

Food-restricted B6-*ob/ob* females may have reduced metabolic rates, as they were fed identical amounts, but weighed 30% more and had three- to fourfold higher body fat percentages than restricted +/+ mice (Table 25.3). Thus beneficial effects of food restriction on obese mice may be mediated through effects on metabolic rates. If metabolic rate is as greatly reduced by food restriction as is

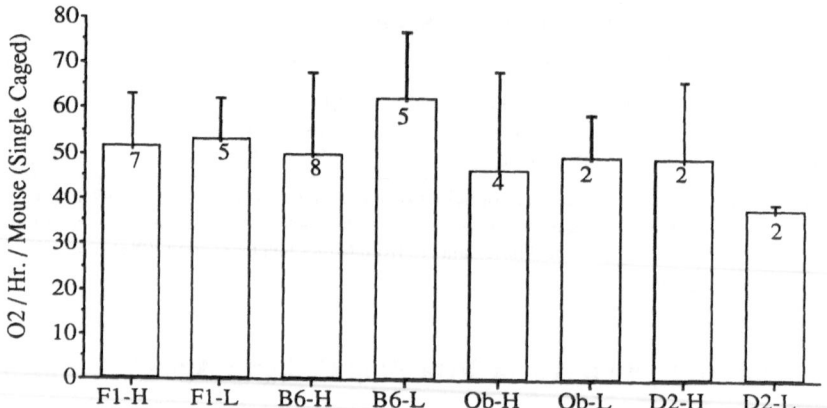

Figure 25.6. Effects of genotype and food restriction on oxygen consumption. Using protocol IIA, mice were fed high- and low-energy diets, shown as $-H$ and $-L$, respectively. The four genotypes were $F1$-B6D2F$_1$, $B6$-C57B1/6J, Ob-B6-ob/ob, and $D2$-DBA/2J. Oxygen consumption was measured over 24 h. *Bar lengths* give means, while *error bars* are SDs and *numbers* of mice tested are *under the bars*. Effects of genotype were not significant (F test 0.64, $p=0.64$) and those of diet were of questionable significance (F test 2.1, $p=0.16$)

implied by the weight and adiposity data, aging processes in which beneficial effects of food restriction result from retarding metabolic rate may be benefitted substantially more in restricted obese than in +/+ mice.

In the studies outlined in Tables 25.3 and 25.5, nutrient levels had been reduced to 70% of normal +/+ consumption during food restriction, perhaps causing deficiencies affecting spleen size or immune response, since food-restricted mice were simply fed 70% of a standard 96WA diet. This problem was corrected by feeding constant levels of nutrients to all mice using the diet in Table 25.1. Figures 25.6 and 25.7 show initial measures of metabolic rates as oxygen consumed per individual when mice were studied in metabolic cages over 1-day periods. Although these mice were studied for only 1 day, there did not appear to be high oxygen consumption due to the stress of recaging, as metabolic rates in B6 mice were the same as those in B6 mice given 2 days to acclimate (data not shown).

Genotype was not important, but numbers of mice of each genotype studied may have been too few to detect biologically meaningful effects. Nevertheless it is surprising that obese mutants consumed about the same amount of oxygen per individual as did normal mice (Fig. 25.6). If this is confirmed, it would suggest that obese mutants use food more efficiently than normal mice to produce fat.

The only significant effect was housing; B6D2F$_1$ mice housed singly consumed more oxygen than those housed four per side (Fig. 25.7). Effects of food restriction using the low-energy diet (protocol IIA) were not significant at $p<.05$; if anything, the mice fed the low-energy diet consumed more oxygen in three of the four genotypes (Figs. 25.6, 25.7). If confirmed in longer-term studies with more

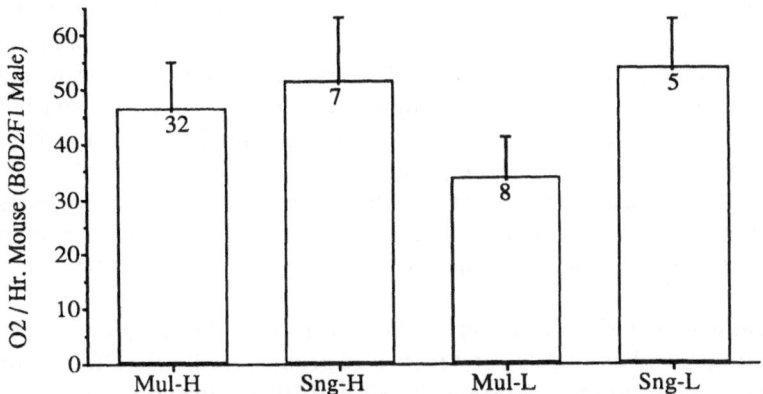

Figure 25.7. Effects of single housing on oxygen consumption. Using protocol IIA, multiply (*Mul*-four per side) and singly (*Sng*) housed mice were fed high (−*H*) and low (−*L*) energy diets. Oxygen consumption was measured over 24 h. *Bar lengths* give means, while *error bars* are SDs, and *numbers* of mice tested are *under the bars*. An ANOVA showed that effects of housing were highly significant, with an F-test of 15.7 for $p=0.0002$. Effects of diet were of questionable significance (F test 2.8, $p=0.10$)

individuals, this would suggest that food-restricted mice have altered metabolic pathways that allow more efficient use of food, perhaps with lower levels of free radical production per oxygen molecule used.

General Discussion

Considerations in Designing Food Restriction Experiments

Although such experiments may be of great value in defining mechanisms of mammalian aging, they are even more difficult to do correctly than are conventional aging studies. All studies of mammalian aging require excellent environmental and genetic definition (Solleveld and Hollander 1984), including pathogen control and pathology, to get repeatable results and to distinguish aging from disease. However, the following considerations are also needed in food restriction studies:

1. There should be controls to distinguish long-term effects of food restriction from short-term effects, since only the former could result from retarding aging processes. The latter might result from reversing aging processes, an exciting possibility but one that should be carefully defined. Controls could be provided as follows: 6 weeks before testing an extra ad libitum group should be fed restricted amounts. Changes should be made gradually to avoid stress, while giving the equivalent of 1 month of full restriction.

2. As mice are food restricted, it is vital to avoid malnutrition. Therefore, either the ad libitum-fed animals must have an excess of all nutrients, or the restricted diet must be enriched by supplementation of each nutrient. In protocols IIA and IIB, we supplement the diet fed to restricted animals, to avoid confounding ad libitum feeding with feeding excessive nutrients. Our restricted mice have an enriched diet, but this should not cause problems because consumption of all nutrients except carbohydrates and fats will be the same among animals fed ad libitum and food restricted, and there will be adequate amounts of carbohydrates and essential fats in all diets.

3. A serious source of error in interpreting effects of food restriction is a different meal time for ad libitum-fed and restricted groups, leading to different diurnal rhythms. To avoid this, in protocols IIA and IIB we feed both ad libitum-fed and restricted groups one meal per day, such that the animals finish their rations in less than 4 h and retain similar diurnal rhythms. Ad libitum-fed mice will be actually meal fed, receiving amounts with nearly the same caloric contents as those consumed by mice of the same genotype and age fed ad libitum with the conventional Jackson Laboratory 96WA diet. They finish their food in 4 h because they do not have food constantly present.

4. To measure food intake and metabolic rates accurately, mice must be housed singly; however, this unnatural condition may stress mice because they lack social interactions and cannot conserve heat by huddling. Even in colonies with excellent pathogen control, longevities of singly housed F344 male rats fed ad libitum (Bertrand et al. 1980) were 6 months less than those of the same strain housed in groups of three (Coleman et al. 1977). Thus multiply housed controls should be used to test if singly housed mice receiving the same diet differ in any changes with age or in longevity, or singly housed controls should be used to test accuracy of measures of food intake and metabolic rate.

5. Changes in efficiency of food use or in body composition as a result of food restriction should be tested. These require long-term metabolic measures and direct measures of body composition.

How the Response to Food Restriction May Have Evolved

We have suggested that the ability to retard aging in response to food restriction evolved due to the selective advantages for females whose reproductive life spans were extended by food restriction (Harrison and Archer 1988a). Consider the enormous selective advantage of mice that could respond to a drought, or other long-term condition that severely reduced food supplies, by delaying reproductive senescence. This would be most important in females, as their reproductive senescence occurs at younger ages than that of males. After a drought outlasting the maximum female reproductive life span, a surviving female that is still able to reproduce could repopulate huge areas once weather conditions and food supplies became favorable. Survivors lacking the food restriction response would have reached female reproductive senescence before food supplies became adequate to raise offspring.

This hypothesis results in a unique and testable prediction: that the beneficial effects of food restriction will be greater in species with shorter female reproductive life spans. This would occur because selective pressure would be greater for mechanisms that prolong life in response to food restriction in species with shorter reproductive life spans. The shorter the reproductive life spans, the more likely are droughts that outlast maximum reproductive life spans.

Our hypothesis could be tested by comparing the two mouse species *Mus musculus* and *Peromyscus leucopus* because the latter live about twice as long as the former, despite their similar sizes and metabolic rates (Sacher and Hart 1978). We predict that food restriction will have significantly less benefit for *Peromyscus* mice because their female reproductive life spans are already twice as long as those of *Mus* mice.

This leads to a related hypothesis: that the same selective pressures favoring prolonged life spans in response to food restriction in *Mus* also caused the extremely long life span potential in *Peromyscus*. Thus there are two ways for short-lived rodents to survive droughts longer than the normal female reproductive period and retain the ability to reproduce and repopulate the area once food supplies improve. One is to increase maximum potential longevities in response to food restriction. The other, taken by *Peromyscus* mice, is to increase longevities in general, whether or not there is food restriction.

If the above hypotheses are correct, they lead to two important consequences: (1) Female reproductive senescence is regulated by at least some of the same mechanisms that control longevity, and thus is a good model for studies of aging; (2) the degree of genetic change required to extend longevity in response to food restriction may be similar to that required to extend longevity in general, since at least one species, *Peromyscus*, adapted using the more general mechanism.

If the beneficial effects of food restriction are greater in species with shorter female reproductive life spans, the specific treatment of food restriction would have little beneficial effect on human longevity. However, studies of the underlying mechanisms by which food restriction retards aging in short-lived species would still be highly relevant to human health, since the same mechanisms might retard aging in man.

Acknowledgments. The authors are grateful to Mrs. K. Davis, Mr. C.J. Bennet, and Mrs. W. Waterman for skillful technical assistance. We thank Mr. G.E. (Bud) Sorensen for imaginative design and painstaking machinist work in helping produce automated feeding systems and oxygen consumption measuring devices. The diets in protocol IIB were designed by Dr. J. Knapka, whose advice has been very important.

References

Astle CM, Harrison DE (1984) Effects of young and old marrow on immune responses: studies on the cellular level. J Immunol 132:673–677

Barrows CH, Kokkonen CG (1977) Relationship between nutrition and aging. Adv Nutr Res 1:253–298

Bertrand HA, Frederick TL, Masoro EJ, Yu BP (1980) Changes in adipose mass and cellularity through adult life of rats fed ad libitum or a life-prolonging restricted diet. J Gerontol 35:827–835

Bieri JG, Stoewsand GS, Briggs GM, Phillips RW, Woodard JC, Knapka JJ (1977) Report of the American Institute of Nutrition ad hoc committee on standards for nutritional studies. J Nutr 107:1340–1348

Brooks GA, White TP (1978) Determination of metabolic and heart rate responses of rats to treadmill exercise. J Appl Physiol: Respirat Environ Exerc Physiol 45: 1009–1015

Campbell BA (1982) Behavioral markers of aging in the Fischer 344 rat. In: Reff ME, Schneider EL (eds) Biological markers of aging. NIH 82-2221, pp 78–86

Cathcart R, Schwiers E, Saul RL, Ames BN (1984) Thymine glycol and thymidine glycol in humans and rat urine. Proc Natl Acad Sci USA 81:5633–5637

Christadoss P, Talal N, Lindsrom J, Fernandez G (1984) Suppression of cellular and humoral immunity to T-dependent antigens by calorie restriction. Cell Immunol 88:1–8

Coleman DL (1985) Increased metabolic efficiency in obese mutant mice. Int J Obes 9:69–73

Coleman GL, Barthold LSW, Osbaldiston GW, Foster SJ, Jonas AM. (1977) Pathological changes during aging in barrier-reared Fischer 344 male rats. J Gerontol 32: 258–278

Cutler RG (1985) Peroxide-producing potential of tissues: inverse correlation with longevity of mammalian species. Proc Natl Acad Sci USA 82:4798–4802

Fernandes GP, Friend P, Yunis EJ, Good RA (1978) Influence of dietary restriction on immunologic function and renal disease in (NZB/NZW)F1 mice. Proc Natl Acad Sci USA 75:1500–1504

Forsum E, Hillman PE, Esheim MC (1981) Effect of energy restriction on total heat production, basal metabolic rate, and specific dynamic action of food in rats. J Nutr 111:1691–1697

Harman D (1981) The aging process. Proc Natl Acad Sci USA 78:7124–7128

Harrison DE, Archer JR (1983) Physiological assays for biological age in mice: relationship of collagen, renal function and longevity. Exp Aging Res 9:245–251

Harrison DE, Archer JR (1987) Genetic effects on responses to food restriction in aging mice. J Nutr 117:376–382

Harrison DE, Archer JR (1988a) Natural selection for extended longevity from food restriction. Growth Dev Aging 52:65

Harrison DE, Archer JR (1988b) Biomarkers of aging: tissue markers. Future research needs, strategies, directions and priorities. Exp Gerontol 23:309–321

Harrison DE, Archer JR, Sacher GA, Boyce FM (1978) Tail collagen aging in mice of thirteen different genotypes and two species: relationship to biological age. Exp Gerontol 13:63–73

Harrison DE, Archer JR, Astle CM (1984) Effects of food restriction on aging: separation of food intake and adiposity. Proc Natl Acad Sci USA 81:1835–1838

Hill JO, Latiff A, Digirolamo M (1985) Effects of variable caloric restriction on utilization of ingested energy in rats. Am J Physiol 248:R549–559

Ingram DK, Archer JR, Harrison DE, Reynolds MA (1982) Physiological and behavioral correlates of life span in aged C57BL/6J mice. Exp Gerontol 17:295–303

Lane PW, Dickie MM (1958) The effect of restricted food intake on the life span of genetically obese mice. J Nutr 64:549–554

Masoro EJ, Yu BP, Bertrand HA (1982) Action of food restriction in delaying the aging process. Proc Natl Acad Sci USA 79:4239–4241

McCarter R, Masoro EJ, Yu BP (1985) Does food restriction retard aging by reducing the metabolic rate? Am J Physiol 248E:488–490

McKay CM, Maynard LA, Sperling G, Barnes LL (1939) Retarded growth, life span, ultimate body size and age changes in the albino rat after feeding diets restricted in calories. J Nutr 18:1–13

Meade CJ, Sheena J (1979) Immunity in genetically obese rodents. In: Festing MFW (ed) Animal models of obesity. Oxford University Press, New York, pp 205–220

Miller RA (1984) Age-associated decline in precursor frequency for different T-cell mediated reactions, with preservation of helper or cytotoxic effect per precursor cell. J Immunol 132:63–68

Miller RA, Harrison DE (1985) Delayed reduction in cell precursor frequencies accompanies diet-induced lifespan extension. J Immunol 134:1426–1429

Nichols WK, Spellman JB, Daynes RA (1978) Immune responses of diabetic animals. Diabetologia 14:343–349

Pearl R (1928) The rate of living. Knopf, New York, pp 1–185

Ross MH (1969) Aging, nutrition and hepatic enzyme activity in the rat. J Nutr 97 (Suppl 11):563–602

Ross MH (1976) Nutrition and longevity in experimental animals. In: Winick M (ed) Nutrition and aging. Wiley, New York, pp 43–57

Rubner M (1908) Das Problem der Lebensdauer und seine Beziehungen zu Wachstum und Ernaehrung. Oldenbourg, Munich

Sacher GA (1977) Life table modification and life prolongation. In: Finch C, Hayflick L (eds) Handbook of biology of aging. Van Nostrand Reinhold, New York, pp 582–638

Sacher GA, Duffy PH (1979) Genetic relation of life span to metabolic rate for inbred mouse strains and their hybrids. Fed Proc 38:184–188

Sacher GA, Hart RW (1978) Longevity, aging and comparative cellular and molecular biology of the house mouse, Mus musculus, and the white-footed mouse, Peromyscus leucopus. Birth Defects 14:71–96

Solleveld HA, Hollander CF (1984) Animals in aging research: requirements, pitfalls and a challenge for the laboratory animal specialist. Vet Q 6:96–101

Sprott RL, Eleftheriou BE (1974) Open-field behavior in aging inbred mice. Gerontology 20:155–162

Walford RL (1969) The immunologic theory of aging. Norton, New York

Walford RL, Liu RK, Delima-Berbasse M, Mathis M, Smith GS (1974) Long-term dietary restriction and immune function in mice: response to sheep red blood cells and to mitogenic agents. Mech Ageing Dev 2:447–454

Weindruch RH, Suffin SC (1980) Quantitative histologic effects on mouse thymus of controlled dietary restriction. J Gerontol 35:525–531

Weindruch R, Gottesman SRS, Walford RL (1982) Modification of age-related immune decline in mice dietarily restricted from or after midadulthood. Proc Natl Acad Sci USA 79:898–902

Weindruch R, Walford RL, Fligiel S, Guthrie D (1986) The retardation of aging in mice by dietary restriction: longevity, cancer, immunity and lifetime energy intake. J Nutr 116:641–654

Young VR (1979) Diet as a modulator of aging and longevity. Fed Proc 38:1994–2000

Yu BP, Masoro EJ, McMahan CA (1985) Nutritional influences on aging of Fischer 344 rats: physical, metabolic and longevity characteristics. J Gerontol 40:657–670

CHAPTER 26

1200-Rat Biosure Study: Design and Overview of Results

F.J.C. Roe[1]

Introduction

It is clear from many earlier studies that restriction of caloric intake is associated
with improved survival and reduced incidence of ageing-associated non-neoplastic
and neoplastic diseases both in mice (Conybeare 1980) and rats (Berg and Simms
1960; Tucker 1979). In the case of rats, an earlier small-scale study (Salmon et al.
1990) suggested that calorie restriction reduces the age-standardized incidence of
fatal and/or potentially fatal cancers of many kinds. This latter observation needed
to be checked in a larger-scale study and a number of other questions merited careful investigation. In the light of these needs scientists in several organizations
agreed to undertake a large-scale study giving their own time and services for minimum fees or without any charge. The individuals and companies involved are
listed in Table 26.1 and the main aims of the study in Table 26.2.

At the time of the preparation of this preliminary report the histopathological
evaluation was virtually complete for 10 of the 12 groups of rats in the study with
only a very few additional tissue sections awaited. Most of the in-life data for
individual animals have been entered into the computer but no attempt has yet
been made to analyze the data for individual animals in relation to the last aim
listed in Table 26.2.

A full and detailed report on the study will be prepared for publication in due
course.

Materials and Methods

Twelve groups consisting of 50 male and 50 female SK&F Wistar weanling rats
aged 3 weeks were constructed using a random allocation procedure. They were
numbered and housed five to a cage in grid-bottomed cages. Sick animals were

---

[1] 19 Marryat Road, Wimbledon Common, London SW19 5BB, UK

Table 26.1. Collaborating scientists and sponsors.

Scientist	Role	Sponsor
David Kelly	Protocol design and organization	Biosure
Geoffrey Conybeare	In-life observations and necropsies	Smith Kline and French
Graham Tobin	Nutritional aspects	Biosure
Peter Donatsch	Hormonal assays	Sandoz
David Prentice, Bernhard Matter, and Peter Stirnimann	Histopathological processing and second opinion on assessment	Sandoz
Francis Roe	Histopathological assessment	Biosure and self
Peter Lee	Data processing and statistical analysis	Sandoz and self[a]

[a]Further sponsors are needed.

isolated to avoid cannibalism. Euthanasia was by carbon dioxide asphyxia. Animals were checked twice daily for general health and clinically examined and weighed once weekly.

Table 26.3 lists the data that were collected during life, the observations made at necropsy, and the tissues examined microscopically.

Study Design

Table 26.4 lists the diets given to the 12 groups of rats. Included in the different dietary formulations in varying proportions were barley, maize meal, oats, oat feed, wheat, wheat feed, dried skimmed milk, single cell (yeast) protein, soya protein, white fish meal, yeast, minerals, vitamins, and essential amino acids. (Although the actual formulae of the diets are commercially sensitive, further information will not be withheld in the event of enquiries from interested scientists.)

The composition of the diets in terms of nutrients and certain minerals, by analysis, is summarized in Table 26.5.

Table 26.2. Main aims of 1200-rat Biosure study.

1. To see whether dietary restriction (80% of ad libitum) reduces the age-standardized incidence of fatal or potentially fatal neoplasia before the age of 30 months
2. To see whether the beneficial effects of diet restriction can be achieved by (a) limiting the daily period of access to food to 6 h, or by (b) limiting the energy value of the diet
3. To see whether reduced calorie intake between weaning and age 4 months influences survival and/or incidence of non-neoplastic and neoplastic diseases
4. To compare effects of food consumption, energy intake and protein intake on survival and disease
5. To study the relationships between body weight at different ages with eventual survival and disease incidence
6. To provide a data base for studying relationships between various in-life measurements and eventual survival and disease incidence in individual animals

Table 26.3. Data collected in life and at necropsy and stored.

In-life

Measure	Time interval
Body weight	Weekly
Food consumption	During 7 days weekly at first, then monthly
Water consumption	During 7 days during week 4, then 12-weekly
Clinical chemistry	20 rats/sex per group 6-monthly
Haematology	20 rats/sex per group 6-monthly
Urinalysis	During 4 h 20 rats/sex per group 6-monthly
Terminal plasma	Samples in store at $-18°C$

At necropsy

Measure	Tissue
Macroscopic examination	Complete except spinal cord
Weight	Liver, kidneys, heart, adrenals, testes, prostate, seminal vesicles, ovaries, pituitary, and brain
Microscopic examination	Liver ($\times 3$), kidneys ($\times 2$), heart, spleen, lungs ($\times 2$), pancreas, adrenal cortex ($\times 2$), adrenal medulla ($\times 2$), thyroid ($\times 2$), parathyroid ($\times 2$), testes ($\times 2$), epididymides ($\times 2$), ovaries ($\times 2$), uterine horns ($\times 2$), skin, mammary gland, and pituitary. *Plus* all tissues thought to be abnormal at necropsy.

N.B. All tissues in formalin fixative.
Terminal blood plasma samples at $-18°C$
Carcasses stored in formalin.

Table 26.4. Study design.

Group (50 males + 50 females)	During first week after weaning	From 1 week after weaning for 12 weeks	13 weeks–30 months
1	SB24	SB24	SM24
2	SB24	SB24	LM24
3	SB24	SB24	SMR
4	SB24	SB24	SM6
5	SB24	SBR	SMR
6	SB24	SBR	LM24
7	SB24	SBR	SM24
8	SB24	SB6	SM6
9	LB24	LB24	SM24
10	LB24	LB24	LM24
11	LM24	LM24	LM24
12	PRD24	PRD24	PRD24

SB, standard breeding diet; SM, standard maintenance diet; LB, low nutrient breeding diet; LM, low nutrient maintenance diet; PRD, Porton Research diet; 24, ad libitum for 24 h/day; 6, ad libitum for 6 h/day; R, restricted to 80% of ad libitum

Table 26.5. Percent composition of diets in terms of nutrients and certain minerals by analysis.

Nutrients	Diet				
	SB	LB	SM	LM	PRD
Crude oil	3.4	3.2	3.2	3.0	3.0
Crude protein	19.7	16.0	14.4	13.4	19.8
Crude fibre	2.3	9.2	6.4	11.6	5.4
Carbohydrate	55	53	57	53	53
Starch	49	32	42	26	31
Calcium	0.71	0.73	0.71	0.85	0.65
Phosphorus	0.74	0.70	0.69	0.77	0.68
Magnesium	0.17	0.24	0.2	0.28	0.17
Ca:P ratio	0.98	1.03	1.02	1.1	0.95
Lignosulphonates	0	1.0	0 up to 1.5	0 up to 1.8	0

SB, standard breeding diet; LB, low nutrient breeding diet; SM, standard maintenance diet; LM, low nutrient maintenance diet; PRD, Porton Research diet

Statistical Analysis

All data were entered onto ROELEE 84, a computer-based system for recording, processing, reporting, and statistically analyzing pathological and other toxicological data. Analyses of data recorded after 13 weeks on test were complicated by the change in dietary regimen so that in assessing the possible effects of different diets prior to changeover, comparison had to be made between pairs of groups fed differently before 13 weeks, but the same after. Thus, standard breeding diet ad libitum for 24 h/day (SB24) and standard breeding diet restricted to 80% of ad libitum (SBR) could be compared based on the pairs of groups 1 and 7 (both given standard maintenance diet ad libitum for 24 h/day, SM24, after week 13), 2 and 6 low-nutrient maintenance diet ad libitum for 24 h/day (LM24), and 3 and 5 standard maintenance diet restricted to 80% of ad libitum (SMR). These three sets of comparisons can then be combined for a more powerful test, "stratified" for diet after changeover. Initial comparisons of diets after changeover were stratified for diet before, but as it became clear that for many endpoints diet before week 13 had little or no effect, many of the analyses presented here are unstratified, thus allowing inclusion of data from all relevant groups and a somewhat more powerful test.

In the case of in-life measurements, the analysis of data at a given time point involved a combination of presentation of distributions, means, and standard errors. Rank tests were used to test statistical significance, stratification being carried out as described by Fry and Lee (1988).

Histopathological data were analyzed by the method of Peto et al. (1980), which compares the observed and expected incidences of a lesion after adjustment for time of death, taking account of whether the condition was classified as having caused or heavily contributed to death (fatal) or having not done so (incidental). The method allows stratification.

Comparisons were normally made with the SB24 or SM24 diets or with group 1, with tests carried out for males, females, and, as appropriate, for sexes com-

Table 26.6. Mean food consumption (g/day), energy intake, (kJ/day), and protein (g/day) during weeks 0–52 as percent of group 1.

	Males			Females		
Group	Food consumption (%)	Energy intake (%)	Protein intake (%)	Food consumption (%)	Energy intake (%)	Protein intake (%)
1	100	100	100	100	100	100
2	110	112	104	108	110	102
3	84	84	85	84	84	85
4	107	107	106	103	103	102
5	80	79	79	80	80	79
6	110	111	102	105	107	98
7	91	91	90	95	95	94
8	100	99	98	94	94	93
10	117	118	105	116	117	101
12	117	119	146	108	111	136

bined (stratified for sex). Both the rank tests and the method of Peto et al. (1980) result in chi-squared statistics from which two-tailed p values were calculated. P values are usually presented as ***$p < 0.001$, **$p < 0.01$, *$p < 0.05$, or (*) $p < 0.1$ for positive differences, minus signs similarly being used for negative differences, and NS (not significant) indicating $p \geq 0.1$.

Results

Food Consumption and Energy and Protein Intake. Table 26.6 summarizes the data for food consumption, energy intake, and protein intake. The mean values for rats in group 1 (SB24 for 13 weeks then SM24) were taken to be 100%. The food consumption and other values for the two groups rationed to 80% of ad libitum from week 13 (groups 3 and 5) approximated to 80% of group 1 as planned. The values for rats fed continuously on the Porton Research diet (PRD) (group 12) were considerably higher than those for group 1 in terms of all three intake parameters. In this study, although not in a previously reported study in rats of the same strain, restricted access to food to a period of 6 h per day throughout life (groups 4 and 8) had little effect on their intakes of food, energy, or protein. Rats fed ad libitum on LM24 from week 13 (groups 2, 6, and 10), by increasing their food consumption, increased their mean energy intake to above that of group 1 and managed to equal group 1 in mean protein intake.

Water Consumption. Compared with rats fed ad libitum on SM24 (group 1), rats restricted to 80% of ad libitum (groups 3 and 5 – SMR) drank significantly less water on all the occasions water consumption was measured, up to and including week 108. Not unexpectedly water consumption by rats in groups 4 and 8 (SM6) which managed to eat as much food in 6 h as group 1 ate in 24 h was similar to

Table 26.7. Survival.

Group	Diet to week 13	Diet after week 13	Males Deaths before week 52	104	130	Females Deaths before week 52	104	130
1	SB24	SM24	0	15	29	1	9	30
2	SB24	LM24	0	12	26	1	5	20
3	SB24	SMR	0	7	15	2	5	12
4	SB24	SM6	0	6	19	1	19	27
5	SBR	SMR	0	6	16	0	6	12
6	SBR	LM24	1	8	23	1	8	20
7	SBR	SM24	1	13	30	3	14	36
8	SB6	SM6	0	8	18	3	13	28
9	LB24	SM24	1	12	30	0	17	31
10	LB24	LM24	1	13	28	0	8	20
11	LM24	LM24	2	13	26	3	9	24
12	PRD24	PRD24	1	18	36	1	12	29

Diet abbreviations as in Table 26.4.

that of group 1 rats. By contrast, rats on the LM diet or the PRD diet drank significantly more water than rats on the SM diet.

Survival. Survival data are summarized in Table 26.7. Compared with rats on the SM diet from week 13 (groups 1, 7, and 9−SM24), rats in both sexes in the two restricted groups (groups 3 and 5−SMR) showed very much improved survival ($p < 0.001$). Survival was also significantly improved ($p < 0.001$) in

Table 26.8. Body weights at start, 1 week, 13 weeks, 12 months, and 18 months.

Group	Diet to week 13	Diet from week 13	Males Start	Week 1	Week 13	Month 12	Month 18	Females Start	Week 1	Week 13	Month 12	Month 18
1	SB24	SM24	81	125	459	545	554	71	104	249	293	337
2	SB24	LM24	82	121	416	460	471	72	105	243	256	268
3	SB24	SMR	84	124	420	410	405	70	103	243	251	256
4	SB24	SM6	82	123	423	484	489	70	102	239	255	273
5	SBR[a]	SMR	83	126	353	391	389	71	103	218	245	254
6	SBR[a]	LM24	83	125	359	459	477	72	104	221	253	269
7	SBR[a]	SM24	81	121	344	505	528	71	103	222	291	323
8	SB6[a]	SM6	82	125	364	460	470	70	102	217	250	272
9	LB24	SM24	83	120	385	525	543	69	97	224	282	316
10	LB24	LM24	80	119	381	453	468	70	99	222	251	266
11	LM24	LM24	82	115	340	457	471	70	96	212	251	268
12	PRD24	PRD24	82	125	419	542	546	70	103	237	299	327

Diet abbreviations same as in Table 26.4.
[a]During the 1st week after weaning, rats in group 5–8 were fed on SB24. Diet restriction began thereafter.

Table 26.9. Liver and kidney weight relative to body weight in terminally killed rats.

			Males				Females		
Group	Diet to 13 weeks	Diet after week 13	BW	BW as % of group 1	LW/BW (%)	KW/BW (%)	BW	BW as % of group 1	LW/BW (%)	KW/BW (%)
1	SB24	SM24	456	100	3.6	0.79	318	100	4.3	0.91
2	SB24	LM24	392	86	4.0	0.78	268	84	4.6	0.92
3	SB24	SMR	405	89	3.3	0.72	258	81	3.8	0.85
4	SB24	SM6	418	92	3.4	0.77	276	87	3.8	0.91
5	SBR	SMR	386	85	3.3	0.72	251	79	3.4	0.84
6	SBR	LM24	371	81	4.0	0.81	264	83	4.5	0.89
7	SBR	SM24	452	99	3.6	0.75	301	95	4.1	0.88
8	SB6	SM6	404	89	3.4	0.78	273	86	3.9	0.86
9	LB24	SM24	433	95	3.9	0.85	291	91	4.3	0.88
10	LB24	LM24	374	82	3.8	0.81	262	82	4.4	0.91
11	LM24	LM24	383	84	3.8	0.81	259	81	4.7	0.94
12	PRD24	PRD24	411	90	4.2	0.98	311	98	4.7	0.89

Diet abbreviations as in Table 26.4.
BW, body weight; LW, liver weight; KW, kidney weight

males in the interrupted groups (groups 4 and 8 – SM6) and in females on the LM diet (groups 2, 6, 10, and 11 – LM24). Survival of rats fed PRD24 throughout the study (group 12) was similar to that of rats fed SM24 from week 13.

Body Weight. Initial mean body weights and mean body weights at 1 week, 13 weeks, 12 months, and 18 months are shown in Table 26.8. The feeding of LB24 or LM24 instead of SB24 during the first 13 weeks of the study reduced body weight gain up to week 13 by 20%–22%. Thereafter the rats of these groups tended to catch up with other groups fed similarly after week 13. Sustained reduction in body weight was seen in the two groups whose food intake was restricted to 80% of ad libitum from 13 weeks (groups 3 and 5). Males of these groups weighed on average 26% less than group 1 males at 12 months and 28% less at 18 months. The corresponding reductions for females were 15% and 14%, respectively. The feeding of LM24 instead of SM24 after week 13 reduced weight gain in males by about 14% and in females by about 20%.

Liver and Kidney Weight Relative to Body Weight in Terminally Killed Rats. In both sexes kidney weight relative to body weight in terminally killed rats was significantly lower in the restricted groups (groups 3 and 5) than in any of the other groups. Relative liver weights in the same two groups were significantly lower than in the group fed SM24 or PRD24 after week 13. They were also low in the SM6 groups. The data are summarized in Table 26.9.

Urinalysis Data. Urine samples were collected over 4 h between 1600 and 2000 hours. Since ad libitum-fed animals both eat and drink mainly during the dark (1800 to 0600 hours) animals in the 24-h per day-fed groups would have taken in little food or water during the 10 h prior to being put in metabolism cages without access to either food or water. By contrast animals in the restricted groups (SBR,

SMR, and SM6) which were provided with food at 1000 hours each day would have been eating and drinking up to, perhaps, 2 h before being put in metabolism cages. Not surprisingly, therefore, the volumes of urine samples collected from restricted animals significantly exceeded those for 24-h per day-fed animals.

The pH of urine collected from SM24 and PRD24 groups ranged from 5 to 7, that from LM24 groups from 5 to 9, that from SM6 groups from 6 to 8, and that from SMR groups from 7 to 9. Throughout the study urine samples from SMR, SM6, and LM24 groups (both sexes) contained less protein than SM24 or PRD24 groups. After 18 months females of the SMR, SM6 and LM24 groups had higher levels of ketones than the SM24 or PRD24 groups. Also after 18 months, urine samples from males in SM24 and PRD24 groups had more blood than males from SMR, SM6, or LM24 groups.

Clinical Chemistry Data. SMR, SM6, and LM24 were associated with lower total serum protein levels and lower serum albumin levels during the first half of the study than SM24. However, a difference in the opposite direction was evident at 30 months. Blood glucose was consistently lower in SMR groups than in SM24 groups.

Haematological Data. White blood cell counts (WBC), red blood cell counts (RBC), and haemoglobin (Hb) were significantly lower in the SMR groups than in the SM24 groups during the first 18 months of the study. The same is true for SM6 in respect of WBC. After 18 months WBC, RBC, and Hb tended to be lower in LM24 groups than SM24 groups.

Tail Necrosis. Tail necrosis was seen significantly more frequently in the two restricted groups (groups 3 and 5). This may be due to the fact that animals in these groups were more active but drank less than animals in other groups during long stretches of the day when their food baskets were empty.

Histopathological Findings: General. At the time of preparing this report the data for groups 9 and 11 are not available.

Corticomedullary and Pelvic Nephrocalcinosis. As shown in Table 26.10, diet restriction (SBR or SB6 prior to week 13, groups 5 to 8) irrespective of diet after 13 weeks was associated with significantly higher incidence of corticomedullary nephrocalcinosis in males and a significantly lower incidence of the same change in females than SB24 prior to week 13 (group 1). LB24 followed by LM24 (group 10) was also associated with less corticomedullary nephrocalcinosis in females. PRD24 gave rise to significantly less corticomedullary nephrocalcinosis (females) and significantly more pelvic nephrocalcinosis (both sexes) than was seen in any other group. It is noteworthy that analysis of the diets indicated that only in the cases of SB and PRD was the calcium-to-phosphorus ratio less than unity (see Table 26.5).

Nephropathy, Myocarditis, Polyarteritis, and Prostatitis. The findings in respect of nephropathy, myocarditis, polyarteritis, and prostatitis are summarized in

Table 26.10. Percent incidences of corticomedullary and pelvic nephrocalcinosis.

Males	\multicolumn{10}{c}{Group}									
	1	2	3	4	5	6	7	8	10	12
Diet to week 13	SB24	SB24	SB24	SB24	SBR	SBR	SBR	SB6	LB24	PRD24
Diet from week 13	SM24	LM24	SMR	SM6	SMR	LM24	SM24	SM6	LM24	PRD24
Corticomedullary										
nephrocalcinosis										
Any	0	0	0	0	24**	24***	30***	14*	0	0
Moderate/severe	0	0	0	0	4	6	6	2	0	0
Severe	0	0	0	0	0	0	2	0	0	0
Pelvic										
nephrocalcinosis										
Any	8	6	8	8	6	10	10	6	2	24*
Moderate/severe	0	0	2	0	2	0	0	0	2	10(*)
Females										
Corticomedullary										
nephrocalcinosis										
Any	100	96(*)	98	96	86	74***	68***	86*	54***	32***
Moderate/severe	80	64(*)	76	56	38**	20***	26***	36***	10***	2***
Severe	16	24	18	22	12	4	4	6	0*	0*
Pelvic										
nephrocalcinosis										
Any	0	0	0	2	0	0	2	4	0	28***
Moderate/severe	0	0	0	0	0	0	0	0	0	4

Diet abbreviations same as in Table 26.4.
Significance values relate to comparisons with group 1.
$(*)p<0.1$; $*p<0.05$; $**p<0.01$; $***p<0.001$

Table 26.11. Compared with SM24, PRD24 was associated with higher incidences of all these changes except prostatitis, while SMR, SM6, and LM24 were associated with lower incidences of all these changes.

Mammary Acinar Hyperplasia and Secretory Activity. Diet restriction (SMR— groups 3 and 5) was associated with significantly less mammary hyperplasia and secretory activity in females and less secretory activity in males than SM24. The same is true for mammary hyperplasia and secretory activity in females in LM24 groups (groups 2, 6, and 10).

Neoplasms of Endocrine Glands and the Exocrine Pancreas. As shown in Table 26.12, SMR and LM24 from 13 weeks were associated with significant ($p<0.01$ or $p<0.001$) reductions in incidence of neoplasms of the anterior and intermediate lobes of the pituitary and islet cell tumours of the pancreas compared with SM24. Adenomas of the exocrine pancreas were encountered less frequently ($p<0.05$) in SMR and SM6 groups. PRD24 was associated with a significantly higher incidence of tumours ($p<0.05$) of the adrenal medulla. Leydig cell tumours were seen in significantly higher ($p<0.05$) incidence in rats fed LM24

Table 26.11. Nephropathy, myocarditis, polyarteritis, and prostatitis.

	Gender	Diet from week 13				
		SM24	SMR	SM6	LM24	PRD24
Group		1,7	3,5	4,8	2,6,10	12
Kidney						
Nephropathy						
Any	M	95	42***	86***	77***	98***
Severe/very severe	M	21	0***	5***	1.3***	46***
Any	F	93	42***	66***	59***	100
Severe/very severe	F	15	0***	3**	0***	16
Heart						
Chronic myocarditis						
Any	M	34	38	45	29.3	66***
Any	F	46	25***	19***	11.3***	58
Fibrosis						
Any	M	23	6***	14(*)	11.3(*)	44*
Any	F	19	1***	3***	2.7***	28
Polyarteritis						
Pancreatic artery	M	8	4	2*	8.7	28***
Pancreatic artery	F	10	0***	3(*)	2.7**	20
Mesenteric artery	M	1	1	1	2.7	18***
Mesenteric artery	F	1	0	0	0.7	2
Prostatitis						
Acute						
Any		19	4***	9*	9.3*	16
Moderate/severe		16	2***	7	2***	6
Chronic						
Any		10	3*	8	14.7	12
Moderate/severe		5	1(*)	5	4.7	2

$*p < 0.05$; $**p < 0.001$; $***p < 0.001$

after week 13. No clear differences between groups were seen in tumours of the adrenal cortex, ovary, parathyroid, thyroid follicular cells, or thyroid C cells.

Neoplasms of Epidermis, Adnexa, Jaw, Subcutaneous Tissue, and Mammary Gland. The incidence of neoplasms of epidermis, adnexa, jaw, subcutaneous tissue, and mammary gland is summarized in Table 26.13. Compared with rats on SM24 from week 13, the two SMR groups (groups 3 and 5) developed significantly fewer epidermal adnexal and jaw tumours (both sexes, $p < 0.05$), significantly fewer subcutaneous tumours ($p < 0.05$) and highly significantly fewer mammary tumours (females, $p < 0.001$). In the SM6-fed and LM24-fed rats (groups 2, 4, 6, 8, and 10), mammary tumour incidence was significantly lower ($p < 0.01$) than in group 1. In PRD24 rats (group 12) the incidence of subcutaneous tumours was significantly higher in females ($p < 0.05$) than in group 1.

Neoplasms of the Lungs. As shown in Table 26.14, diet restriction (SMR—groups 3 and 5) significantly reduced the risk of developing an adenoma or primary adenocarcinoma of the lungs.

Table 26.12. Neoplasms of endocrine glands.

	Diet from week 13				
	SM24	SMR	SM6	LM24	PRD24
Group	1,7	3,5	4,8	2,6,10	12
Adrenal cortex					
B or M (males)	0	1	1	0.7	2
B or M (females)	2	3	2	0.7	2
Adrenal medulla					
B or M (males)	3	4	7	6.7	8
B or M (females)	3	1	4	0.7	10
Ovary					
B or M	4	3	0	3.3	0
Endocrine pancreas (islet cell)					
B or M (males)	10	1**	3(*)	2.7*	4
B or M (females)	2	1	0	0.7	2
Exocrine pancreas (males only)					
B	5	0*	0*	1.3	4
Parathyroid					
B (males)	5	1	2	4	4
B (females)	2	1	4	4	2
Pituitary: anterior lobe					
B or M (males)	30	14***	29	21.3	24
B or M (females)	62	46**	47	45.3***	52
Pituitary: intermediate lobe					
B or M (males)	13	9	8(*)	4.7**	8
B or M (females)	6	0**	5	0.7**	10
Testis/Leydig cell					
B	23	30	29	38.7*	36
Thyroid/follicular					
B or M (males)	1	1	1	1.3	4
B or M (females)	2	0	2	0.7	4
Thyroid/C cell					
B or M (males)	2	2	8	6(*)	2
B or M (females)	8	4	2(*)	2.7*	0(*)

Diet abbreviations same as in Table 26.4.

B, benign; M, malignant

(*)$p<0.1$; *$p<0.05$; **$p<0.01$; ***$p<0.001$

Haemangiomas and Angiosarcomas of the Mesenteric Lymph Node. A particularly surprising finding in the study was a significantly higher incidence of haemangiomas and angiosarcomas of the mesenteric lymph node ($p<0.01$ for the sexes combined) in the three groups fed the LM24 diet from week 13 (groups 2, 6, and 10) than in the groups fed SM24 from week 13 (groups 1 and 7). By contrast, SMR significantly reduced the incidence of tumours ($p<0.05$ for the sexes combined) of these kinds. The data are summarized in Table 26.15.

Neoplasms of the Uterus. Table 26.16 summarizes the data for neoplasms of the uterus. At this site diet restriction was without significant effect. However, rats fed LM24 after week 13 (groups 2, 6, and 10) developed significantly more

Table 26.13. Neoplasms of epidermis, adnexa, subcutaneous tissue, and mammary gland.

Diet from week 13	SM24	SMR	SM6	LM24	PRD24
Group	1,7	3,5	4,8	2,6,10	12
Males					
Epidermis/adnexa					
B or M	14	7*	10	8	10
M	7	3	7	5.3	8
Subcutaneous tissue					
B or M	18	6*	10	10.7	18
M	6	4	5	7.3	16
Mammary gland					
B or M	0	0	1	0	2
M	0	0	0	0	0
Females					
Epidermis/adnexa					
B or M	5	0*	2	2.7	6
M	5	0*	2	2	6
Subcutaneous tissue					
B or M	1	2	7	3.3	10*
M	1	2	6	3.7	8(*)
Mammary gland					
B or M	37	9***	19**	9.3***	22
M	5	1(*)	5	2	4

Diet abbreviations same as in Table 26.4.
B, benign; M, malignant
(*)$p<0.1$; *$p<0.05$; **$p<0.01$; ***$p<0.001$

uterine tumours ($p<0.01$), most of which were adenomatous polyps, adenocarcinomas, or anaplastic carcinomas.

Neoplasms of All Sites. As shown in Table 26.17, SMR (groups 3 and 5) significantly reduced the incidence of rats developing one or more benign or malignant tumours ($p<0.001$) before the termination of the study as compared with SM24.

Table 26.14. Neoplasms of the lungs.

Diet from week 13	SM24	SMR	SM6	LM24	PRD24
Group	1,7	3,5	4,8	2,6,10	12
Primary tumours					
B + M (males)	6	0*	2	4.7	2
M (males)	2	0	0	0.7	0
B + M (females)	4	0*	0(*)	1.3	2
M (females)	2	0	0	0.7	0
B + M (males and females)	5	0***	1*	3	2
M (males and females)	2	0*	0(*)	0.7	0

Diet abbreviations same as in Table 26.4.
B, benign; M, malignant
(*)$p<0.1$; *$p<0.05$; ***$p<0.001$

Table 26.15. Haemangiomas and angiosarcomas of the mesenteric lymph node.

Diet from week 13	SM24	SMR	SM6	LM24	PRD24
Group	1,7	3,5	4,8	2,6,10	12
Mesenteric lymph node					
B or M (males)	15	7(*)	11	34**	12
M (males)	6	0**	1*	15.3(*)	0
B or M (females)	8	2	2	14.7	0
M (females)	1	0	0	5.3	0
B or M (males and females)	11.5	4.5*	6.5(*)	24.3**	6
M (males and females)	3.5	0**	0.5*	10.3*	0

Diet abbreviations same as in Table 26.4.
B, benign; M, malignant
$(*)p<0.1$; $*p<0.05$; $**p<0.01$

This was true for both sexes and for malignant neoplasms. The incidence of rats bearing more than one tumour was also significantly reduced ($p<0.001$) in both sexes, and in males the diet restriction significantly reduced the risk of an animal developing two malignant tumours of different kinds ($p<0.01$). PRD24 was associated with a significantly higher incidence ($p<0.05$) of male rats bearing one or more tumours while significantly fewer female rats ($p<0.001$) fed on LM24 developed one or more benign or malignant neoplasms or malignant neoplasms of more than one site.

Table 26.16. Neoplasms of the uterus.

Diet from week 13	SM24	SMR	SM6	LM24	PRD24
Groups	1,7	3,5	4,8	2,6,10	12
Total no. of rats	100	100	100	150	50
Benign tumours					
Papillary adenoma	0	1	0	0	0
Adenomatous polyp	2	2	5	8	1
Squamous polyp	0	0	0	2	0
Fibromyoma	0	0	0	1	0
Total(%)	2(2)	3(3)	5(5)	11(7.3)	1(2)
Malignant tumours					
Adenocarcinoma	3	6	7	20	0
Anaplastic carcinoma	1	1	0	6	0
Squamous carcinoma	0	0	0	1	0
Sarcoma	1	1	6	4	1
Haemangiosarcoma	0	0	0	1	0
Total (%)	5(5)	8(8)	13(13)	32**(21.3)	1(2)
Total tumours (%)	7(7)	11(11)	18(18)	43**(28.7)	2(4)

Diet abbreviations as in Table 26.4
$**p<0.01$

Table 26.17. Neoplasms of all sites.

Diet from week 13	SM24	SMR	SM6	LM24	PRD24
Group	1,7	3,5	4,8	2,6,10	12
One or more sites					
B or M (males)	78	69***	83	84.7	88*
B or M (females)	86	66***	76	78.7***	80
M (males)	30	13***	33	41.3	44(*)
M (females)	34	18***	33	34	40
More than one site					
B or M (males)	52	26***	44*	57.3	46
B or M (females)	55	16***	38*	30***	52
M (males)	7	0**	2(*)	6	0
M (females)	4	1	4	4	6

Diet abbreviations same as in Table 26.4.
B, benign; M, malignant
(*)$p<0.1$; *$p<0.05$; **$p<0.01$; ***$p<0.001$

Discussion

Although some of the results of this study are not yet available, several important conclusions may be confidently drawn in relation to the main aims of the study whilst the associations between the LM24 diet and increased risks of mesenteric node and uterine carcinomas merit further research.

Effect of Diet Restriction (80% of Ad Libitum) on the Incidence of Malignant Neoplasia of All Sites. The finding reported earlier that restriction of calorie intake to about 80% of ad libitum reduces the risk of premature death from malignant neoplasm has been confirmed (see Table 26.17). This beneficial effect appeared to apply to all kinds of malignant neoplasms for which there were sufficient data for comparisons to be made. There was no example of a significantly higher incidence of malignant neoplasia at any particular site in the SMR groups as compared with the SM24 groups.

Effect of Limiting Daily Access to Food to a 6-h Period. Although in the study reported earlier (Salmon et al. 1990), diet restriction to about 80% of ad libitum was achieved by limiting the access of rats to food to 6.5 h per day, this stratagem did not work in the present study. For reasons that are not clear, in the present experiment rats given access to food for only 6 h per day (between 1000 hours and 1600 hours) managed to consume almost as much food as rats given access to food throughout the 24 h of each day. It is not surprising therefore that relatively few differences were seen between rats fed SM6 and SM24 after 13 weeks. Among the differences recorded, however, were significantly lower incidences of nephropathy and myocarditis (Table 26.11), mammary hyperplasia, and mammary tumours (see Table 26.13), and mesenteric lymph node tumours (see Table 26.15). By contrast a significantly higher incidence of subcutaneous neoplasms was seen in SM6 females than in SM24 females (see Table 26.13).

Effects of Limiting the Energy Value of the Diet. Reduction of the energy values of diets fed from weaning to week 13 or from week 13 to the termination of the study led to increased consumption of food and water and to increased energy intake (see Table 26.6). The intake of protein of rats on LM24 was similar to that of rats on SM24. However, like diet-restricted rats (SMR), rats on LM24 had lower serum protein levels and lower urinary protein levels than SM24 rats. The incidences of nephropathy, myocarditis, polyarteritis, acute prostatitis, mammary hyperplasia, and secretory activity were also markedly lower in LM24 rats than in SM24 rats (see Table 26.11). Beneficial effects of LM24 as compared with SM24 were seen in the cases of tumours of the anterior and intermediate lobes of the pituitary gland and islet cell tumours of the pancreas (see Table 26.12) and mammary tumours (see Table 26.13). However, as discussed below, three kinds of tumour—Leydig cell tumours, tumours of the mesenteric lymph node, and uterine tumours—arose significantly more frequently in LM24-fed rats than in SM24- or PRD24-fed rats. Overall benign and malignant tumour incidence was lower in females fed LM24 than in females fed SM24 or PRD24 (see Table 26.17).

Effect of Diet Restriction During the 13 Weeks After Weaning. The effects of diet restriction during the first 13 weeks of the study could be assessed by comparing the findings in groups 5, 6, and 7 with those in groups 3, 2, and 1, respectively. These comparisons revealed very few differences. The most obvious of these was the enhancement of corticomedullary nephrocalcinosis in males and its reduction in females (see Table 26.10). Diet restriction confined to the first 13 weeks of the study had little or no effect on survival, on the incidences of ageing-associated non-neoplastic diseases, or on the incidence of neoplasms of any kind.

Prediction of Risk of Tumour Development from Food, Energy, or Protein Intake During the First 12 Months of the Study. Groups 3 and 5 (SMR from week 13) consumed less food, less energy and less protein than any of the other groups. The same two groups experienced the best survival and the lowest incidences of nonneoplastic and neoplastic diseases. Group 12 (PRD24 throughout) had the highest consumption of food energy and protein and exhibited the highest incidences of all the groups of nephropathy, myocarditis, polyarteritis and of one or more tumours of all sites. The data from the study do not enable one to distinguish between the predictive values of the three forms of intake.

Prediction of Risk of Tumour Development (All Sites) from Body Weight Gain Earlier in the Study. Tumour incidences were highest in the two groups which put on most weight during the first 18 months of the study (groups 1 and 12) and lowest in the two groups which put on least weight during the same period (groups 3 and 5). Restricting access to food to 6 h per day (SM6) reduced body weight gain, nephropathy and myocarditis but had little or no effect on tumour incidence. Survival was improved by SM6 in males but not in females. These results, which confirm and extend those of Turnbull et al. (1985), are consistent with there being a moderately good correlation between body weight gain and tumour risk.

Relationships Between In-Life Measurements, Survival, and Necropsy Findings in Individual Animals. The accumulating data base derived from the study will permit such relationships to be studied. However, no analyses of this kind have yet been undertaken, partly because further data are awaited and partly because further funding is needed.

Adrenal Medullary Tumours and Pelvic Nephrocalcinosis. Earlier we pointed to there being associations between adrenal medullary proliferative disease, enhanced calcium absorption from the gut, and pelvic nephrocalcinosis in rats (Roe and Baer 1985). In the present study, PRD24 gave rise both to the highest incidence of pelvic nephrocalcinosis and the highest incidence of adrenal medullary tumours (see Tables 26.10 and 26.12). The difference in incidence between PRD24 (group 12) and SM24 (groups 1 and 7) was significant ($p < 0.05$), and a similar difference was seen in the incidence of adrenal medullary hyperplasia between the same groups.

The Association Between the LM24 Diet and Tumours of the Mesenteric Lymph Node. The finding of an association between LM24 diet and tumours of the mesenteric lymph node is, as far as we know, an entirely new one and, as such, was wholly unexpected. The mechanism remains to be elucidated.

In the present study high incidences of mesenteric lymph node tumours were seen in all groups, the incidence being higher in males than in females. Thus 89 out of the 500 males (18%) and 34 out of the 500 females (7%) for which histopathological evaluation is complete had haemangiomatous tumours of the mesenteric lymph node. This suggests that the Wistar strain of rats used for the study may have been genetically prone to develop such tumours. Nevertheless an explanation is needed of why, in animals fed on LM24 from week 13, the incidences of these tumours reached 34% in males and 14.7% in females compared with 15% in males and 8% in females fed on SM24 (groups 1 and 7).

Samples of the LM diet preserved in cold storage will now be analyzed for known carcinogens.

The Association Between the LM Diet and Carcinomata of the Uterus. The significantly higher incidence ($p < 0.01$) of uterine tumours in groups 2, 6 and 10 (21.3%) than in groups 1 and 7 (5%) poses the same questions as the higher incidence of mesenteric lymph node tumours in these same groups. The fact that benign uterine tumours – particularly adenomatous and squamous polyps – also occurred in highest incidence in the same groups (7.3% as compared with 2% in groups 1 and 7) reinforces the need to seek answers to the questions posed.

Conclusions

It has long been recognized that genetic and environmental factors interact in the determination of incidence of most diseases. In the past there has been a tendency for experimentalists to assume that the pattern of diseases seen in control groups is determined mainly, or solely, by the genetic constitution of the strain of animals used. The results of the present study illustrate the extent to which diet and food consumption may influence survival and the incidence of degenerative and neoplastic diseases in rats. Both the composition of diets and the amounts consumed are important determinants of disease incidence. However, diet restriction by rationing animals to 80% of what 24 h/day ad libitum-fed animals eat from the age of 4 months seems to be more effective in reducing degenerative disease and neoplasia than any of the other stratagems tried in this study.

The beneficial effects of calorie restriction on the incidence of pituitary tumours, mammary tumours, nephropathy, myocarditis, and polyarteritis are well known. Less well known, although previously reported, is the effect of calorie restriction in reducing the incidence of prostatitis. We believe that the present study provides the first evidence for an effect of calorie restriction on lung tumour incidence in rats, although such an effect has previously been described in mice (Conybeare 1980).

The confirmation provided by the present study that calorie restriction nonspecifically reduces the risk of fatal or potentially fatal malignant neoplasia of seemingly all sites provides substantial support for the view (Gensler and Bernstein 1981; Yu 1989; Roe 1989) that endogenously generated electrophiles (e.g., during lipid peroxidation) may be important determinants both of ageing-associated degenerative diseases and of neoplasia. The underlying theory is that higher calorie intake is associated with higher rates of generation of electrophiles as a consequence of normal metabolic processes and that these electrophiles increase the risks both of ageing-related diseases and cancer.

The high incidences of mesenteric lymph node tumours and uterine carcinomas in rats fed on a low nutrient, high-fibre diet in the present study demonstrates a need for further research.

References

Berg BM, Simms HS (1960) Nutrition and longevity in the rat. II Longevity onset of disease with different levels of food intake. J Nutr 71:255–263

Conybeare G (1980) Effect of quality and quantity of diet on survival and tumour incidence in outbred Swiss mice. Food Cosmet Toxicol 18:65–75

Fry JS, Lee PN (1988) Stratified rank tests. Appl Stat 37:264–266

Gensler HL, Bernstein H (1981) DNA damage as the primary cause of aging. Q Rev Biol 56:279–301

Peto R, Pike MC, Day NE, et al. (1980) Guidelines for simple, sensitive significance tests for carcinogenic effects in long term animal experiments. In: Long-term and short-term screening assays for carcinogens: a critical appraisal. IARC Monogr Eval Carcinog Risk Chem Hum Suppl 2:311–426

Roe FJC (1989) Non-genotoxic carcinogenesis: implications for testing and extrapolation to man. Mutagenesis 4:407–411

Roe FJC, Baer A (1985) Enzootic and epizootic adrenal medullary proliferative diseases of rats: influence of dietary factors which affect calcium absorption. Hum Toxicol 4:27–52

Salmon GK, Leslie G, Roe FJC, Lee PN (1990) Influence of food intake and sexual segre-
gation on the incidence of non-neoplastic and neoplastic diseases in rats. Food Chem
Toxicol 28:39–48

Tucker MJ (1979) The effect of long-term food restriction on tumours in rodents. Int J
Cancer 23:803–807

Turnbull GJ, Lee PN, Roe FJC (1985) Relationship of body weight gain to longevity and
risk of nephropathy and neoplasia in Sprague-Dawley rats. Food Chem Toxicol
23:355–361

Yu BP (1989) Why dietary restriction may extend life: a hypothesis. Geriatrics 44:87–90

CHAPTER 27

Primate Models for Dietary Restriction Research

G.S. Roth[1], D.K. Ingram[1], and R.G. Cutler[1]

Introduction

During the six-odd decades in which reduced dietary intake has been established as the only definitive intervention for extension of maximum life span in mammals, two burning questions have persisted. First, what is the mechanism responsible for the life-prolongation effect, and second, can this effect be achieved in humans (or for that matter, any animal higher on the evolutionary scale than rodents)?

Although many studies have focused on the former question (for a review see Weindruch and Walford 1988), the latter has been largely ignored for a number of reasons. Foremost among these are the high costs associated with maintaining statistically significant numbers of longer-lived species for periods of time sufficient to evaluate the effects of long-term reduced feeding. For the same reason, many mammals with maximal life spans beyond a few years have been poorly characterized with respect to mortality, disease, and various functional and structural indices of aging. Ideally, it would be desirable to conduct such studies on human populations directly, but for both ethical and economic reasons, properly controlled experiments are out of the question at this time. Although "accidents of history" have frequently created undernourished groups of people over the centuries, in essentially all cases proper controls were unavailable for comparisons, health conditions were suboptimal, and malnutrition rather than undernutrition was the rule.

The challenge, therefore, has long been to select an appropriate animal model, with adequate characterization with respect to aging, to most closely mimic the human situation, and to design a study with sufficient numbers of animals, proper environmental and health conditions, economic feasibility, and undernutrition without malnutrition. We seriously began to consider initiation of such a study in the mid-1980s. Among the mammals evolutionarily superior to rodents, dogs have probably been most used for aging studies. Since both their maximal

[1]Molecular Physiology and Genetics Section, Laboratory of Cellular and Molecular Biology, The Nathan W. Shock Laboratories of the Gerontology Research Center, National Institute on Aging, National Institutes of Health, Francis Scott Key Medical Center, Baltimore, MD 21224, USA

Table 27.1. Biological indices of aging in *Macaca nemestrina* (from Short et al. 1987).

Variables	Female Y to M[a]	Female M to O[b]	Male Y to M[a]	Male M to O[b]
Immunoglobulin A	↑	↑	↑	↑
Triglycerides (L)	↑	−	↑	−
Testosterone			↑	↓
Tactile corpuscles	↓	↓	↓	−
Activated suppressor T cells	−	↑	↑	↑
Fingernail growth rates	↓	↓	↑	↓
Heart rate	−	↓	↑	↓
Follicle stimulating hormone	−	↑	−	−
Blood chloride	↓	↓	−	−
Blood calcium	↓	−	↓	↓
Bone thickness	↓	↓	−	↓
Blood albumin	↓	↓	↓	↑
Blood platelets	−	↓	−	↓
Rectal body temperature	↓	↓	↓	↓
Estradiol	−	↓		
Hemoglobin	−	−	−	↑
Blood glucose	↑	↑	−	↑
Blood potassium	↓	↑	−	−
Apolipoprotein-A1 (H)	↓	↑	↑	−
Very low-density lipoprotein	↑	−	−	−

↑ and ↓ indicate, respectively, an increase or decrease in age-group means with increasing age; − indicates no apparent difference, and blanks indicated not tested; H, high-density fraction; L, low-density fraction

[a]Y to M stands for between the young and middle age groups.

[b]M to O stands for between the middle and old age groups.

life spans and housing costs approach those of some shorter-lived primates, our selection was quickly narrowed to the latter. After 2 years of investigating the suitability of several dozen primate species according to the criteria established above, our final choices were rhesus and squirrel monkeys.

Rhesus (*Macaca mulatta*) and squirrel (*Saimiri sciurius*) monkeys have estimated maximal life spans of approximately 40 and 20 years, respectively (Jones 1968). They have been partially characterized with respect to aging (see next section), are available in reasonable quantities, and can be housed under environmental conditions that do not require exorbitant expenditures. However, at the time of our decision little information was available as to the short-term effects of reduced feeding on the general health status of these monkeys. For this, and a number of other reasons, a pilot study was indicated (see further below).

Primate Aging

Due to high costs and relatively long life span, the use of primates as aging models has lagged considerably behind that of rodents and lower species. Only a few colonies contain sufficient numbers of aged and genetically/environmentally

Table 27.2. Assessment of primate aging: effects of caloric modification design for study.

Species	Group label	Estimated age when incorporated into study (years)	Control (*n*)	Experimental animals (*n*)
Squirrel monkeys,	Juvenile	1–4	7	5
Saimiri sciurius	Adult	5–9	7	6
	Old	> 10	4 (8)[a]	–
Rhesus monkeys,	Juvenile	1–4	6 (10)[b]	6 (10)
Macaca mulatta	Adult	5–9	6 (10)	6 (10)
	Old	> 20	6 (10)	– (10)

[a]Four additional old squirrel monkeys were purchased in November 1988. Numbers in parentheses reflect the total.
[b]Numbers in parentheses reflect total monkeys in respective groups after June 1988 expansion of rhesus population.

comparable younger primates for cross-sectional age analyses, let alone longitudinal studies over periods of time long enough to be meaningful. In addition, invasive studies or those requiring killing are limited even more.

Nevertheless, several recent reports have revealed the very tempting potential of nonhuman primates as aging models with a great degree of relevance for humans. One of the most ambitious and comprehensive projects has employed the pigtailed macaque, *Macaca nemestrina* (Short et al. 1987). Approximately 30 animals of each sex ranging from 4 to 28 years were examined over a 4- to 5-year period. Seventy-two different anatomical, biochemical, and physiological measurements were performed. Of these, 23 exhibited significant main effects of age. Some of these are listed in Table 27.1. Although sex differences in age effects can be noted in a few cases, most exhibit fairly consistent patterns across gender. Particularly robust age changes have been reported for rectal body temperature, blood albumin, immunoglobulin A, fingernail growth rate, bone thickness, and blood calcium (Short et al. 1987). It must be emphasized, however, that not all of these cross-sectional age differences prove predictive of longitudinal changes and vice versa (D. Bowden, personal communication).

It is interesting to note that T-lymphocyte mitogenic stimulation failed to show significant age effects in pilot studies and was eliminated from the above investigation. However, a very recent examination of 92 rhesus monkeys (*Macaca mulatta*) of both sexes ranging from 2 to more than 20 years of age indicates that lymphocyte stimulation by concanavalin A, phytohemagglutinin, and pokeweed mitogen all significantly decline with age (Ershler et al. 1988). Greatest decrement in these immunological responses occurs between midlife (11–17 years) and senescence (> 20 years) in all cases. In addition, a roughly progressive 70%–80% decline in natural killer cell function was observed in the same study.

Changes in bone thickness, particularly in females, are another important manifestation of aging reported in *Macaca nemestrina* (DeRousseau 1985b). Complementary observations of age-related osteopenia (DeRousseau 1985b) and degenerative joint disease (DeRousseau 1985a; Kessler et al. 1986) have been

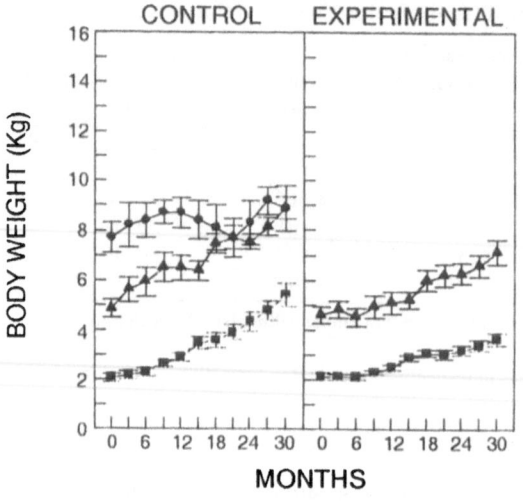

Figure 27.1. Body weights of juvenile (*squares*), adult (*triangles*), and old (*circles*) rhesus monkeys (group 1) during the first 30 months of the study

noted for *Macaca mulatta*. Taken together, these findings further emphasize the similarity of skeletal aging in monkeys and humans.

Neurological degeneration represents still another area in which nonhuman primates may serve as useful models for human aging. For example, squirrel monkeys (*Saimiri sciurius*) were recently reported to exhibit senile plaques in the brain that are quite similar to those occurring in aged rhesus monkeys and

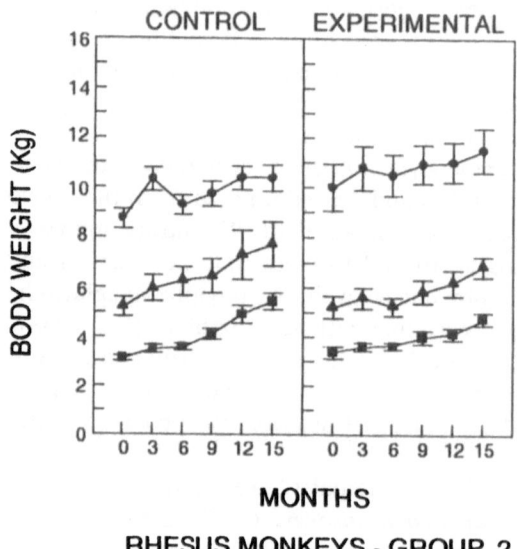

Figure 27.2. Body weights of juvenile (*squares*), adult (*triangles*), and old (*circles*) rhesus monkeys (group 2) during the 15 months after June 1988

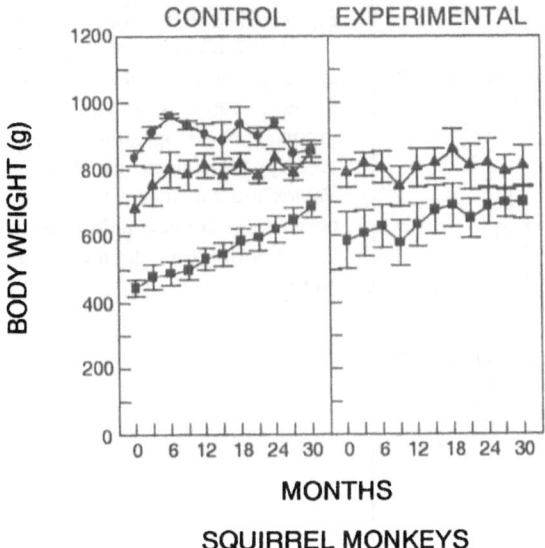

Figure 27.3. Body weights of juvenile (*squares*), adult (*triangles*), and old (*circles*) squirrel monkeys during the first 30 months of the study

humans (Walker et al. 1987). Although this phenomenon may be more representative of age-associated disease than "normal" aging, the parallels with the human situation are impressive. If it is indeed possible to separate pathological and nonpathological aging, a striking marker appears to be the loss of striatal dopamine receptors. This phenomenon has now been observed in rats, mice, rabbits, humans, and, most recently, rhesus monkeys (for reviews see Roth et al. 1986 and Lai et al. 1987).

It should be recognized that several of the age-related changes reported for subhuman primates are similar to those which have been affected by reduced feeding in rodents. Perhaps best known are the reductions in immunological function, body temperature, and striatal dopamine receptor concentrations. Thus, these primate models not only exhibit many parallels with human aging processes, but also may be ideal systems with which to compare dietary modulation of aging rate with studies performed in rodents.

Dietary Restriction in Primates

A number of studies have examined primate nutrition, but to our knowledge none have studied the effects of reduced intake on aging in a manner analogous to that carried out in rodents (e.g., Short et al. 1987).

Therefore, in our initial experiment 30 males of each species were housed at the NIH Animal Center at Poolesville, MD, and divided into the age and diet groups

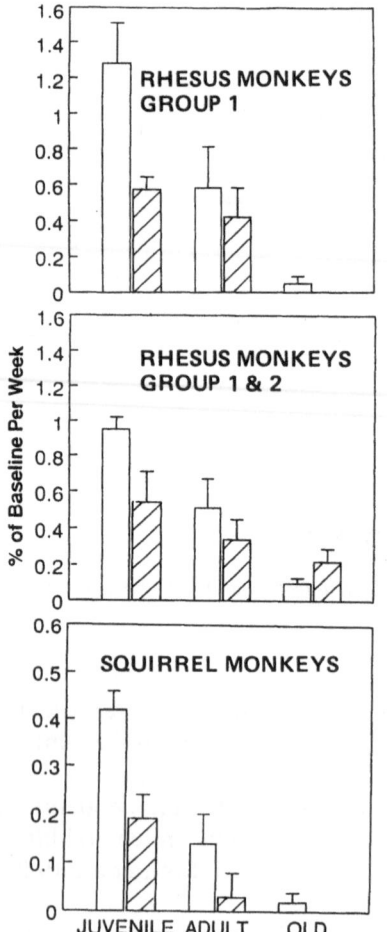

Figure 27.4. Effect of 30% restricted food intake on relative body weight gain in rhesus and squirrel monkeys. Body weight data were obtained through January 1990. Data for the combined rhesus groups include only the initial 19 months of full dietary restriction for the respective groups. Values are the means ± standard errors for the numbers of animals indicated in Table 27.2. *Open bars*, control animals; *hatched bars*, experimental animals

shown in Table 27.2. Semisynthetic diets based on established nutritional requirements for the two species (NIH open formula nonhuman primate diets) were formulated (Ingram et al., to be published) and supplemented with 40% greater vitamin and mineral contents to allow for a 30% reduction in overall intake without reduction in trace nutrients. All animals were fed ad libitum for the first month at Poolesville, then experimental groups sequentially received 10% less food (based on ad libitum intake of body weight matched controls) each month for 3 months, at which point (30% reduction) they have been maintained to date. In June of 1988 additional rhesus monkeys (group 2) were added to the study and in November four additional old squirrel monkeys were added as shown in Table 27.2. In some cases, data from these animals are treated separately.

Figures 27.1–27.3 show the mean body weights for all groups over the course of the study. Despite some heterogeneity within groups, juvenile and adult experimental animals appear to exhibit slower weight gain than their correspond-

Table 27.3. Food consumption in rhesus and squirrel monkeys.

	February 1988		
	% Total consumed	% Control total consumed	% Reduction in intake
Rhesus group 1			
JC	90	90	
JE	100	70	23
AC	91	91	
AE	100	70	24
OC	78	78	
OE[a]	—	—	—
Squirrel monkeys			
JC	83	83	
JE	93	65	22
AC	82	82	
AE	88	62	24
OC	66	66	
	February–March 1989		
Rhesus group 1			
JC	100	100	
JE	100	70	30
AC	100	100	
AE	100	70	30
OC	66	66	
OE*	81	56	15
Rhesus group 2[b]			
JC	88	88	
JE	98	68	23
AC	91	91	
AE	89	62	32
OC	78	78	
OE	89	62	21
Squirrel monkeys			
JC	90	90	
JE	97	67	26
AC	85	85	
AE	91	63	26
OC	80	80	
	December 1989–January 1990		
Rhesus group 1			
JC	98	98	
JE	100	70	29
AC	84	84	
AE	100	70	17
OC	60	60	
OE*	72	50	16
Rhesus group 2[b]			
JC	96	96	
JE	99	69	29
AC	98	98	
AE	97	67	32
OC	77	77	
OE	83	58	25

Continued on next page

Table 27.3. *Continued.*

	December 1989–January 1990		
	% Total consumed	% Control total consumed	% Reduction in intake
Squirrel monkeys			
JC	99	99	
JE	100	70	30
AC	99	99	
AE	100	70	30
OC	88	88	

JC, juvenile control; JE, juvenile experimental; AC, adult control; AE, adult experimental; OC, old control; OE, old experimental
[a]OE formed from half OC in June 1988.
[b]There were 30 additional rhesus monkeys added in June 1988.

ing controls in all cases. Differences between control and experimental old rhesus monkeys are not yet apparent.

Data in Fig. 27.4 indicate that dietary restriction successfully reduced the absolute relative body weight gain in juvenile animals of all groups as well as adult squirrel monkeys. A trend toward reduction in the adult animals of the rhesus group is also evident. No significant differences are apparent for the old experimental group.

Over the same period, food intake was periodically checked to insure that experimental animals were indeed ingesting as close to 30% less than ad libitum-fed controls as possible. Table 27.3 shows that restricted animals in most cases actually consumed 20%–30% less food than their respective controls. The most notable exception is for aged rhesus monkeys since ad libitum-fed animals left the most food in these groups. Consequently, the food allotment for these aged animals has recently been reduced for both control and experimental monkeys. Restricted animals in the juvenile and adult groups averaged between 90% and 100% consumption of their respective allotments (63%–70% of the control allotment), less than complete consumption usually being accounted for by mechanical factors.

In addition to monitoring rates of body weight gain and food intake, a number of biological indices of aging were examined. Some of these have been catalogued in previous reports (Ingram et al., to be published; Roth et al., to be published). Several representative examples will be discussed here.

The monkeys under study are being characterized with respect to a number of endocrine parameters (Blackman et al., manuscript in preparation). Figure 27.5 shows progressive decrease with age in the serum concentration of dehydroepian-drosterone (DHEA) in both control and restricted rhesus monkeys. This situation is roughly comparable to that described for humans (Orentreich et al. 1984). Surprisingly, squirrel monkeys exhibit an opposite phenomenon, progressive increase in concentration with age. Restricted animals are not significantly

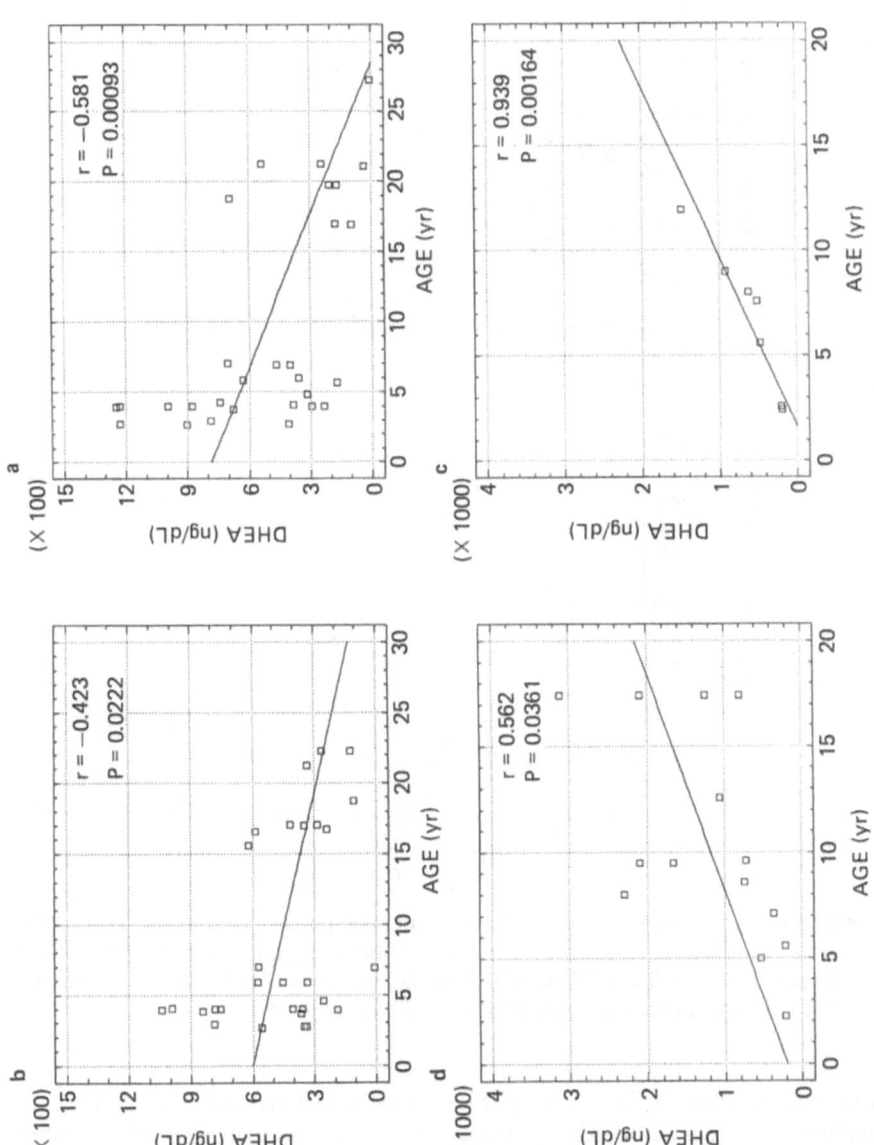

Figure 27.5a–d. Effect of age and diet on dehydroepiandrosterone (DHEA) levels in sera of dietary-restricted rhesus (**a**) and squirrel (**b**) monkeys and their controls (**c** and **d**, respectively)

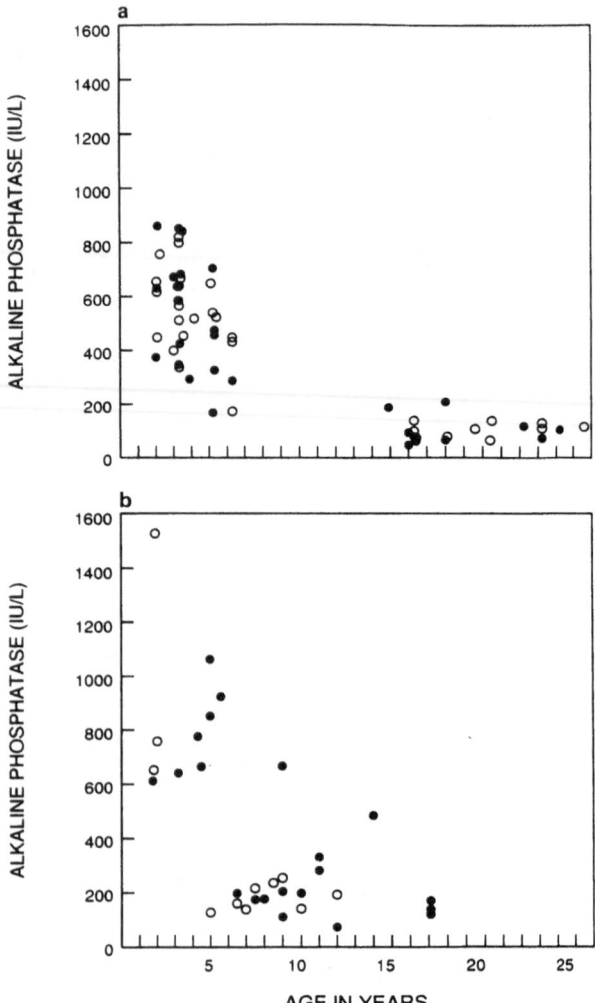

Figure 27.6a,b. Effect of age and diet on serum alkaline phosphatase levels in rhesus (**a**) and squirrel (**b**) monkeys. Value are the means of three separate determinations over a 6-month period between 9 and 15 months after the initiation of full 30% dietary restriction. *Open circles*, experimental animals; *solid circles*, controls

different from controls in either species, thus data have been combined. An explanation for the contrasting patterns of DHEA change will await further study, but the robust age effects may render this an extremely useful parameter for assessing diet effects if they occur.

Various blood chemistry parameters have also been examined. Figures 27.6–27.8 illustrate mean levels of serum alkaline phosphatase, calcium, and globulin over a 6-month period between 9 and 15 months after the initiation of

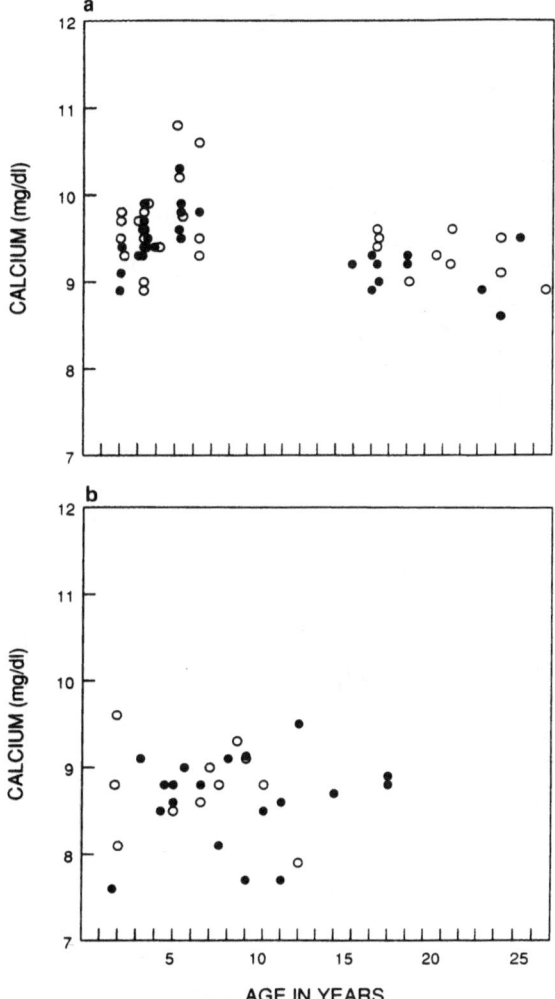

Figure 27.7a,b. Effect of age and diet on serum calcium levels in rhesus (**a**) and squirrel (**b**) monkeys. Values are the means of the separate determinations over a 6-month period between 9 and 15 months after the initiation of full 30% restriction. *Open circles*, experimental animals; *solid circles*, controls

full dietary restriction as a function of individual animal age. Marked cross-sectional reductions with increasing age are observed for alkaline phosphatase levels in both species, along with a slight decrease for serum calcium in the Rhesus monkeys. Trends toward an increase in globulin concentrations occur with increasing age in both species. No striking differences between control and experimental animals are apparent for any of these parameters at the sample times depicted here.

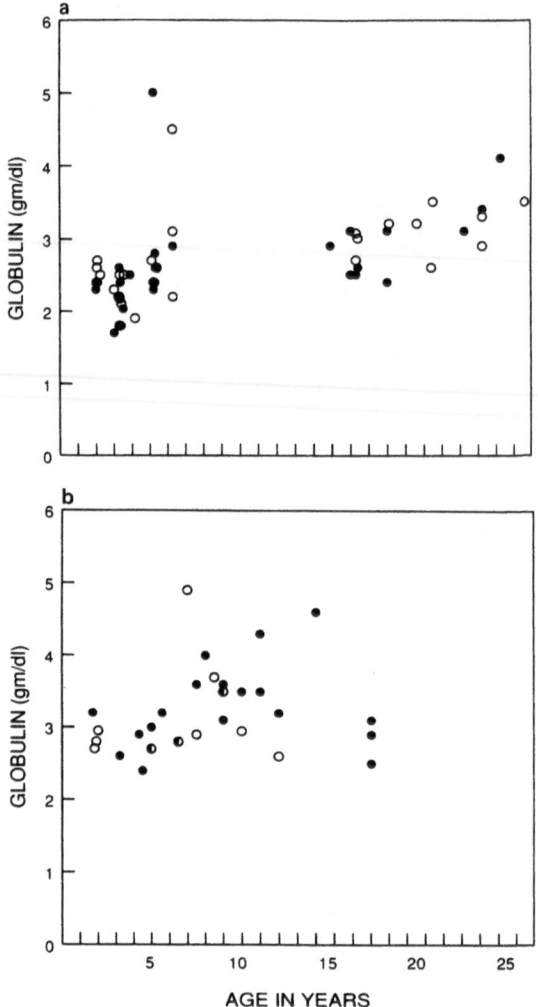

Figure 27.8a,b. Effect of age and diet on serum globulin levels in rhesus (**a**) and squirrel (**b**) monkeys. Values are the means of three separate determinations over a 6-month period between 9 and 15 months after the initiation of full 30% dietary restriction. *Open circles*, experimental animals; *solid circles*, controls

Summary and Conclusions

Rhesus and squirrel monkeys have been administered a semisynthetic, ad libitum or 30% reduced intake, diet for approximately 2½ years and continue to be in good health as determined by physical examinations, hematology, and blood chemistry. Rates of body weight gain in the restricted animals have been substantially reduced. At the same time, a number of anatomical, biochemical, and physiological measurements of parameters that are correlated with age have been identified. Although significant cross-sectional age differences have been documented for many of these measurements, no significant effects of diet on parameters other than body weight have been detected to date. Nevertheless, with the expansion of the study to 90 monkeys in 1988 and the relatively robust age differences already documented, it should be possible within the next 3–5 years to detect dietary effects if they do indeed occur.

Acknowledgments. The authors wish to thank the staff of the NIH Animal Center in Poolesville, MD, for their excellent assistance in the current study, Dr. Joseph Knapka for his valuable nutritional advice, Dr. Bernadette Marriott for her superb contributions to numerous aspects of the study, Ms. Sharon Davison for statistical assistance, and Ms. Rita Wolferman for typing the manuscript.

References

DeRousseau CJ (1985a) Aging in the musculoskeletal system of rhesus monkeys: II Degenerative joint disease. Am J Phys Anthropol 67:177–184

DeRousseau CJ (1985b) Aging in the musculoskeletal system of rhesus monkeys. III. Bone loss. Am J Phys Anthropol 68:157–167

Ershler WB, Coe CL, Gravenstein S, Schultz KT, Klopp RG, Meyer M, Houser WD (1988) Aging and immunity in nonhuman primates. I. Effects of age and gender on cellular immune function in rhesus monkeys (*Macaca mulatta*). Am J Primatol 15:181–188

Ingram DK, Cutler RG, Weindruch R, Renquist DM, Knapka JJ, April M, Belcher CT, Clark MA, Hatcherson CD, Marriott BM, Roth GS (1990) Dietary restriction and aging: the initiation of a primate study. J Gerontol 45:B148–163

Jones ML (1968) Longevity of primates in captivity. Int Zool Yearbook

Kessler MJ, Rawlins RG, London WT (1986) The hemogram, serum biochemistry, and electrolyte profile of aged Rhesus monkeys (*Macaca mulatta*). J Med Primatol 12:184–191

Lai H, Bowden DM, Horita A (1987) Age-related decreases in dopamine receptors in the caudate nucleus and putamen of the rhesus monkey (*Macaca mulatta*). Neurobiol Aging 8:45–49

Orentreich N, Brind JL, Vogelman JH (1984) Age changes and sex differences in serum dehydroepiandrosterone sulfate concentrations throughout adulthood. J Clin Endocrinol Metab 59:551–555

Roth GS, Henry JM, Joseph JA (1986) The striatal dopaminergic system as a model for modulation of altered neurotransmitter action during aging: effects of dietary and neuroendocrine manipulations. Prog Brain Res 70:473–484

Roth GS, Ingram DK, Cutler RG (to be published) Effects of caloric modification on aging rate in nonhuman primates: a progress report. In: Baker GT, Ingram DK, Livingston GE, Shock NW (eds) The potential for nutritional modulation of the aging process. Food and Nutrition, New York

Short R, Williams DD, Bowden DM (1987) Cross-sectional evaluation of potential biological markers of aging in pigtailed-macaques. J Gerontol 42:644–654

Walker LC, Kitt CA, Schrvam E, Buckwald B, Garcia F, Sepinwall J, Price D (1987) Senile plaques in aged squirrel monkeys. Neurobiol Aging 8:291–296

Weindruch R, Walford RL (1988) The retardation of aging and disease by dietary restriction. Thomas, Springfield

Part VI
Impact of Dietary Restriction on Bioassays and Recommendations for Future Research

CHAPTER 28

Impact of Dietary Restriction on Bioassays and Recommendations for Future Research: Panel Discussion

Chair: C.J. Henry[1]
Panelists: D.B. Clayson[2], G.N. Rao[3], F.J.C. Roe[4],
R.J. Scheuplein[5], and D.E. Stevenson[6]

Dr. Henry: One of the things that we all have learned is that diet is tremendously important not only for human health and aging, but for a lot of the animal work on which we make a fair number of public policy decisions. We have assembled a panel of individuals with very distinguished backgrounds in animal toxicology or regulatory issues. Each panelist will give his views on the work on caloric restriction and on how it might have an impact on bioassays.

By bioassays, toxicologists mean those studies in which animal models are used to assess the toxicity of chemicals that enter our society either through industrial chemicals or food additives, ingredients, and drugs, whatever those substances might be. The term bioassay has a lot of meaning for toxicologists but may not be quite as familiar to the nutritionists and the gerontologists who are here. We have assembled three very different groups of scientists, and one of the things that will be beneficial is the interchange and information exchange and conversations that will go on among these three groups toward advancing the fields of our respective interests.

The panelists are Dr. Ghanta Rao from the National Toxicology Program (NTP); Dr. Bob Scheuplein from the Food and Drug Administration (FDA); Dr. David Clayson from the Health Protection Branch in Canada; Dr. Francis Roe, a consultant from the United Kingdom; and Dr. Donald Stevenson from Shell Oil Company.

Dr. RAO: NTP is the National Toxicology Program of the Public Health Service under the Department of Health and Human Services. It is located in the National Institute of Environmental Health Sciences (NIEHS) in Research

[1]ILSI Risk Science Institute, 1126 Sixteenth Street, N.W., Washington, D.C. 20036, USA
[2]Toxicology Research Division, Bureau of Chemical Safety, Food Directorate, Health Protection Branch, Health and Welfare Canada, Ottawa, Ontario, Canada KIA 0L2
[3]National Toxicology Program, National Institute of Environmental Health Sciences, Research Triangle Park, NC 27709, USA
[4]19 Marryat Road, Wimbledon Common, London SW19 5BB, UK
[5]Toxicology Sciences/OTS, Center for Food Safety and Applied Nutrition, U.S. Food and Drug Administration, 200 C Street, S.W. (HFF-152), Washington, D.C. 20204, USA
[6]Shell Oil Company, P.O. Box 4320, Houston, TX 77210, USA

Triangle Park, North Carolina. NIEHS is an institute of the National Institutes of Health under the Public Health Service.

The purpose of the NTP, when it was conceived in 1978, was to centralize the testing capability of chemicals to aid in safety assessment for human health. For this purpose, chemicals are nominated by various regulatory agencies, departments of the federal and state governments, and the public. These chemicals are prioritized on the basis of their importance for public health and then tested by various procedures.

As Dr. Henry mentioned, the purpose of the chemical-toxicity studies is to understand the full toxic and carcinogenic potential of chemicals and not to decrease tumors in rats or to prolong the life of the rats. The goal of chemical-carcinogenicity studies in rats and mice is a lot different, probably the opposite of the objectives of many of the dietary restriction studies reported at this meeting.

During this conference you heard about the effects of dietary restriction on levels ranging from whole animal to molecules. I would like to concentrate on a few of those.

Dietary restriction resulting in a decrease in body weight of about 20% adversely affects the fertility and reproduction of animals. I hope you had a chance to review the poster by Chapin et al. on mice. A similar but abbreviated study was done by Hamm et al. at the Chemical Industry Institute of Toxicology (CIIT) on Fischer 344 rats, with similar results.

At 90% of the ad libitum diet there are no substantial effects on fertility, resorptions, and litter size. At 80% of the ad libitum diet there is a tremendous effect on fertility and litter size. This is a serious effect on the basic physiologic processes of fertility and reproduction.

In the 90-day studies we observed an adverse effect on the growth of long bones. There is a partial or a complete loss of metaphyseal bone trabeculae of the long bones in 13-week studies, and it was consistently associated with marked reduction ($\leq 80\%$) in relative weight gain.

Reducing body weight by 25% or more by dietary restriction adversely affects the carcinogenic potential of chemicals requiring metabolic activation. In the last 50 years there have been a lot of reports regarding the effect of dietary restriction on carcinogenic potential.

The table shown by Dr. Ip presented a summary of those reports. Methylazoxymethanol is a carcinogen but requires metabolic activation to be carcinogenic. In the ad libitum-fed animals, 90% will get tumors of the small intestine. However, if you restrict their diet so that body weight is reduced to 75%, the tumor rate goes down. This is not the primary goal of the carcinogenicity study.

To summarize, lowering body weight by dietary restriction or other means adversely affects the basic physiologic process of reproduction, adversely affects the growth of the support structures such as bone, and decreases the carcinogenic potential of some classes of chemicals, which is not a desirable modification of the animal model to assess the carcinogenic potential of chemicals.

However, we have a practical problem. Those of you working with rats and mice know that they are getting heavier each year. A male Sprague-Dawley rat used to weigh 700 g and now weighs 1000 g or more. Male Fischer 344 rats had a

20%–25% weight increase and female Fischer 344 rats had a 15% increase over the 16-year period from 1971 to 1987 in NTP–National Cancer Institute (NCI) studies.

This higher body weight is causing problems. It is resulting in increased incidence of endocrine tumors, especially the pituitary tumors, in both sexes and endocrine-related tumors, such as the mammary tumors, in females. In addition, we are seeing a substantial decrease in survival at the end of the 24-month study. That means that the life span is being decreased. We have to find a way to lower this weight gain. I do not think we want to go back to the 1971 level, but how do we accomplish this?

I see at least two choices. One is dietary restriction. That is, daily feeding with a predetermined amount of diet every day. This is labor intensive, cumbersome, and expensive and will be subject to a lot of errors because during the 2-year study there are plenty of weekends, holidays, bad-weather days, emergency days, etc. In a chronic study this may not be a manageable procedure.

Another choice is feeding animals at night, 4 P.M. to 8 A.M. Rodents are nocturnal animals and do not have to be fed during the day. Daytime feeding is like giving a midnight snack. This choice is a little more manageable than the above procedure, but it could have problems as well.

A third choice is to decrease the caloric density so that we can slow the growth and have a lower body weight at the end of the study. One way to accomplish this is to increase the nonnutrient type of ingredients, probably the fiber in the diet.

Because of this trend of increasing body weights seen in the last 15 years, we designed a small study in which we modified the NIH-07 diet, called the NTP 88 diet, by decreasing the protein to 15% from 23%. We added about 3% more fiber and decreased the fat by a small amount, and then we fed some animals only at night.

At 6 months, comparing animals fed the NIH-07 ad libitum with animals feed NTP 88 at night, there was a 10% lower body weight in males and a 8%–9% lower body weight in females. At 12 months a similar difference existed, but the relative proportions changed. At 18 months there was a little more narrowing of the relative difference.

We were encouraged by these results. The modifications appeared to be working, but further modifications are needed so that we can stabilize the procedure. Night feeding will become a problem if we run too many studies in different places. There could be days when animals will be fasted, and if we fast too many times during the course of a 2-year study, the validity of the study will be in question.

We may have to modify the diet further to decrease the caloric density, but if we add nonnutrient ingredients such as fiber, there may be changes in the gut flora, and gut flora is very important in the metabolism of the chemicals we are testing. There could be a decrease in the absorption of nutrients and of chemicals as well.

Even with these drawbacks, I believe that modification of the diet will be a simple procedure that can be implemented to lower body weight gain.

Dr. Scheuplein: Diet has an enormous impact on toxicologic endpoints, particularly tumor responses. At the FDA we do risk assessment on the basis of bioassays. After you have done about a dozen of these bioassays, you wish that they were better. You wish that you had a way of understanding what their relevance is for humans.

Incidentally, the FDA does not regulate calories. Maybe it should, but at the time the Food and Drugs Act was passed in 1906, food was considered wholesome if you cooked it at home, and the only thing that was making it possibly unsafe was the additives that the manufacturers of processed food were putting into it. We have come a long way since those days and we realize now that the diet probably has a lot more to do with any manifested toxicology or toxicity later on in life than any additives or contaminants. In a few years we may be doing risk assessment differently. Perhaps we will assume that everyone is exposed to various carcinogenic initiators, and then we will look at how the diet modulates those.

Are we all fat rats? If we are fat, what kind of rats are we? How much dietary restriction is appropriate for an animal test model?

We have heard that there are enormous numbers of effects, everything from changing DNA repair to changing adducts, to lengthening life, to reducing tumor rates, to changing endocrine balances. These effects are clearly important. What we need to do now is to design studies that help us get to the human relevance of these issues.

I am very critical of our current bioassay system: it needs to be improved. You heard Dr. Rao speak of the effects of dietary restriction of up to 20% on reproduction in rats and Dr. Rao talk about the endocrine-related tumors. These issues will be brought up as reasons for going a little slowly in adopting an animal model on a restricted diet. These are the questions I think we have to debate. There is an effect of the diet and there is an effect on tumor responses. What does a high-dose feeding study mean when we do risk assessment on the basis of it and do not consider the diet at all?

Dr. Clayson: I am one of the people who believes very strongly that dietary restriction is not the way to try to sort out the problems of the carcinogenesis bioassay. In my opinion, there are good scientific and economic reasons why this should not be done.

We can start from the fact that if you restrict your animals, you will most usually and most economically give the diet once a day, and the animals will consume it as a bolus. This can affect the metabolism, the toxicity, and the tissue dose of the compounds you are giving. You can get over this by spending a little bit more money and giving the animal food in three or four portions during the day, but it does not help very much.

If you restrict animals, we have been told innumerable times at this meeting, you prolong life span. The carcinogenesis bioassay is designed to cover the majority of the life span of the test animal. If we increase life span, we should increase the observation period. In other words, we shall increase our costs again in keeping the animals alive for a bit longer.

We have again been told by several people that reduction in the amount of diet you feed leads to the diminution in the number of both spontaneously occurring and induced tumors. That this slight increase in the longevity of the animal in the course of the bioassay may possibly restore the tumor yields, which we are concerned about, to the same level is not very likely. In other words, the bioassay will not give the public the same degree of protection. Dietary restriction may necessitate doubling or even quadrupling the size of the bioassays to make sure we are getting the same degree of protection.

When we have done all of this, we have perhaps suppressed some of the endocrine tumors in rats and mammary tumors in mice, and we find out that the test agent has brought those tumors back again. Are we any further forward? I think not. We are still puzzled as to whether that agent has worked by some chance mechanism, by some sort of promoting action on these tumors, or has worked as a real carcinogen.

I think that the cards are very stacked against a useful reduction in diet during the carcinogenesis bioassay. This is not to say that I believe that the cancer bioassay is fine as it is. I think we only have to look at Haseman and Huff's figures: in the NTP/NCI bioassay program there is agreement between rats and mice for only 74% of the considerable number of chemicals that they considered. Goodness only knows what the predictive rate for humans is based on that; it is probably about the same.

We are going to have to consider mechanisms first, and we are going to have to consider them with sufficient data and with sufficient strength so that we can sell them to the regulators and to the public as well.

We should begin by dividing carcinogens into those that directly affect DNA, the genotoxic carcinogens, and those that do not. I think we have a much better chance of genotoxic carcinogens showing some effect in humans.

The work done at the CIIT on the male rat kidney tumor has shown the mechanism by which this acts. This effect probably has a threshold, and it appears that agents like 2,4,4-trimethylpentane are unlikely to have an effect on humans. In Canada we have taken part in a large international effort that has reached a similar conclusion with the food antioxidant BHA: at reasonable doses humans are very unlikely to be affected. The U.S. Environmental Protection Agency (EPA) has considered the evidence on thyroid tumors. It has concluded that the effects of goitrogens may have a threshold. This type of work may sort out some of the difficulties we are having with the bioassay rather than addressing the difficulties by restricting the diet.

Dr. Roe: At this conference we are a mixture of nutritionists, toxicologists, and gerontologists, and a very interesting mix it is. Traditionally, nutritionists have been disinterested in toxicology. They have designed diets for growth. It has been said that they are obsessed with the risk of malnutrition and that they have not been interested in, or are only perhaps now beginning to be interested in, designing diets and dietary regimens for maintaining animals in good physiologic health into old age. This is really what is needed.

Toxicologists for their part have been disastrously disinterested in diets. They have assumed that their responsibilities end when they check that the diet that they are using conforms to the protocol for the study. They have taken no interest in the diet that is given to purchased animals before they arrive. Yet, as we know, slight differences in diet make enormous differences in the endpoints, which toxicologists wax poetic about at the end of studies.

As far as I am concerned, Walford and Weindruch are gods. Their work, fortunately, has at last begun to have an impact, but are they doing enough? Are they and other gerontologists doing enough to have an impact on the bioassay program?

I would point out to you, although I would not want to belittle the dangers of deaths from cancer, that the proportion of people who die from aging-related diseases is very much greater. Perhaps the problem is that we have come to look upon the bioassay as an assay for cancer and perhaps we should be designing tests for "gerontoxicity," of which cancer is perhaps no more than an important part. If we started designing tests for all aspects of aging, then perhaps we would get a different view about what we should be doing.

We have heard some very interesting reports here of mechanistic studies. I think that the mechanistic study that David Clayson described and the effects on cell turnover are extremely important. I think that all the studies on oxidative damage and DNA repair are extremely important. I think Dr. Good's ideas as to where we might go with short-lived strains are very important. I am sure that in the longer run the studies on chronic biology are going to be interesting, but I am not sure that I quite understand how they have an impact now.

What about bioassay testing? As David Clayson said, one really has to start off by saying that what may be true for genotoxic carcinogens may not be true for nongenotoxic carcinogens. On the other hand, one often cannot make that distinction before one has done the test because very often one does not really know. In any event common sense dictates that the way we are conducting bioassay tests at present is nonsensical.

We know that there is not a population of humans that has a 100% incidence of pituitary tumors as do many strains of rat. We do not have humans, nor indeed any subgroup of humans, that has high incidences of liver tumors, lymphoma, or lung tumors, as found in many strains of mice. The use of such rat and mouse models makes no sense and could not be justified to any member of the general public who has a modicum of common sense.

Time is running out. Unless industry gets its act together, unless governments get their acts together, the antivivisectionists are really going to win the day and one is not going to be able to do the tests we are talking about because they will be seen to be generating more nonsense than sense. I believe that this is a very, very serious problem.

There are three interrelated problems: one is overfeeding, another is use of the maximum tolerated dose (MTD), and the third is the practice of looking only at the tumor incidences that increase, ignoring those that decrease and, more importantly, ignoring all the nonneoplastic pathology, which can tell one so much. Again, this is why we should be thinking in terms of tests of "gerontoxicity" and not simply tests for carcinogenesis.

I am not saying that one has enough knowledge now to start designing a sensible diet-restricted regimen for all bioassays. I do not think we are there by any means. I think there are a lot of problems. What I am saying is that we should be putting money and resources into finding out how to maintain untreated control animals in normal physiological status into old age.

Dr. Stevenson: I agree with much of what has been said, but I think we have to recognize that what we are looking at is part of an evolutionary change in our thought process. I think we all agree that the bioassay has a very limited use. We go back to what I call the Epsteinian view of cancer. In the 1960s and 1970s it was suggested that the majority of human cancer was due to man-made chemicals and, furthermore, that only perhaps 10% of chemicals were going to be carcinogenic; if we could rout those out we could solve the cancer problem. Dr. Roe wrote a very elegant chapter on that in a book called *Theories of Carcinogenesis*.

We now realize that this is not the true situation. If we are interested in public health and toxicology, our approach should now be very different.

There is a lot we can learn from the animal studies. The NTP itself has changed direction very radically in the last 2 or 3 years. The initial program, which was actually designed by NCI, was designed specifically to have qualitative studies to identify carcinogens, and therefore the studies had only one or two dose groups, and therefore these had to be high doses. This design was carried on somewhat when the NTP took over the program, but in the last 2 or 3 years there has been a very significant change toward looking at mechanisms of carcinogenicity; this is a change in direction that we thoroughly applaud.

One of my personal interests is risk assessment, and I feel that one of the tragedies we have right now is that we are trying to use a bioassay in a risk assessment context where it is by no means an optimal design. For instance, the question of thresholds is being raised. I can give or not give you a threshold in a study, depending on the length of the study, because tumors are time related. One of the problems in cutting a study off at 2 years is that sometimes you create a threshold in the study.

To me the dietary-restriction angle is only one part of a multidimensional problem. When we were developing our program in England, for instance, we used different types of diets and found that we could get the same kind of effects as you get with dietary restriction by using, for instance, pelleted diets, particularly hard-pelleted diets instead of powdered diets. So it is not only a question of the composition of the diet but also the way in which the diet is presented.

Another interest of mine is in the bacterial flora of the gut, and, again, this has been mentioned in passing, but I think this is a tremendously important issue that has not been looked at to any degree. The pattern of feeding and the amount of food have a major impact on the gut flora. When we consider that many of the chemicals in which we have an interest are metabolized at least in part by that flora, clearly it is a practical issue.

The practicalities have been mentioned: cost, life expectancy, and so forth. For example, it was once stated in one of the guidelines that if Long-Evans rats were used in a bioassay, because traditionally they do not start dying until well after 2 years, then the study should be a 3-year study instead of a 2-year study. That could increase the cost by perhaps 30% and would probably kill the use of that particular strain.

So where do I come down in all this? I think that I would not necessarily come down on the side of saying you have to use diet restriction or you do not have to. I think you really have to take into account that this is one of the variables that you use to interpret the study along with a lot of other information.

I would like to see the bioassays altered to include the time element. I feel that we could gain a lot if we had partial-lifetime studies, that is, that you try a regimen for a certain period in the beginning or a certain period at the end of the life span of an animal. This would give you a tremendous amount of information.

One dichotomy between toxicologists and nutritionists is the question of whether a useful experiment can be done with one dose. I would like to see some of the data expanded to include one or two other levels of dietary restriction.

If you are not careful, you make the assumption, as the toxicologists have tragically done, that when you only have one or two dose-response points, the response is linear between those two points. I suspect that in many cases you may see a biphasic response. Small amounts of dietary restriction may have one effect, and I would not be surprised in some cases to see that reversed as the effect is made more severe.

One small point that was raised was the question of the increasing body weights of the Fischer 344 rats. We had a similar problem with Charles River rats, and we found that what was happening was that the animal technicians choosing the breeding animals preferred larger animals and therefore were unconsciously selecting them. I just wonder if the same is happening in Fischer 344 rats.

This is really part of a very complex situation. The regulatory policy of this country is somewhat misguided and, therefore, not only do we have to educate the public, we also have to educate Congress into thinking that if we take all the nutrition information and all the toxicology information into account, we would be spending our monies in very different ways.

As Dr. Hart is fond of saying, the cost or the consequences of what we are doing runs into maybe $50 to $100 billion a year, and when we consider what Congress is not able to do in terms of providing primary health care to people in this country, what we are talking about is a real moral issue.

Dr. Henry: What you are hearing is a reflection of the frustration of the panel members, individually and as representatives of their agencies or organizations, with how the bioassay is used for assessing human health risks and with the sort of regulatory policies which must be imposed for certain chemicals.

The frustration level is extremely high. I think the common denominator for all five of our panel members is that the bioassay could greatly benefit from an understanding or interpretation of the results of the body of work that we have heard discussed here.

I think you must also appreciate that this information exchange needs to continue. I think it was very well stated that the toxicologists have not worried about diet and the nutritionists have not necessarily been aware of all that is going on in toxicology and, perhaps more importantly, in risk assessment, which is where a great amount of emotion is being displayed.

Dr. Rao: The question about hard pellets was raised. I believe that using hard pellets is a good idea. It is true to some extent that chewing satisfaction from hard pellets may decrease feed consumption. In toxicology studies, however, we have to incorporate chemicals in the diet. When we start repelletizing to obtain extruded pellets, we have the problem of effects on the chemicals under study. If a chemical has an effect on animals, such as a sedative or other effect, the animals may not be able to chew the hard pellets.

Second, there is the question of why body weights are increasing. Honestly, how many of you order animals by age? Ninety percent of the rats and 90% of the outbred mice were ordered by weight. The investigator could not care less if they are 6 or 8 weeks old. The breeders let animals grow fast and propagate to enhance their business.

Third, you have to realize that between 1970 and 1990, the rats and mice have gone through 60 generations. That is more than 1000 years in human terms. There are genetic changes and selections. You cannot say the rat of 1970 is the same as the rat of 1990.

In addition, we eliminated all infectious diseases. Animals must be growing because we eliminated all those stresses. In toxicology studies and animal experimentation these days, because of animal welfare, we have to control all aspects, especially the environment, which adds to feed efficiency and metabolic efficiency, and this may be why the animals keep growing.

Dr. Henry: I think that scientists across disciplines are very well able to say how much we do not know and why we need more research. I would like to try and guide us into a more positive aspect, because otherwise it seems that we really do not know a lot, that we cannot decide what to do. This is not to say that that is not reflecting a lot of frustration that individuals have to deal with in their own work.

Dr. Meyer: Otto Meyer from the Institute of Toxicology in Copenhagen. I share the frustration shown by members of the panel. One thing I have learned here is that you should take into consideration the ethics we have heard about when you interpret your studies. I do not think that dietary restriction is a good idea because as you make a shift, you may extend the longevity of the animal, but the time to occurrence of the tumors is shifted in the same way. Perhaps, therefore, you do not see more.

I agree that you can reduce the background noise in pathology, but you also may reduce the sensitivity of the model.

I think we should not disregard what we have heard, but perhaps we should better define the model. We may not have looked in enough detail at the caloric need, for example. Perhaps we should reduce the calories because it is unphysiologic to give the animal a diet too high in calories when using a purified diet.

Dr. Roe: I think we are falling rapidly into a deep trap by suggesting that all that diet restriction does is to delay the onset of the same spectrum of aging-related changes that non-diet-restricted animals develop. This is just not true.

It is not true for nephropathy. One of the stupidest things that we do is to use overfed animals that have grossly abnormal kidneys. Now, this must have enormous metabolic and other consequences. How can one meaningfully use an ordinary overfed rat to test for renal toxicity or renal carcinogenesis? It is simply not valid to do so.

The fact is that one can, by means of simple diet restriction, maintain these animals with pretty normal kidneys into old age. Thus, tests involving a kidney endpoint can give qualitatively different results in rats of only 12 to 18 months under conditions of overfeeding compared with diet-restricted rats of more than twice that age. The same is true for many other end points.

Dr. Clayson: I think that we should look beyond just plain dietary restriction. Our preliminary studies suggested that fibers and fats from different sources have different effects on cellular proliferation, each in different tissues. Maybe we should first look at the composition of our diets to see if we can standardize things a bit more. Second, we should forswear the use of diets in which the fat and everything else comes from different sources depending on cost. This could well be putting a fairly big spanner in the works of any biological experiment, let alone the bioassay.

Dr. Walford: Dr. Roe questioned whether the effect is linear biphasic, and that is an important question. Most of the published studies on dietary restriction compare ad libitum animals with severely restricted animals. However, in essentially all the work from my laboratory, there are three groups of animals: one is fed ad libitum, one is mildly restricted, and one is severely restricted. We consider the ad libitum and the mildly restricted animals as two control groups for the severely restricted animals. In all the tests that I remember, the mildly restricted group gives less, but similar response to the severely restricted, so the effect seems to be linear.

I suggest that animals used in toxicology studies might fit into our category of mildly restricted animals. They eat all of the food you give them and they do not live a lot longer; they live a little bit longer, but it seems to be a linear effect.

Dr. Hart: Dr. Roe is correct. The shifts in tumor incidence and in other degenerative diseases are not uniform and proportional to the extension of maximum life span. That surprised me. So your point is well taken, and because we have the largest data bases, I would assume that we have the data to justify that statement in both cases.

I have been associated with toxicology for some time, and the predominant approach of counting lumps and bumps has certain limitations. Toxicology is far more than simply testing for chronic bioassay results. We use toxicologic testing for setting guidelines for pharmaceuticals, poison control, safety of products, and so on.

The importance of caloric restriction to the toxicologist, from my point of view, is something far greater than simply its impact upon the occurrence or lack of occurrence of tumors. Its importance has to do with drug metabolism, intermediary metabolism, excretion, detoxification, and transport.

The point of what is a functional kidney is very important with respect to these endpoints as well. To limit this discussion only to the counting of lumps and bumps is, I think, a total disregard for the field of toxicology and its broad importance in pharmaceutics, poison control, and the other areas that we were very interested in until we started doing one chronic bioassay after another.

Anonymous: I think it is not true that we may be delaying diseases. Some diseases simply do not occur; good examples are chronic nephropathy and cardiomyopathy. They simply do not occur in food-restricted animals.

By simply delaying the occurrence of the restriction, one can avoid the failure or the marked retardation of development of the reproductive system and of development of the bone structure and still completely avoid kidney disease and cardiomyopathy and eventually get the tumors. So if you maintain the animals ad libitum, for instance, until 6 months of age and then restrict the animal, you get rid of all of these other disease processes and have left only the issue of tumors that eventually occur.

Dr. Rao: That is a good idea, but we have to realize that we start with a MTD of chemical and if we change at 6 months to dietary restriction, we have to be sure that the dose will not end the study by death of animals, or we may end up changing the dose during the course of the study.

Dr. Henry: That suggests a very interesting pilot or validation study to see how best to incorporate some of this information into changing the bioassay.

Dr. Rao: Right, but then you end up with pilot studies of 1-year duration.

Dr. Henry: I think we should all recognize there are not any magic bullets here. I do think, even among the people who have spoken most strongly for changing the bioassay, that we cannot go out and do it tomorrow. However, I think we better start to make some first steps.

Dr. Harrison: It would be foolish to take steps until you know what you want to do. What you are looking for is very different, of course, from what the gerontologist is looking for. It seems to me that you are looking for the highest possible signal-to-noise ratio. A part of that, of course, is reducing the variability and the noise, which is why you have often gone to a single inbred strain. You reduce the variability and noise further with food restriction, but if you reduce the signal proportionately, the question, apart from considerations of time and cost, is what is that going to do to the signal (i.e., the response to the carcinogen relative to the noise)? I do not think gerontologists have specifically addressed this in a quantitative enough way to give you any more than hints.

You also have to consider genotype, because genotype is going to interact with the signal-to-noise ratio. I would like to suggest a syllogism. You cannot appropriately model a multigenic population, such as people, with a monogenetic model, such as a single inbred strain of mouse or rat or even a single F_1 hybrid, because an individual is going to have individual characteristics. It might be a wonderful model for those people who are analogous, but it is not going to be a model for all people in the multigenic population. Of course, you cannot have a multigenic population because that is going to have far too high a noise ratio and the variability is going to spoil your signal. I would like to suggest that you might consider the possibility of adding to your long-lived healthy F_1 hybrid with a minimum amount of background tumor, a tumor-vulnerable-type individual who was studied for a shorter period of time. This way at least you attempt to model people who are particularly genetically susceptible as well as the long-lived strain that is particularly genetically resistant.

Dr. Roe: Genes are important, but in the toxicological literature – for instance in the classified lists of inbred strains – one finds statements to the effect that particular strains and sublines are susceptible or resistant to the development of various kinds of neoplasms or other diseases. There is, however, a lot of nonsense here because many of the diseases and lesions listed are actually of dietary origin. There are in the literature, for instance, statements to the effect that some strains of rat are especially liable to develop adrenal medullary tumors while other strains are not. The truth is that one can change a so-called low adrenal-medullary-tumor strain to a high one just by relatively simple dietary intervention.

Many observations in the literature attributed to genetic differences have little or nothing to do with genes. That is not to say that I do not agree that everything that we see is an interaction between genes and environment, but the environment plays a very large part.

Dr. Ruben: Zadok Ruben from Hoffmann-La Roche. I would like to restrict my comments to two areas. The first one is to complement or to add to what Dr. Hart just said, and that is with regard to drugs and metabolism.

In the pharmaceutical industry we are not only facing the issue of metabolism in the effect of the drugs, but we are dealing with chemicals that we are choosing to be biologically potent and we hardly know anything about them. They are novel. They are not like the aflatoxins or the PCBs or other compounds for which there is a tremendous amount of information in the literature, and within 5–10 years, because of business considerations and competition, we need to get them marketed if possible.

The second thing is about the positive points that were just raised. Probably one of the most important things that I learned from this conference is limitations. Every model and every system has limitations, and if we take these limitations into account, it is going to help us in the interpretations that we make, or at least are attempting to make, and in applications, in this case, to humans.

With regard to optimization, there may be no optimum. It may just be that we know what the assay is offering us, which allows us, within limitations, to make the best interpretation.

The best interpretation does not only mean scientific interpretation. Responsibility towards the society and business considerations must be kept in mind. In the past sometimes we did not recognize these limitations and we condemned a lot of compounds, as has been mentioned before. A lot of chemicals that may have been useful and possibly also preferable were condemned, just because we did not consider the limitations of what we have.

Maybe we can move to something better from here. It does not mean that what we did before or what we have been doing is entirely bad.

Dr. Knapka: We had some discussion today that nutritionists were not interested in some of this toxicology work. I really do not think that is true. Many people think that the diet that produces the fastest growing animal is the best diet. This comes from our agricultural background, has always been this way, and nobody has really questioned it.

When you begin a toxicology study with one diet, you have to continue it throughout the study. It is a tradeoff, I am sure, but I think we are ignoring one fact, a very important fact, that the nutrient requirements for animals change as animals age.

We have a requirement for growth, we have a requirement for reproduction, and we have a requirement for maintenance. When we start with a growth diet, as in many toxicology studies, and continue it throughout the study, we are putting a lot of stress on animals. This is where we get obese animals. We have got to determine the maintenance requirement. If we have long-term studies, we should be feeding these animals at maintenance requirements, not at growth requirements.

The American Institute of Nutrition (AIN) diet was not made for long-term study. Some people say that they used the AIN for 16 months and lost every one of the animals. Other people said they did not have any problems. I think that you need to look at who prepares the diet. Preparing a purified diet is a little different from preparing a natural-ingredient diet. You have a lot of problems. You have interactions and you have more decay of nutrients. If you mix the minerals and vitamins, you are going to get vitamin decay. Purified ingredients do change.

At one of our nearby universities, one of the people from the biochemistry department sent me two samples of the AIN diet that they made in the laboratory's diet kitchen and wanted me to analyze it because one had a little different color than the other. So I analyzed it, and I found a 4% difference in protein between the two samples. It turned out that they had used two different kinds of casein: one was 90% protein, the other was 80% protein. You might as well be feeding the natural diet. So, I caution that before getting too critical about the AIN diet, you need to watch who is preparing it and how it is being prepared.

Dr. Henry: I think there have been at least two unified calls for understanding maintenance requirements, which really does get back to what Dr. Hart was suggesting—understanding mechanisms. Dr. Stevenson may be known for coming down with "definite maybes," and I come down for "it depends." We have to understand what question we are asking, where we want to go, and how to design not only the diet but the study or your own research.

Dr. Bechtel: Dave Bechtel with Best Foods. Dr. Clayson warned us that because dietary restriction decreases chemical potency, especially carcinogenic potency, that dietary restriction might lead to a problem. Dr. Rao's data showed a decrease in tumor incidence. I suggest that the dose levels used in the bioassays are determined from shorter-term studies, and if dietary restriction—if that is what we want to call a caloric restriction—is part of the whole regimen, if there is a change, it will be reflected in a change in the potency at the lower levels which will be translated into a change in the estimated MTD.

Dr. Henry: Again, a suggestion for pilot studies.

Dr. Bechtel: Regarding Dr. Rao's comment about sedation preventing animals perhaps from eating hard pellets, I submit to you that if the animals are sedated to the point where they cannot chew pellets, you are exceeding the MTD.

That is an issue that we have to look at in some of these studies. To me as a toxicologist, when I hear the gerontologists talking about a 40% caloric deficit in feeding the animals, I think that they must really be confused.

Well, I do not think that a physiologist would view MTDs the way that they seem to be applied as any different in extreme than in what I view 40% caloric restriction to be; that we are pushing the MTDs beyond the physiologic limits. And if we are going to justify doing that for any chemical, we have to look at the metabolism and pharmacokinetics and demonstrate that we are within some sensible physiologic boundary. Otherwise one cannot interpret the data.

Dr. Henry: I think the conference has done much to help us further define what might be a more reasonable definition of MTD.

Dr. Bechtel: If I might make one additional comment along the same lines. We need that kind of data. I think Dr. Hart is right, besides counting lumps and bumps, I do not know how you can correlate dose and response unless you know what is going on mechanistically.

In fact, as I understand from one of Dr. Scheuplein's comments about the problems of doing risk assessment, bioassays were never intended to be used for risk assessment. If you are going to use dose-response data from a two-level study, I think you are in trouble already.

At their peer reviews the NTP goes to great lengths to say that it only decides whether something is potentially carcinogenic or not and does not tell the world anything about its hazards to man. The FDA takes not only the bioassay data but a whole lot of other data. I think we should request the FDA to redouble its efforts to not rely on the bioassay alone and that if it does not have sufficient ancillary data to interpret it, then the bioassay data ought to be considered insufficient.

Dr. Clayson: I think you slightly misinterpreted one point I made. I expect many of you know that in the early 1940s, Dr. Tannenbaum demonstrated that mammary tumors, presumably induced by the mammary-tumor virus, in the strain of mouse he was using could be completely suppressed by going to a restricted diet as late as 9 months of age, which was about half the animal's life in those days.

If the same sort of thing happens with chemicals, you are just not going to see that these chemicals in fact are toxic and carcinogenic agents. Why carry out the bioassay at all if you want to reach that status?

Dr. Scheuplein: May I respond to the question of whether we should ignore bioassays or reinterpret them?

The NTP program only exists because Congress thought it would be relevant to humans. What NTP does for us is to provide bioassays. Regardless of what they write down on their data sheets, regardless of the special qualified conditions of the NTP experiments, Congress expects us to use that data to provide regulations and regulate food additives and anything else that needs to be regulated, whether under FDA authority or under the Toxic Substances Control Act or Federal Insecticide, Fungicide and Rodenticide Act.

We are all aware of the limitations of those bioassays, and it would be nice if we could interpret them, but interpret what? All we have are lumps and bumps, very little mechanistic data, almost no pharmacokinetic data. Certainly no dietary information accompanies those bioassays. Until it does, we are going to have to rely on those bioassays as best we can. We cannot ignore a carcinogen, even though we might want to, that has a clear dose response, despite our problems with the extrapolation to low dose. You need data to interpret those studies. You need the kinds of data that we talked about here. We cannot go to court and say well, it is a bioassay, you know how bad they are, nobody believes them anyway. That simply will not work.

Dr. Rao: I would like to go back to my colleague's comments here on NTP study reports. You will find all the information you need on diets. If you want more, we will be glad to provide it. In addition there are mechanistic data and short-term toxicity data as well.

Anonymous: I would like to make a positive comment concerning the design of most of the experiments we have heard these last two days. All the comparisons were made between ad libitum and restricted animals, and restricted animals were fed some number of hours per day. We know that the change in the pattern of food intake has major effects on the metabolism of these animals. For example, we have restricted diets by 25%. What we found was an increase in insulin and an increase in lipoprotein. These are effects of changing the feeding pattern.

We need to have control ad libitum animals fed for a limited time, like the 6-h restricted animals you had this morning, to have the animals eating at exactly the same time per day. Another way is to use automatic feeders. There are many laboratories using automatic feeders that give the animals the desired quantity of feed. In this way, you can have restricted animals eating the same number of meals as control animals, which should be important if you want to understand the consequences of calorie intake.

Dr. Fernandes: It is interesting that we have been hearing this information from gerontologists and immunologists. I did work in nutrition with Dr. Good almost 20 years ago, then with Dr. Masoro. All these groups here were sort of promoting food restriction in prolongation of life and looking into what exactly the mechanisms were, at least from an immunological point of view.

We learned so much about it, although the restriction was very severe, and many areas were opened up. For example, what exactly is the subpopulation of cells affected? Are they susceptible? What kind of lymphokines are they producing? We learned about doubling or tripling the life span of short-lived animals. So many things have come out in the last few years.

From a nutritionist's point of view, we are learning more about dietary lipids. We have been listening to what people said about saturated and polyunsaturated fatty acids: we were told to increase polyunsaturated fatty acids and decrease saturated fatty acids. A trickling of information is coming that polyunsaturated fatty acids are not good for certain mammary tumors.

There has been a lot of criticism of n−3 fish oil, but I think that from n−3 fish oil we can learn how to prevent peroxidation. You can have prolongation of life without restricting calories, but if you restrict the calories and consume fish oil, you can have much more. We have been learning many things about approaches to restriction.

Anonymous: I see a data gap in tying together the existing cancer models and the dietary restriction model. The mechanism for prolonging life by diet restriction is simply the dose-response relationship.

If we had an adequate data base for carcinogens with diet restriction in the range of up to 20%, we might be able to take a close look at the historical data. For aging, with regards to mechanisms, it seems you could do shorter-term studies by using good dose-response relationships. The pharmacologist in me should give historical perspective on this. For 30 years physiologists missed the fact that there were two different types of adrenergic receptors because they did not do adequate dose-response relationships with different agonists. It was not until Dr. Alquist came along with four different agonists and did dose-response curves that anybody was able to point out the simple fact that there were two different types of adrenergic receptors.

The same kind of analogy should be used to caution the researchers to do that type of dose-response study to evaluate mechanisms.

Dr. Hathcock: John Hathcock from the FDA. The first thing we should do is recognize that there is no such thing as a perfect diet. The perfect diet depends on what the purpose is. You have to decide what it is you are trying to achieve before you look for the correct diet. We can look for the maximum growth rate, we can look for maximum longevity, we can look for the optimum, whatever that is; I am not about to define it. We have all heard reproduction mentioned. With the world's population being what it is, is anyone suggesting that maximum reproduction is optimum?

The purified diets that nutritionists have used have achieved some things. They have allowed us to vary one nutrient at a time and to see what its effects are. Some of the difficulties with using diets that give better performance in terms of longevity and reproduction also confound the nutritional interpretations by making multiple nutrient variations at the same time, that is, if we change the ratios of corn, soybean, alfalfa, and so forth.

We have to look at the title of this conference, and it is dietary restriction. Some of us have taken that to mean caloric restriction, but then we also should look at its biological effects. It is not longevity, nor is it carcinogenesis or anti-carcinogenesis. We have to define our interest on each side of the spectrum, both in nutrition and toxicology, before we select a diet or toxicologic endpoints.

Dr. Henry: I thank everyone for their comments. I would like to especially thank the panel. May I close by saying that I believe that this has provided us with more insight so that we can sharpen the questions we address, understanding a bit how we are going to use the answers that we get.

There has been a tremendous amount of progress in the area of dietary restriction in the last 20 years, and I think we should go forward from here to see how we can improve it, remembering that most of our jobs are associated with trying to improve public health.

Index